FLIGHT

THE EPIC HISTORY OF THE QUEST FOR THE SKIES

Text by
Riccardo Niccoli

whitestar

Contents

Editorial project
VALERIA MANFERTO DE FABIANIS
LAURA ACCOMAZZO

Layout and graphic design
MARIA CUCCHI

Image research
CLAUDIA ZANERA

Piazzale Luigi Cadorna, 6
20123 Milan, Italy
www.whitestar.it

New expanded and updated edition

Translations
Neil Frazer Davenport - Hugh Swainston - John Venerella

ISBN 978-88-544-2225-4
1 2 3 4 5 6   30 29 28 27 26

Printed in China

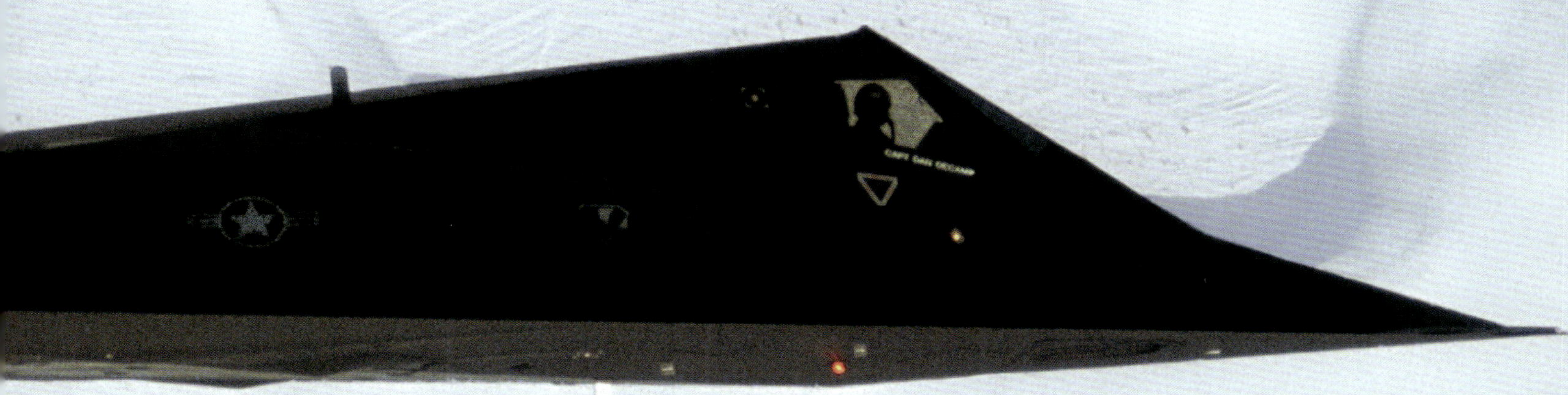

1 top The NASA Space Shuttle "Challenger" in orbit with its loading bay open, seen during its STS-7 mission.

1 bottom A reinforced wing designed by Leonardo da Vinci (1452-1519), who proposed the first technical approach to the problem of human flight.

2-3 One of the RAF's heavily armed Panavia Tornado F.3 fighters in high-speed flight. The Tornado was the first example of large-scale multinational cooperation in European aircraft design and manufacturing.

4-5 The Stealth F-117A fighter-bomber's futuristic form highlighted during a low-altitude turn over the White Sands desert region of New Mexico.

5 bottom An impracticable airship design as imagined in the nineteenth century.

6-7 In the early days of flight, getting airborne called for desperate measures. Here a horse and cart help launch a Caudron biplane glider into the air.

Preface

Nine hundred years. So much has happened since that cold winter day of 17 December 1903 when a heavier-than-air engine-powered aircraft first demonstrated that controllable human flight was possible. Rather than a wavering flight, it was more a series of long hops, the first, 39 yards in length, and the longest about 282 yards, but in substance, mankind, human progress had actually taken a much greater leap. Probably few of those watching Orville and Wilbur Wright on the sand dunes at Kitty Hawk in North Carolina could have imagined that, within just a few years, man would begin his mastery of the air, building smoother, faster, more powerful, and increasingly manageable airplanes.

In comparison with the rate of invention in previous epochs, technological progress in the 20th century and at the beginning of the 21st century went lightning fast, and much of this was due to the efforts invested in research and development in aeronautics and space. A little more than a decade after the Wright brothers' flight, planes were able to exceed 125 mph and perform breathtaking acrobatics. In the 1920s, just two decades later, the first commercial transport planes appeared, able to convey passengers over long distances in times absurdly shorter than those for ships or trains. The advent of World War II accelerated development dramatically, producing conventional, propeller-driven planes capable of doubling their predecessors' power, speed, range, and load capabilities, but more importantly, it brought about the invention of the jet engine and the first usable helicopters. Only 44 years after the Wright brother's flight of 12 mph, a jet engine broke the sound barrier. A mere 60 years later, man flew in space. We could go on with the list of achievements, but this would be pointless: the results of over 100 years of flight are before our eyes every day.

Thanks to air travel, we can now reach every corner of the globe in just a few hours at costs affordable for almost everyone. Goods and materials can be transported with the same ease. Lives can be saved as a result of rapid search and rescue facilities. The sick and injured can be carried to distant hospitals where the finest care facilities or organ transplant capabilities are available. Populations struck by natural disasters or war can be reached and aided, and life-threatening fires can be accessed and extinguished.

Aircraft are vital in making land and sea surveys and in monitoring the environment and pollution. They are important in geography, topography, geology, archaeology, and zoology. For our safety, aircraft can scan, detect, and effectively monitor illicit actions and trafficking of a criminal or terrorist nature, often succeeding in preventing crimes.

In the military, aviation has revolutionized the rules of engagement and become the decisive instrument for victory not only for land battles, but even for the conflicts themselves. The recent wars in the Persian Gulf, Kosovo, Afghanistan, and Libya have shown how air power can decide alone the outcome of an entire regional crisis. As a result of specialized optical and electronic recognition capabilities, aircraft and satellites have made the world a safer and more peaceful place, rendering the great powers more knowledgeable about one another. Space travel has opened new horizons for the scientific world and brought benefits even to the man in the street. Shuttle flights and space station programs have increased human scientific knowledge in many fields, while orbiting satellites have spurred enormous advances in telecommunications and precision navigation.

This book pays tribute to the History of Flight. Through 27 chapters dedicated to its most renowned protagonists, events, and epochs, the principal engine-powered aircraft (planes and helicopters) marking the history of flight since 1903 are briefly examined. In spite of their technological interest, however, experimental aircraft have unfortunately had to be excluded from our inquiry: the purpose of the book is to draw attention to those craft that have really and effectively operated on a broad basis, whether for the civilian or military sectors. The history of flight is immensely wide and deep in scope, and regrettably the limitations of a single volume have meant that dozens and dozens of aircraft deserving inclusion in both the text and photographs have been omitted.

Nor should we forget, beyond the many words, the impulse that has always urged man in his attempt to fly. By design, this book is also a celebration of the realization of one of man's most ancient and yearned-for dreams: that of overcoming his physical limitations and the constraints of earthbound living, so that he might dare to hover within a more spiritual, a more divine dimension, that of flight.

Chapter 1

The Earliest Endeavors

Man has felt the urge to fly from the earliest of times. As ancient legends recount, even the most powerful of men yearned for the sensation of potency and freedom that they imagined that the ability to launch themselves into the sky would provide. One of these legends, originating in Persia, refers to King Kai Kawus' flight in a chariot drawn by eagles in around 1500 BC. Another, this time from Greece, recounts that Alexander the Great saw the skies from a wickerwork chest lifted by griffins. Moreover,

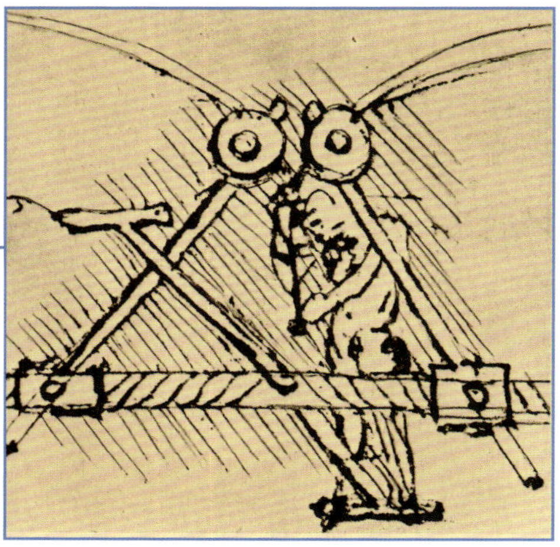

eagles and hawks, birds of prey that dominate the skies, have always been seen as symbols of power. During antiquity, flight was considered to be the prerogative of the gods. In fact, in the legend of Daedalus and Icarus and their feathers-and-wax wings, Daedalus escaped death only by flying low. Icarus, intoxicated by the thrill of flight, flew ever higher toward the sun, the symbol of divine power, until its heat melted the wax, with his sacrilegious act receiving appropriate punishment. Alongside these imaginative accounts are others, some thousands of years old, that testify to man's practical efforts to conquer the third dimension. It appears that as early as two thousand years before Christ, the Chinese were capable of using kites to lift lookouts into the sky for military purposes. In 852 AD at Cordoba, an Arab savant was killed during an attempt at free flight, while in 1020 an English Benedictine monk named Oliver of Malmesbury threw himself from the top of a tower, wearing a pair of feathered wings; unfortunately he crashed to the ground breaking both legs. The years that followed were punctuated by a series of similar failures. In 1496 it was the turn of Senecio of Nuremberg, in 1503 the mathematician Giovan Battista Danti of Perugia, and in 1628 Paolo Guidotto, also from Perugia. For these and for others like them, the same irresistible temptation led to the same disastrous fate. In the meantime, studies into the phenomenon of flight were increasingly elaborate. The chronicles of 1630 report that a certain Hezarfen Ahmet Celebi succeeded in gliding from the top of the Galata Tower in Istanbul to the far side of the Bosporus, thanks to a pair of rudimentary wings. In 1670, Francesco Lana Terzi, a Jesuit, described a flying machine composed of four copper spheres in which a vacuum was to be created and a sail provided for the directional control of the flight. Early in the eighteenth century, another Jesuit, Laurenào de Gusmão of Brazil, designed a bird-like glider that he used

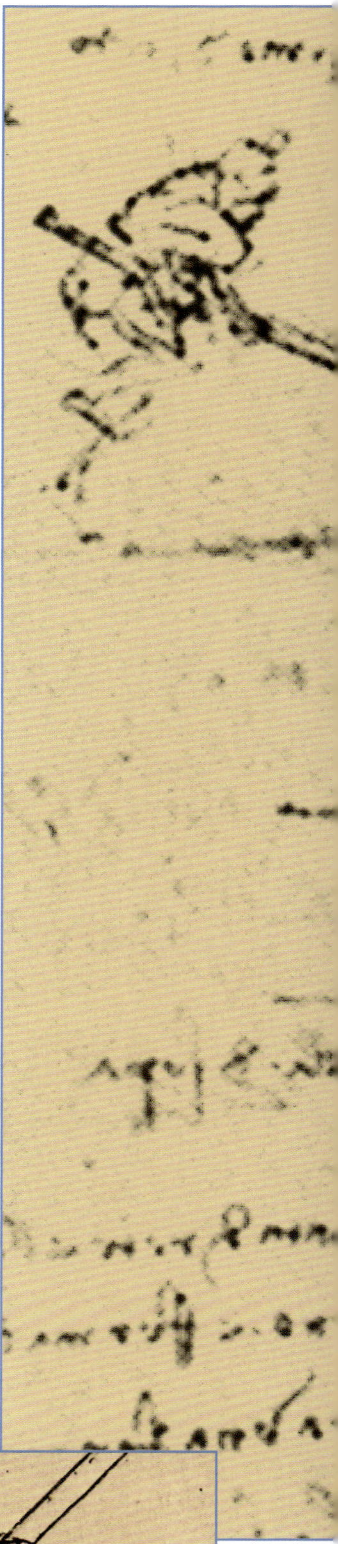

8 Leonardo da Vinci was the first scientist to consider flight from a rational viewpoint. He transferred the workings of his imagination to sketches and drawings that inspired many other thinkers in later centuries. The two small sketches provide examples.

8-9 Surrounded by Leonardo's characteristic mirror-writing is a sketch showing an experiment to prove the lift of a wing. Published for the first time in 1797, Leonardo's famous notebooks were a great revelation.

for experiments: he also built what may be considered a primitive balloon, which he successfully demonstrated to the King of Portugal in 1709. In 1764 and 1781, the German Bauer and the Frenchman Blanchard designed and tested flying machines. In earlier times it appears that balloons and other lighter-than-air vehicles were also flown in China and Cambodia. It was not, however, until 1797 that Leonardo da Vinci's codices were published for the first time, opening an incredible new window onto the science of aeronautics.

9 right Leonardo da Vinci: a self-portrait. The brilliant master, who excelled in the figurative arts, architecture and engineering, was born at Anchiano, outside Vinci in Tuscany in 1452. He died at Amboise in France in 1519.

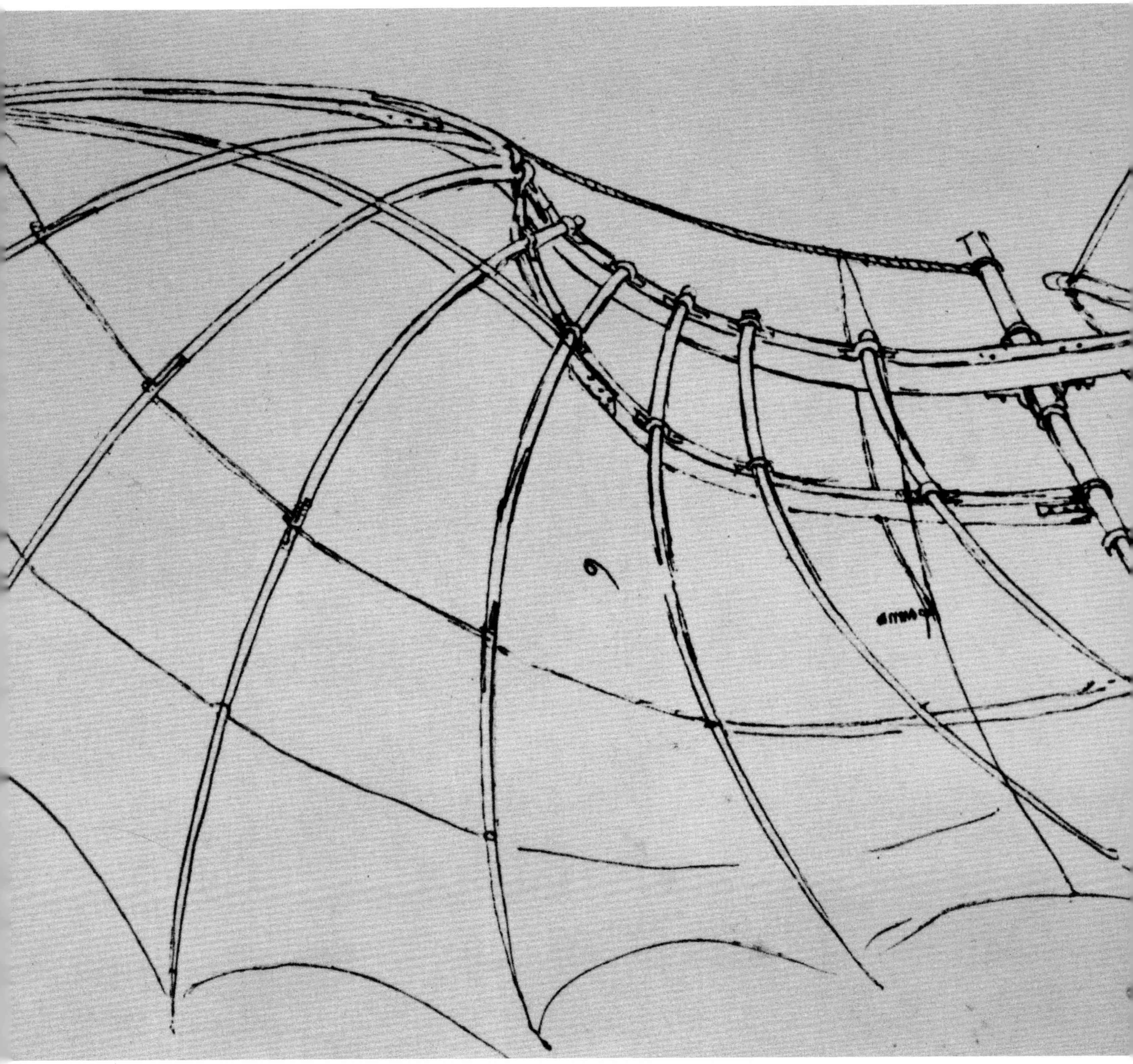

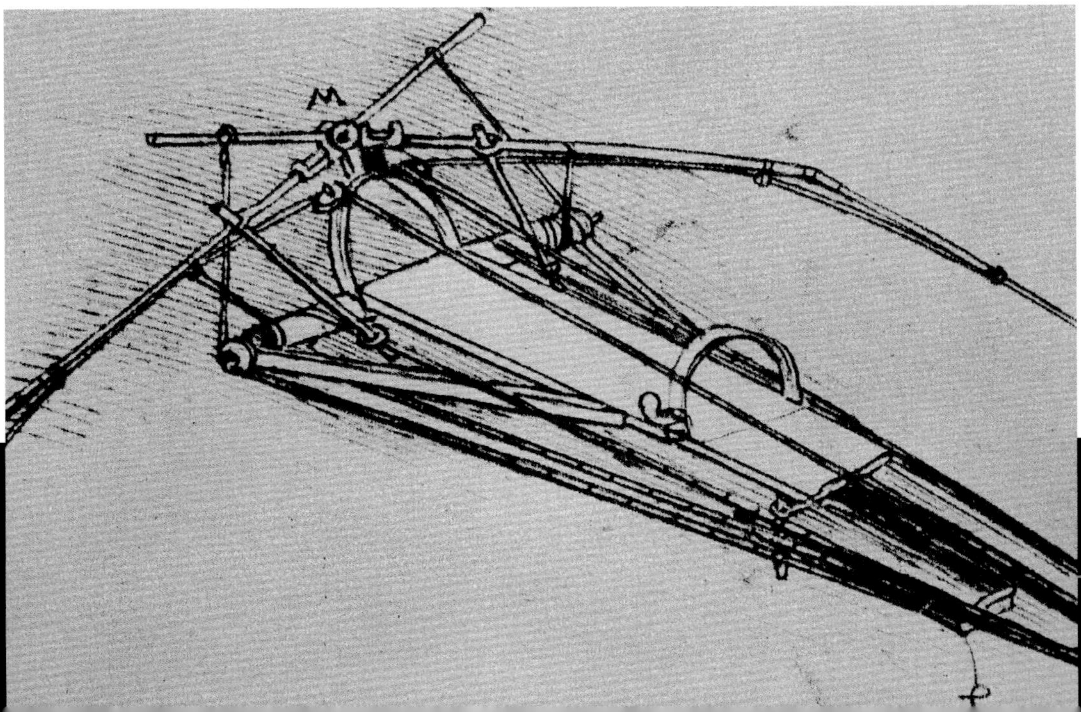

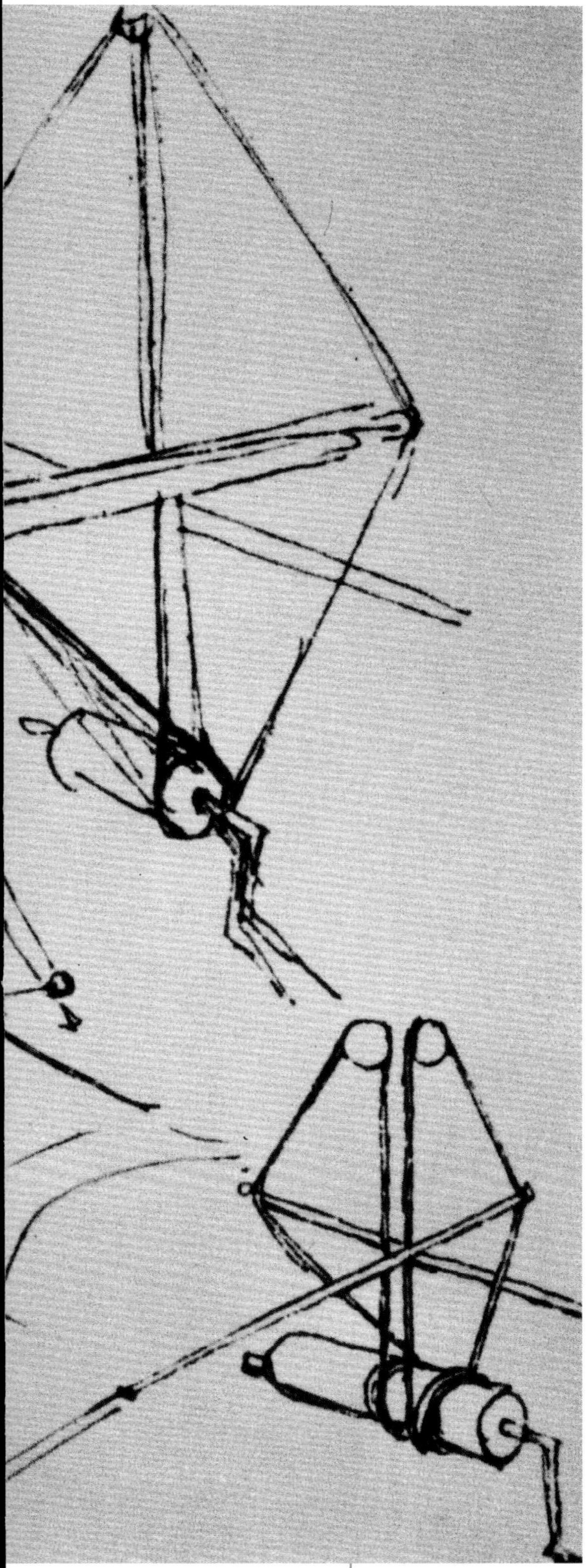

10-11 One of Leonardo's many drawings dedicated to flight: a crank-powered wing clearly inspired by the movement of a bird's wing, which it copies both in form and action.

11 bottom In this drawing, Leonardo drew what today we would call the fuselage of a flying machine, complete with cables and tie-rods for the movement of the wings.

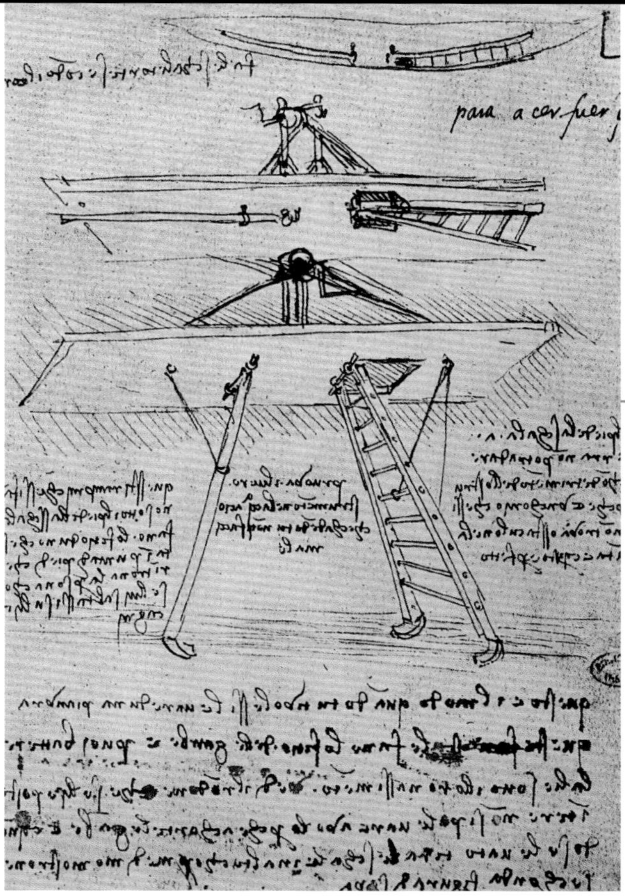

11 Two more mechanisms designed by Leonardo da Vinci. Above, a pulley-driven propulsion mechanism operated by cables. Below, a detail of the model of a flying machine with four wings powered by a man standing inside the framework. The complicated arrangement of pulleys and tie-rods that moved the wings is clearly shown.

The Tuscan genius, who was born in 1452 at Anchiano, outside Vinci, and died at the Manoir du Cloux in Amboise in the Loire region of France in 1519, was the first person known to have taken a scientific, rational approach to the problem of flight. The codices contain no fewer than 150 designs for devices, machines, or parts of machines of various types, including a parachute, the "aerial screw" (the first example of a helicopter) and the propeller. However, even Leonardo long remained anchored to the concept of imitating the flight of birds and designed the so-called ornithopter in which the weight of the pilot was to be supported by the movement of a pair of mechanical wings. The great defect in this concept was that the power required for flight was far greater than that calculated by Leonardo, who was convinced that muscles alone could supply it. Toward the end of his life, however, he recognized these limitations and intuited the need for a means of multiplying power and for a fixed wing, innovations that were to revolutionize his research. While Leonardo did not survive to put these ideas into practice, his studies and drawings reveal the exceptional clarity of his thinking and observations in all fields. Today many consider him to be the spiritual father of manned flight.

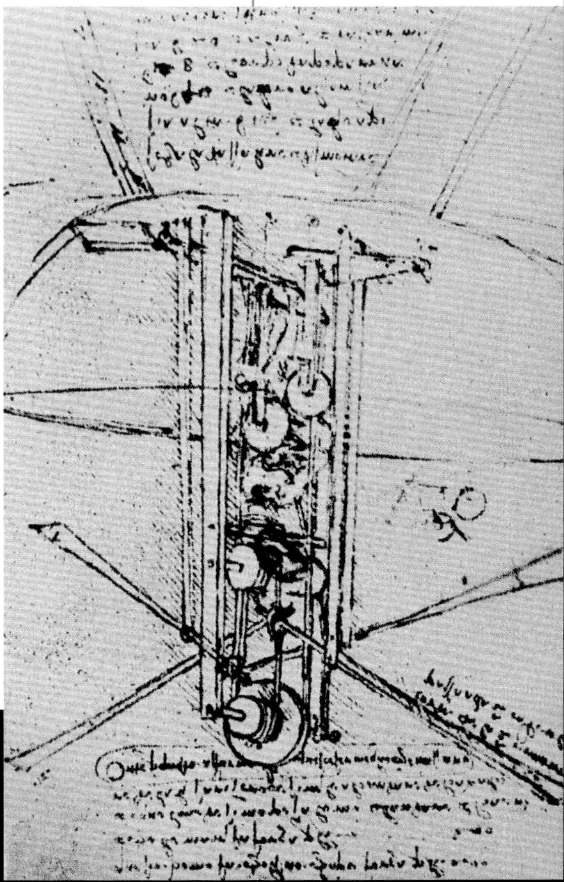

12 left A contemporary engraving of the balloon the Italian Vincenzo Lunardi used in the ascent he made on 13 May 1785.

12 right and 13 left These late 18th-century drawings show two anonymous designs: an aerostatic balloon pulled by birds, and a sail-powered balloon. Their ingenuity and the approximation to known concepts is evident. The two balloons were still based on imagination rather than on technical studies.

As is frequently the case, the first efficient solution to the problem of flight came from an unexpected source and utilized simple, natural technology, that of the balloon. The intuition that made this conquest possible came from two French brothers, Joseph Michel and Jacques Etienne Montgolfier, proprietors of a paper-mill near Lyon. Their observations concerned the lifting force of the hot air above a fire and how it was capable of lifting upturned paper bags. They set to work and in 1782 conducted a series of experiments with large, light envelopes filled with hot air. On 4 June 1783, a large balloon with a diameter of over 32 feet was constructed and tested at Versailles with a sheep, a goose and a cockerel aboard. The attempt was successful and the "passengers"

13 center This print illustrates the flight Jean-Pierre Blanchard and John Jeffries made on 7 January 1785. They crossed the English Channel (from Dover to the forest of Guines, near Andrei) in a historic "first."

13 right The Mongolfier brothers first balloon project, successfully tested at Annonay, on 5 June 1783. The balloon, which had no passengers, rose to just over 6,500 feet.

returned safely by balloon to the ground after a flight of almost two miles. The time was now ripe for the first manned flight. The historic event took place in Paris on 21 November 1783. Jean François Pilâtre de Rozier (a physicist) and François d'Arlandes (an army officer) boarded a balloon 72 feet high and 49 feet wide and were lifted into the air to great acclaim at 1:54 p.m.: man was flying!

13 bottom Portraits of the Montgolfier brothers from the Bibliothèque Nationale, Paris. Left, Etienne de Montgolfier (1745-1799); right, Joseph de Montgolfier (1740-1810).

14 top This colorful and imaginative drawing shows an anonymous project for an aerostatic balloon measuring 120 feet in diameter (Carnavalet Museum, Paris).

14 bottom The military potential of balloons was appreciated by many armies. Here we see a military balloon ascending during the siege of Paris, in the Franco-Prussian War of 1870.

The course had now been set and experimentation flourished.

The hot air balloon (known as the montgolfière in honor of its inventor) was joined on 1 December 1783 by the hydrogen balloon invented by the Frenchman, Jacques César Charles (and consequently named the charlière). He lifted off from the Tuileries in Paris and completed a flight of no less than 26.5 miles. Immense curiosity and desire flared up, and flights were made in many countries. In Italy there were those of Andreani and the Gerlis at Milan on 25 February 1784. In London that of Vincenzo Lunardi with a charlière hydrogen balloon on 15 September 1784 and at Philadelphia on 9 January 1793, that of the Frenchman Blanchard. On 5 January 1785, a passenger-carrying balloon crossed the English Channel for the first time.

In 1794 the maturity of ballooning technology was demonstrated by its adoption by the French army for artillery range-finding purposes. Its first use occurred during the battle of Fleurus in Belgium on 26 June 1794, when the French defeated the Austrian forces. Five years later, however, Napoleon Bonaparte disbanded the world's first air force, and military use of the balloon was not revived until the second half of the nineteenth century.

14-15 and 15 bottom Manufacturing a gas-tight balloon was a delicate and arduous task. The lower picture, from 1794, shows Nicolas Jacques Cont's cutting technique (1755-1805) in making a military balloon. The cutting was done outdoors, using primitive methods. Above: Cont painting the balloon, an indoor process.

16 top left The steam-powered flying machine of William Henson (1805-1888) was the first heavier-than-air craft. Henson decided to patent his 'steam-powered flying carriage' and founded the world's first air company in 1842. However, the machine never flew.

16 bottom left German engineer Otto Lilienthal (1848-1896) was one of the greatest C19th investigators of human flight, which he dreamed about from the age 15. This drawing of a glider is from a patent he took out in 1895.

The Englishman Sir George Cayley (1773-1857), who was active in the late eighteenth and early nineteenth centuries, had a great influence on the future development of aviation. As early as 1799 he had put forward the hypothesis that a flying machine could succeed in taking off only if it had separate means of ensuring lift and propulsion. This meant abandoning the idea of the ornithopter and the adoption of the heavier-than-air concept, a remarkable step at that time, given that it was highly unlikely that Cayley had had an opportunity to examine Leonardo's codices. In 1804, Cayley designed a dirigible powered by a propeller and continued further with his studies into aerodynamics. In 1809, he published his research into manned flight in a paper entitled On Aerial Navigation. In this work he codified a number of suppositions regarding the governability of an aircraft in flight and even suggested the use of internal combustion engines for propulsion. He wrote, "The whole problem [of powered flight] is confined within these limits, viz. To make a surface support a given weight by the application of power to the resistance of air." Cayley's extremely advanced ideas were not put into practice immediately, but during the nineteenth century there were numerous other attempts at flight as imaginative as they were ineffective. In 1842 another Englishman, William Samuel Henson, designed and patented an interesting machine, the Aerial Steam Carriage. It never flew, but it was a fixed-wing aircraft designed to be propelled by propellers driven by a 25-hp steam engine. Other machines were designed along similar lines but it became clear that while the concept of powered flight was undoubtedly valid, a means of propulsion more suitable than the steam engine was required along with a truly efficient wing. Until the introduction of the internal combustion engine that would

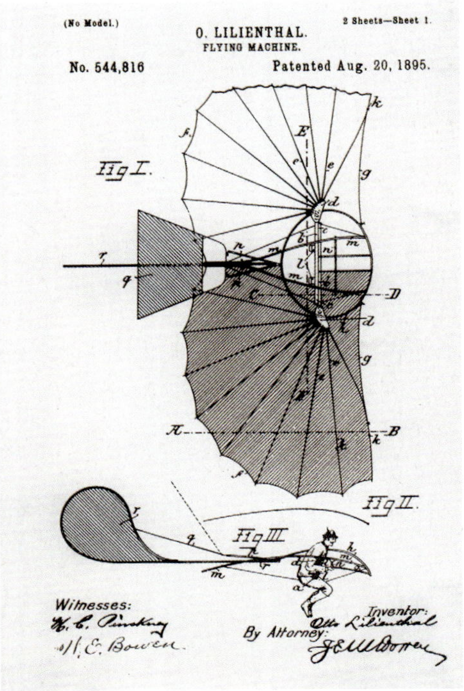

16 right The first three photographs portray Otto Lilienthal in Fliegeberg in August 1894. It is clear that he still believed in gliders that used bird-shaped wings. As shown by the fixed (no longer beating) wing and from the large photograph, the first tail units were appearing. In the last picture, Lilienthal is shown with a two-wing glider in 1896. Lilienthal was killed using one of these, after more than 2,000 flight attempts, in a glide during which his craft overturned at a height of 50 feet.

17 top Lilienthal's 1894 flight attempt.

17 center A two-wing glider built by Otto Lilienthal in 1895. His flight trials of October 1895 at Fliegeberg shows he had abandoned the primitive natural-wing form and moved toward more sophisticated aerodynamic designs, with a tail unit and dual wings.

17 bottom left George Cayley, an Englishman, made this sketch of a glider in 1804. The aircraft became a reality 49 years later when it made a short flight near Scarborough.

17 bottom right The notion of "heavier than air" – an aircraft fitted with means to provide it with lift and power – was George Cayley's.

eventually power a true aircraft, studies continued on the problem of lift.

One of the most celebrated researchers was the German Otto Lilienthal (1848-1896). At the age of fifteen he began to take an interest in flight and together with his brother built a six-and-a-half-foot ornithopter that never actually managed to take off. The determined and technically minded Lilienthal graduated in mechanical engineering and retained a passionate interest in flight, even though his work prevented him from devoting himself to research in the field. Once he had retired he published his studies in 1889 in the book The Flight of Birds as the Basis for the Art of Flying (Der Vogelflug als Grundlage des Fliegkunst) and dedicated himself to putting his theories into practice. In the absence of an appropriate engine, Lilienthal constructed gliders that made use of the force of gravity and he became the first man in world to launch himself into the air and fly with such devices. In 1892, having abandoned wings based on those of birds, he began to build wings with radial structures and then biplanes that were easier to control. On 9 August 1896, however, one of his gliders overturned almost fifty feet above the ground, and on the following day Lilienthal died of his injuries.

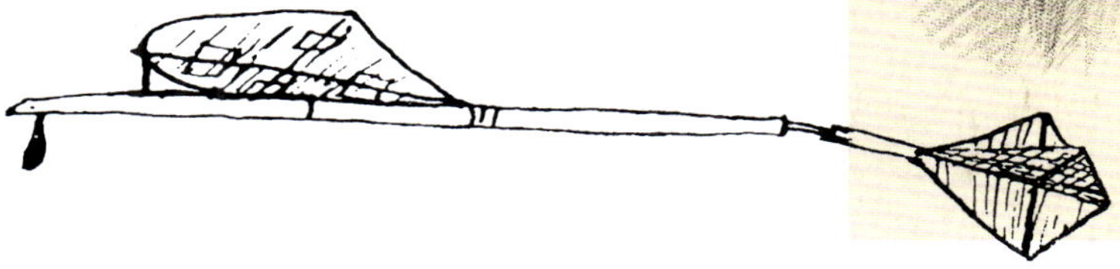

His more than two thousand glider flights were nonetheless a great inspiration for other pioneers such as Rumpler and Octave Chanute (1832-1910). The latter, born in Paris, emigrated to the United States at the age of six with his father, a man of science who had been appointed vice-president of Jefferson College. After qualifying as an engineer, Chanute embarked on what was to be a brilliant railway career, completing great civil engineering projects and earning an enviable reputation. From 1874 he began to take an interest in the problems of flight but had to abandon all serious study because it would have interfered with his professional commitments. He put his projects to one side and only revived them in 1889, when he began to complete his documentation, drawing on the experiments of Lilienthal and others. In 1894 he published the book Progress in Flying Machines, in which he reviewed all the experiments conducted to date in the field heavier-than-air craft. He then began to work on his own designs in which the lifting areas of his gliders were automatically balanced to bring the center of pressure onto the same axis as the center of gravity. He also continued Lilienthal's research into gliders and recognized the inherent dangers of his machines. Unfortunately, this recognition came just a week before Lilienthal's fatal accident: he could not be informed of the danger he faced and saved from his tragic fate. Chanute's glider designs and the publication of his works in America and France in 1897 attracted the attention of the American brothers Orville and Wilbur Wright. Within three years they had established contact with the French engineer in search of greater detail and information. Chanute thus effectively became the Wright brothers' mentor and helped them demonstrate the feasibility of controlled, powered flight in 1903. Octave Chanute lived to celebrate the Wrights' success with them, but in 1910 died in Chicago after an illness.

This review of the most important pioneers could hardly be complete without mention of the Frenchman Clément Ader (1841-1925) and the American Samuel Pierpoint Langley (1834-1906). Ader designed and built two aircraft powered by a steam engine driving a propeller, the Eole and the Avion III, while Langley constructed the Aerodrome, a monoplane with tandem wings powered by a combustion engine that drove two propellers. The Eole attempted take-off in 1890 and the Aerodrome in 1903, but both projects were unsuccessful. They marked the end of the period of improvised attempts and solutions. A new era dawned with the work of the Wright brothers.

18 left The Avion III, built by Clément Ader, was a steam-powered, twin-engined craft driven by two large four-blade propellers. Ader was not successful as a constructor but his theories of military aviation were valid, even if appreciated at the time.

18 top right Octave Chanute designed and tested several gliders, like this five-winged craft. Chanute's work inspired the Wright brothers, who turned to him for information that contributed to their own success.

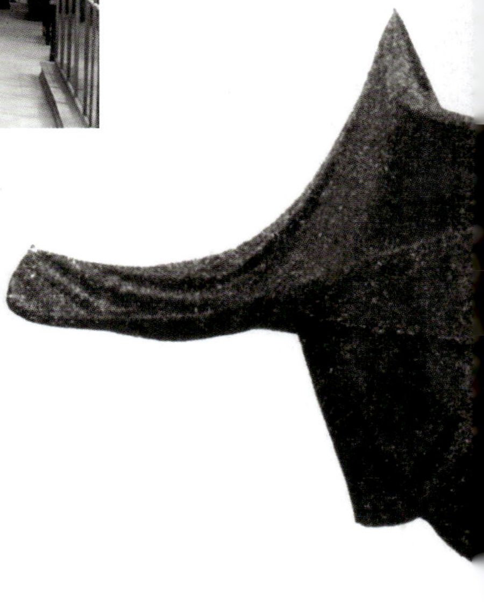

18-19 Clément Ader's Avion III, shown here with its wings folded. The plane failed at the demonstration flight before a military commission in October 1897 due to a strong wind. The French Defense Ministry rejected the design.

19 top Léon Delagrange, to the left in the photo, was one of aviation's pioneers. In the March of 1908 he became the first pilot in Europe to fly with a passenger, Henri Farman. In the May of that year he was also at the controls of the first aircraft to take off from Italian soil.

20-21 Balloons were still successfully used even at the start of the 20th century. In this photograph of 15 May 1913, a balloon is being maneuvered by the "Eclaireurs de France" at St.-Cyr, France.

20 bottom Two press pictures showing the world record for a balloon ascent, achieved by the Belgian scientist Auguste Piccard on 27 May 1931. With his colleague Paul Kipfer, Piccard reached an altitude of 51,762 feet after taking off from Augusta in Germany, and landing in the Tyrolean Alps in Austria. The record caused a large public stir even though balloons were by then an obsolete form of air transport.

21 top right In the early 1900s, balloons were still highly regarded. Shown are examples exhibited at the First International Fair of Air Navigation, held in the Grand Palais, Paris, in September 1909.

21 bottom The starting line for the Gordon Bennett Cup of 1907. Six different types of balloon wait for the start signal.

LE PETIT JOURNAL

ILLUSTRÉ

HEBDOMADAIRE - 42e Année
61, rue Lafayette, Paris

7 Juin 1931 - N° 2111
PRIX : 50 CENTIMES

LE MAGNIFIQUE EXPLOIT DU PROFESSEUR PICCARD

ILLUSTRAZIONE DEL POPOLO

Supplemento della "GAZZETTA DEL POPOLO„

Domenica 14 giugno 1931 (I.

Anno XI - Numero 24

L'ascensione del prof. Piccard. — Dopo essere rimasto per molte ore nella stratosfera, ad un'altezza non mai raggiunta dall'uomo, il pallone dell'audace scienziato è disceso sul ghiacciaio del Gurgler, nelle Alpi tirolesi, dove la navicella sferica dell'aerostato rimarrà a ricordo della memorabile impresa.

(Disegno di Aldo Molinari).

In questo numero: Confidenze di Virgilio Brocchi

CHELIN

RESS
ENC

Chapter 2

Dirigibles

The first means of aerial transport was the balloon, which since the end of the eighteenth century had been used for both civil and military purposes. The balloon alone was inadequate, being subject more to the whims of the elements, especially the wind, than to the will of its crew. However, after a few decades, progress made in propulsion devices led to the birth of a more complete and versatile aircraft, the dirigible.

The first of these craft was an "airship" powered by a small steam engine. Henri Giffard, the Frenchman who designed and constructed it, flew it for the first time on 24 September 1852. This type of engine proved to be technically unsuitable and, above all, incapable of providing sufficient power for this kind of application, just as it had done with heavier-than-air machines. It was only the advent of the internal combustion engine that allowed successful dirigibles to be developed. The No. 1 of the Brazilian Alberto Santos-Dumont (1873-1932) was the first of these craft. It made its maiden flight in Paris on 28 November 1898. The wealthy Santos-Dumont settled in France and, gripped by a passion for flying, began to construct balloons and dirigibles. He designed no fewer than 18 different examples of the latter before turning his attention to powered airplanes in 1907.

Three major categories of dirigibles were developed. They were the non-rigid type, in which the pressure of the contained gas alone shaped the external envelope; the semi-rigid type, in the pressure of the gas and a longitudinal metal keel maintained the form; and the rigid type, in which a complex metal framework guaranteed external form. This last structure was the only one suitable for large craft.

22-23 top Alberto Santos-Dumont (center) in front of one of his airships. In October 1902, he won the Deutsche de la Meurthe prize for the first aeronaut to complete the St. Cloud – Eiffel Tower – St. Cloud route in less than half an hour.

22 center The first modern airship was built by Alberto Santos-Dumont, who linked a balloon to an internal combustion engine. Shown here is the Santos-Dumont 9, with a large gondola.

22-23 bottom In 1906, Santos-Dumont moved away from constructing airships to concentrate on airplanes. Shown here is the 14bis, an airplane with which he participated in the French Grand Prix of 1906.

14
14bis

The country that did most to develop and perfect the dirigible was Germany, where in 1875 Paul Haenlein built the nation's first powered rigid example. In 1888, Kurt Wolfert built a dirigible equipped with the first Daimler-Benz aircraft engine, a 2-hp unit. The man generally considered to be the father of the dirigible, Count Ferdinand Adolf August Heinrich von Zeppelin, was at this time still a cavalry officer in the German imperial army. Born in Konstanz in 1838, von Zeppelin took part in the American Civil War and made his first balloon flight in America. He fought in the Austro-Prussian War (1866) and the Franco-Prussian War (1870), reaching the rank of general, but in 1890 abandoned his military career in order to devote himself completely to the development of lighter-than-air vehicles. In 1898 he founded the Zeppelin Company at Friedrichshaven. From the very beginning von Zeppelin tried to interest the German army in his craft but was faced with competition from two other pioneers, Major von Parseval (who proposed a non-rigid dirigible) and Captain Gross, an advocate of a semi-rigid type. These two models were adopted before the Zeppelin airships because the army considered them to be better suited to military operations as they could be easily dismantled and transported. In effect, the key to the eventual success of Zeppelin's designs was the fact that for first time large rigid dirigibles could be constructed. The Count's first airship, the LZ-1 (Luftschiff Zeppelin Eins) was no less than 420 feet long, over 38 feet wide and had a volume of about 400,000 cubic feet. The LZ-1's most serious problem was a lack of power from its two 16-hp Daimler-Benz engines, which provided a top speed of only just over 17 mph. The LZ-1 completed its maiden flight on 2 July 1900, over the Bodensee. Despite the design's evident quality, eight years passed before the German army ordered its first Zeppelins, the LZ-3 and the LZ-5. In 1909, the German navy also ordered its first four dirigibles, to be used for reconnaissance purposes. In that year Zeppelin and Hamburg-Amerika Reederei founded Delag, the first commercial airline equipped with dirigibles. On 28 June 1910, the company inaugurated its service with the LZ-7 *Deutschland*, capable of carrying 24 passengers. In Delag's four years in service prior to the

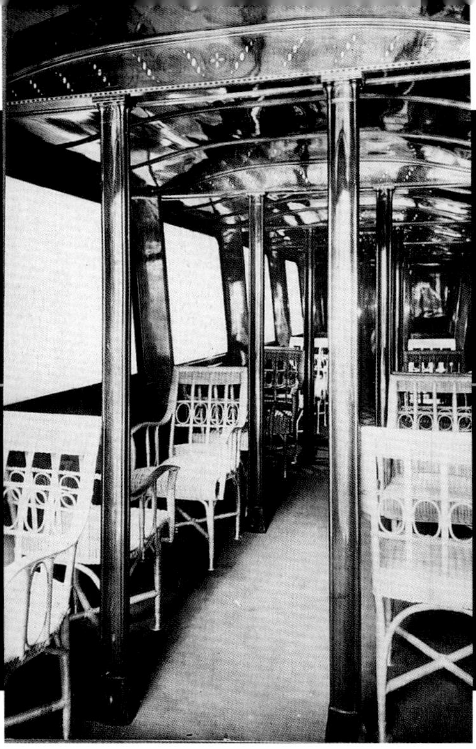

24 top and bottom Count Ferdinand von Zeppelin, a German, is generally considered to be the genius behind airships. He founded the Zeppelin company in 1898. Shown are two of his airships; the lower one is a military L2, used by the German army during the First World War.

25 top The passenger cabin of the LZ-7 Deutschland of 1910, a commercial dirigible that could carry 24 passengers. The airship was operated by the DELAG company, which continued in business until 1914.

outbreak of the Great War, the company carried over 10,000 fare-paying passengers and completed 3,193 hours aloft, covering over 100,000 miles.

When World War I broke out in August 1914, the German army was equipped with twelve dirigibles while the navy had one. Having received authorization from the Kaiser, in 1915 Zeppelins began to be used to bomb London and other targets in Great Britain. However, as the conflict wore on, anti-aircraft defenses became increasingly effective and Zeppelin losses began to become significant. On 25 September 1916, the German army's dirigible service flew its last mission over London. The German navy instead acquired no fewer than 74 dirigibles during the course of the war and continued to use them until the armistice of 1918.

25 center The gondola of the first Zeppelin accepted by the German army. This was the LZ-3, which entered service as the Z1 in 1908. In 1909, it took off from Metz and remained continuously in the air for four days.

25 bottom A large Zeppelin dirigible, surrounded by a curious crowd, landing in Constance in April 1909. Note the many control surfaces at the rear of the airship. The Zeppelin company was based in Friedrichshafen on Lake Constance.

In the post-war period, Delag resumed operations with two airships, and Count Zeppelin continued to develop ever larger and more capacious rigid dirigibles.

On September 18, 1928, the LZ-127 *Graf Zeppelin,* an immense airship 774 feet long and containing 3,920,000 cubic feet of hydrogen, made its maiden flight. The ship, which was powered by four 550-hp engines, had a top speed of 79.5 mph. The manufacturers intended that this craft should become the principal means of intercontinental transport: it was faster than existing ships and safer than the airplanes of the day, which were handicapped by severely restricted range and could carry only a third of the passengers the *Graf Zeppelin* could. In August 1929, this airship, which could carry up to 39 passengers in luxurious accommodations, inaugurated transatlantic service to North and South America. It also completed a round-the-world trip in 29 days with intermediate stops at New York, Los Angeles and Tokyo only.

26 Three photographs of the Graf Zeppelin from 1928. Designated the LZ-127, the Graf Zeppelin, the largest airship of the period, was an enormous intercontinental transport dirigible, 774 feet long and containing 3.9 million cubic feet of hydrogen. Above, the control panel; below, phases of construction building process in the company's largest hangars.

27 top The Graf Zeppelin during trials over Lake Constance. It is flanked and escorted by two Swiss army fighter planes.

27 bottom Two photographs showing construction of the Hindenburg, the largest dirigible that the Zeppelin company built in Friedrichshafen. Its maiden flight was on 1 May 1936. Left, the construction of the large metal rings of the framework; right, the airship is almost complete; it already flies the Third Reich's swastika flag on its tail unit.

28-29 The LZ-129 Hindenburg shown flying over Manhattan in 1936. This dirigible was the pride of Nazi Germany and was fitted out for intercontinental passenger transport. It was 804 feet long and the fuselage enclosed a volume of over 5.5 million cubic feet.

28 bottom left The Hindenburg's radio cabin. Thanks to radio, the Zeppelin could receive meteorological information and stay in touch with radio stations everywhere.

28 bottom right Passenger life on the Hindenburg. It had large inclined windows for observation of the countryside below, and various services and entertainments for the 36 passengers it could carry.

The largest and most famous Zeppelin dirigible was the LZ-129 *Hindenburg,* which was 804 feet long, had a top speed of 84 mph, and a range of about 9,000 miles, while passenger space remained unchanged. The LZ-129 made its maiden flight on 1 May 1936, and completed a further 63 (of which 37 were transatlantic trips) before a tragic accident while docking at Lakehurst, New Jersey, destroyed it completely on 6 May 1937. The *Hindenburg* burst into flames for reasons still unknown, and 12 of the 36 passengers and 22 of the 61 crew members were killed in a catastrophe caught on film by the media cameras. That tragedy was the swan song of the great dirigibles and marked their decline as a means of transport, a function for which even then technical advances in airplane design had made them obsolete. The last Zeppelin was the LZ-130 *Graf Zeppelin II,* which first flew in September 1939. The following year it was broken up along with the LZ-127 on the orders of the Luftwaffe commander, Hermann Goering.

29 Three pictures of the Hindenburg disaster. On 6 May 1937, the LZ-129 mysteriously caught fire as it was docking at Lakehurst, New Jersey. A huge blaze immediately followed, caught by film cameras on the airfield. In the fire, whose causes remain unknown, 12 passengers and 22 of the 61 crew members lost their lives.

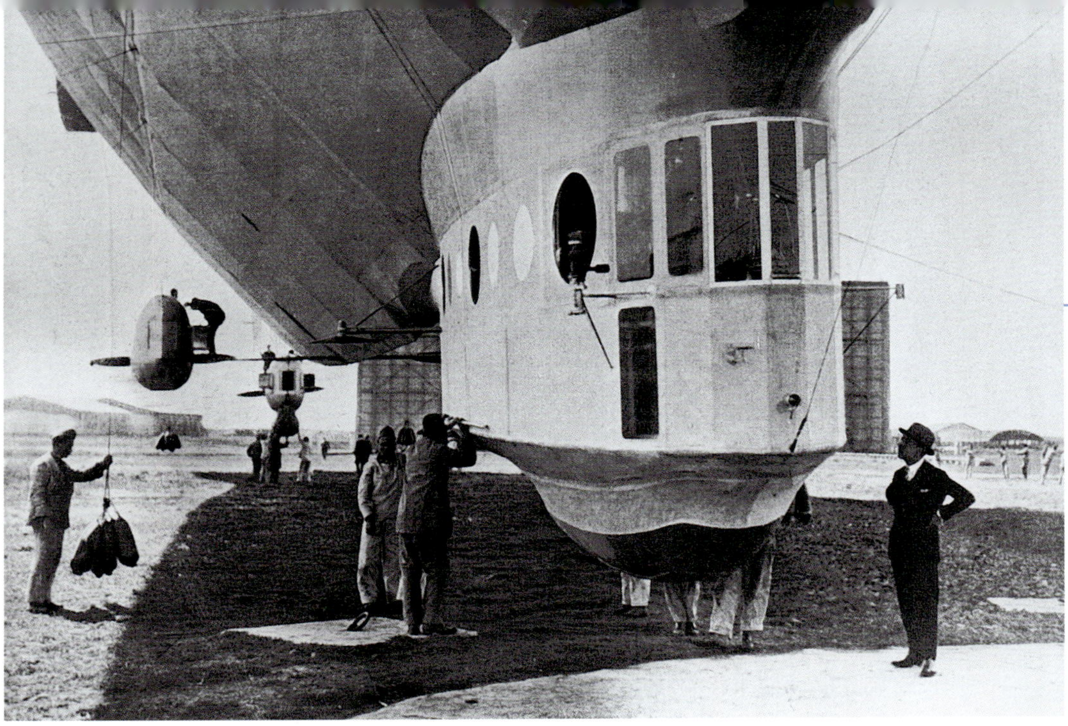

30-31 and 31 top
The Norge expedition was followed throughout the world, even though the airship was beaten by an airplane: Richard Byrd flew over the North Pole in a Fokker two days before the Norge. General Nobile with Governor Cremonesi arriving in Rome on 3 August 1926. The large photograph shows the Norge docking. It was 348 feet long and had an overall volume of 633,000 cubic feet.

30 bottom right 11 May 1926. General Nobile (left) with another member of the Norge gondola crew, shortly before taking off from King's Bay for the historic flight over the North Pole.

30 top, center and bottom left The first polar expedition by dirigible was made in 1926 by Norway. The N-1 airship, designed by Umberto Nobile, was renamed Norge for the occasion. The expedition was led by Norwegian explorer Roald Amundsen, who placed the Norge under Nobile's command. Above, the airship during the expedition; below, photographed at Pulham, Great Britain, Nobile with Crown Prince Olav of Norway, the British Air Minister, Sir Samuel Hoare, and the dirigible's crew.

In the 1920s, however, Italy played a major role among the countries involved in the development of the modern dirigible, as the work of General Umberto Nobile, an Italian Royal Airforce officer, confirms. The Italian army and navy had already been using dirigibles for a number of years when Nobile, born in the province of Avellino in 1885, was assigned in 1915 to the technical office of the Aeronautical Factory at Rome. There he contributed to various airship projects, including the Type O built for the navy and also sold abroad. In 1918 Nobile was promoted to director of the Aeronautical Factory, where he continued with his design work. In collaboration with others he developed the T34, a semi-rigid dirigible that was to remain the largest ever built in Italy. The *Roma,* as it was named, was 410 feet long, had a volume of 1.2 million cubic feet, and could raise a payload of 18.7 tons.

In 1921 the *Roma* was sold to the United States, where it aroused considerable interest. In 1923 Nobile designed his first airship, the N-1, a semi-rigid dirigible powered by three engines. It was 351 feet long, with a volume of 579,000 cubic feet. Three years later it was acquired by Norway and renamed *Norge.* Under Nobile's command and with the Norwegian explorer Roald Amundsen aboard, this airship made the first crossing of the polar ice cap, from Spitzbergen to Alaska, actually flying over the North Pole.

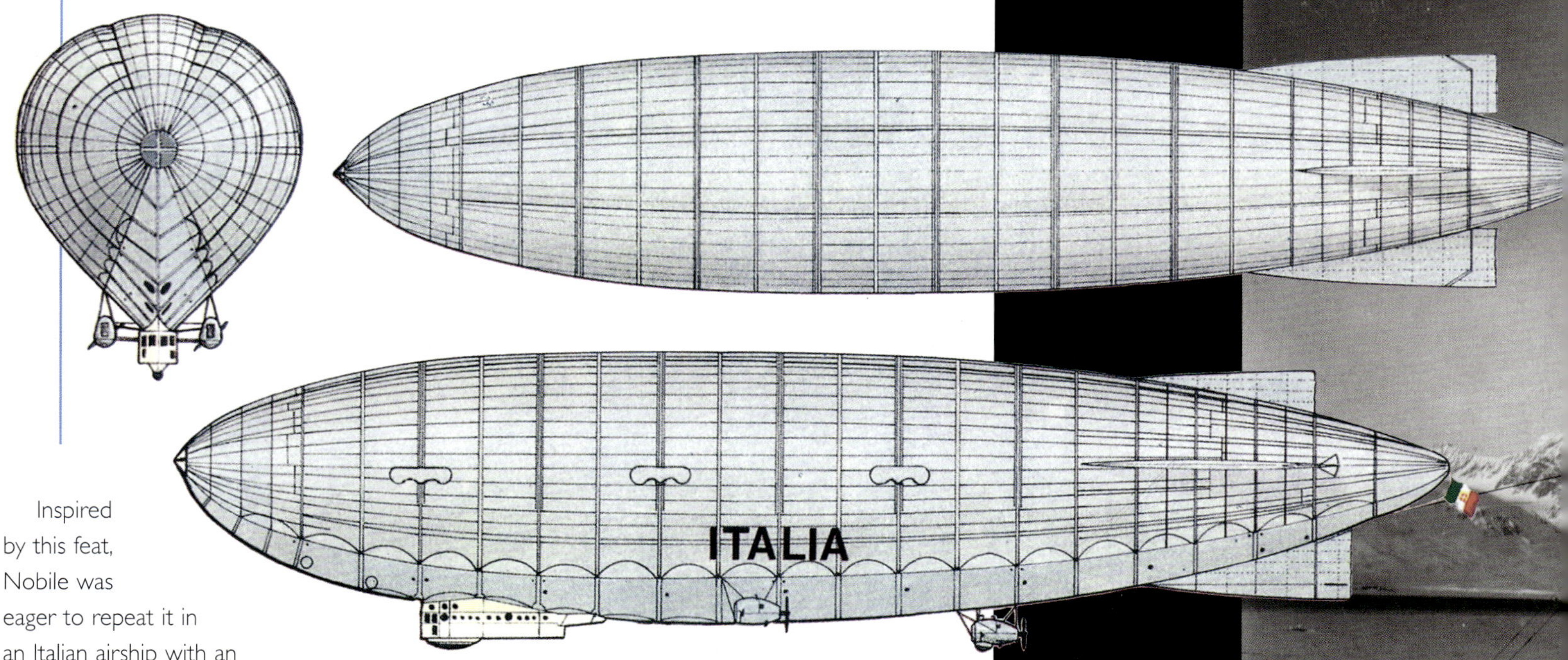

Inspired by this feat, Nobile was eager to repeat it in an Italian airship with an Italian crew, introducing the added attraction of a landing at the North Pole. The expedition received little support from the Fascist regime and its preparation was probably not as thorough as it should have been. The N-4 Italia, an airship almost identical to the Norge, departed from Milan on 15 April 1928. Technical and meteorological problems were encountered, but things seemed to have taken a turn for the better after 11 May, when the Italia began a series of explorations of the pack ice. On 23 May, the dirigible set out for the North Pole, reaching its destination at about midnight. Weather conditions were worsening, however, and the landing at the Pole was not attempted. On its return flight the Italia encountered strong headwinds and a severe storm. On 25 May, burdened by ice, the dirigible struck the pack ice, and the gondola with its nine occupants was torn from the semirigid structure. This, with the remaining six crew members, regained altitude and disappeared, never to be seen again. Six of the nine survivors, whose only shelter was provided by a small tent (the famous red tent whose color was intended to increase its visibility), held on for about a month before rescuers arrived. The other three had set out on foot, heading toward Nordaustlandet, and were eventually found, although only two survived. During the search the explorer Amundsen was killed when his seaplane was lost. The enterprise and its disastrous conclusion were followed by a drawn-out dispute, in part because Nobile had been the first to be carried to safety by the rescuing seaplane. In 1929, an embittered Nobile resigned from the Italian Royal Air Force. He moved to Moscow in 1932 and worked on Soviet dirigibles until 1936. Between 1939 and 1943 he taught aeronautical engineering in the United States, and then returned to Italy in the post-war period and was elected as a member of the Constituent Assembly. He died in Rome in 1978.

32 The dirigible Italia, used on Nobile's second expedition to the North Pole in 1928, is shown docking at Slupsk, Poland, a stop on the way. The Italia, designated N-4, was almost identical to the Norge.

32-33 The Italia shown docking in King's Bay. The airship left from Milan on 15 April 1928 and reached the North Pole on 23 May after several stops and exploratory flights over the ice-pack. Shortly afterwards, the tragedy occurred.

33 center Some members of the Italia's crew during a pause. The expedition had 15 members of which only two, the meteorologist (Finn Malmgren, Swedish) and the physicist (Franz Behounek, Czech), were not Italian.

33 bottom Holding a megaphone, General Umberto Nobile, Italy's greatest expert on dirigibles, directs the Italia during an exercise before departure.

Another country had been conducting interesting experiments with dirigibles, transforming them into flying aircraft carriers. In the early 1930s, the US Navy had two airships, the USS *Akron* and the USS *Macon*. The latter had a top speed of 84 mph and could carry four Curtiss Sparrowhawk monoplane fighters slung below its belly. The fighters could be launched and recovered in flight. Unfortunately, the Akron crashed into the sea in 1933, as did the *Macon* in February 1935, following a structural failure.

However, the 1930s saw the decline of the dirigibles as airplanes surpassed them in speed, robustness and reliability. The tragedies of the *Italia* and the *Hindenburg* only accelerated the end. During the Second World War, the Goodyear Company constructed a series of semi-rigid dirigibles that proved their worth in anti-submarine reconnaissance work.

In the post-war period, the United States and the Soviet Union maintained a number of airships in service with secondary maritime reconnaissance duties. In more recent times the only civil dirigibles

34 top The US Navy was also interested in dirigibles and at the start of the 1930s introduced two that could carry aircraft. The photograph shows the USS Macon inside its hangar, with the crew lined up in front of the gondola.

35 bottom A Curtiss Sparrowhawk in its dock on the airship USS Macon. This type of airship could accommodate five such planes, and launch and reattach them in flight.

were the Goodyear craft used for publicity purposes in the United States and Europe.

The advent of new technologies (such as electronics and composite materials) brought dirigibles back into the limelight in the 1990s. It has become clear that the dirigible is still a valid option for touristic, surveillance and advertising purposes, as well as for camera platforms and for lifting bulky loads. The year 1993 saw the foundation of Zeppelin Luftschifftechnik GmbH, a company dedicated to the development of semi-rigid airships using non-inflammable helium gas. The new NT series dirigibles have overcome the primary problem of safety during take-off and landing thanks to computer technology and the adoption of three 200-hp piston engines with tri-blade propellers, two swiveling through 90° and one at the tail that can be used to control yawing. On 18 September 1997, the Zeppelin NT LZ N-07 flew for the first time, and the first production model was completed in 2000. Other companies such as the German Cargolifter, the American ABC and the British ATG are also currently working on the design of new dirigibles.

34 bottom A spectacular formation of 11 US Navy dirigibles planes in flight shortly after the outbreak of World War II.

35 top The USS Macon over San Francisco Bay. This dirigible could transport four Curtiss Sparrowhawk fighters attached to its belly. It was lost at sea in February 1935.

35 center The great difference in size between the intercontinental Graf Zeppelin and a small American Goodyear dirigible is clear from this late-1920s photograph.

Chapter 3

Heavier-than-Air Machines

By the end of the nineteenth century, man's dream of flight had in a certain sense been realized, thanks to balloons and dirigibles. However, this was not the kind of flying that inspired and truly interested aviation pioneers. What they really wanted was to fly controllable aircraft equipped with wings and an engine; to this end, research carried out with gliders proved to be invaluable and finally made it possible to fit them with a power unit suitable for aeronautical purposes. This was the air-cooled internal combustion engine. Its inventor was the German engineer Nicklaus Otto, who in 1877 created a four-stroke thermodynamic cycle, now known as the Otto cycle. Gottlieb Daimler further developed this concept, building the first high-speed (800 rpm) fixed engine in 1883. The internal combustion engine proved to be suitable for automobiles and subsequently for aviation, thanks primarily to progress made in ignition. This advanced first from glow plugs (used until circa 1900), to coil ignition, and then to high-tension electrical ignition using magnetos, which Robert August Bosch introduced in 1902.

The American brothers Wilbur and Orville Wright were the first men to successfully combine a suitable airframe and an internal combustion engine to create a controllable airplane. Wilbur was born in Millville, Indiana in 1867 and Orville in Dayton, Ohio in 1871. Their father, a Protestant pastor, one day gave them a toy helicopter powered by elastic. For the two brothers, then eleven and seven years old, this was a revelation that pushed them toward the worlds of mechanics and flight. After opening a bicycle repair shop, in 1896 the pair became inspired by the work of Otto Lilienthal (he died that year) and attempted to emulate his experiments with gliders.

They considered these to be an intermediate step on the path toward creation of a powered airplane. The brothers began by

36-37 The Wright brothers performed many experiments with gliders. In this photograph from 1903, Orville Wright makes a heavy landing while his brother Wilbur runs to help him.

36 bottom Wilbur Wright at the controls of The Flyer, the first heavier-than-air plane to fly successfully with an internal combustion engine. The date: 17 December 1903.

building a working model, a biplane controlled from the ground via four cables that warped the wings. Then, in 1900, thanks in part to the advice and friendship of Octave Chanute, they began to build and test small gliders. The results were somewhat disappointing, however, and encouraged the Wrights to revise their theories about flight. They built a kind of wind tunnel at their workshop in Dayton, coupling a two-yard long, square-section tube with a fan. Using this rudimentary structure they managed to design and create efficient airfoil shapes and in 1902 they built a glider with satisfactory flight characteristics. Their next step was the design of the first powered airplane.

The first hurdle to be overcome was finding a suitable engine. After fruitless investigation of the automobile industry, the Wrights turned to a mechanic, Charles Taylor, and together with him they designed an in-line four-cylinder internal combustion engine. It produced a maximum power output of 12 hp. Centrally mounted, this engine drove two contra-rotating pusher propellers via a dual chain transmission system. The airframe was a biplane structure with canard-type forward stabilizers, a wingspan of 40 feet, a length of 21 feet, and a lifting area of 510 square feet. The brothers named it the "Flyer."

37 top left and right Two pictures of the Wrights' experiments with gliders. Left, a photo of Wilbur Wright at the controls of the Wright No.1 in October 1902 during an attempt at Kill Devil Hills, near Kitty Hawk, North Carolina. Right, a flight by Orville Wright on the same site, in a more sophisticated machine.

37 center and bottom Two pictures from the start of the century showing attempts – often naive and approximate – by the first pioneers of motor-powered flight. Center, a twin-engined Pean monoplane. Bottom, a curious seven-winged aircraft with an oval frame. Neither were successful.

38 top During the early 1900s, the Wrights' Flyer remained a reference point for the rest of the nascent world of aviation. Here the airplane is on Auvours airfield in France for a demonstration flight in September 1908.

38 bottom The US Army was also interested in The Flyer. In 1908 Orville Wright made a series of demonstrations at Fort Meyer, Virginia and, the following year, the Model A Flyer entered service as the world's first military airplane.

38-39 In 1908-1909, Wilbur Wright took The Flyer to Europe for a demonstration and promotional tour. Here the airplane is in France during a demonstration flight organized to mark the visit of the king of Spain.

39 bottom The amphibious version of The Flyer! Wilbur Wright is shown during a flight over New York Bay. Below the craft, a canoe was attached in place of landing gear in the hope of saving the plane should it come down in the water.

The assembly work was completed in December and, impatient to test the airplane despite the cold weather, the Wrights decided to make an initial attempt on 14 December 1903, close to the beaches at Kitty Hawk in North Carolina. This test was unsuccessful: immediately after take-off the Flyer, piloted by Wilbur, and still equipped with skids, stalled and crashed to the ground.

Three days later, on 17 December, events took a different course. Having been repaired, the Flyer was taken back to Kitty Hawk and set up on the launch rails. At the controls this time was Orville. At 10:35 that historic morning, the Flyer took off at a speed of 10 mph, immortalized by a photographer present for the occasion. The airplane flew for just 12 seconds and covered just under 120 feet, but the two brothers completed further flights that day, the longest of which lasted all of 59 seconds.

After this success, the two brothers quietly set to work once again to improve their airplane, trying to maintain a veil of secrecy in order to obtain maximum benefit from their efforts. That first model was followed by the Flyer II, powered by a 15-hp engine and then, in 1915, by the Flyer III, which on 5 October proved capable of flying for over 39 minutes and covering a distance of over 23 miles. However, news of the Wrights' work had spread, in part thanks to the enthusiastic Chanute, and eventually they received a French offer of 500,000 francs to provide a two-hour demonstration flight. The brothers then divided their forces, with Orville conducting a series of demonstrations for the American army and Wilbur travelling to France in June 1908. Following the success of his demonstration flights, Wilbur founded the world's first flying school at Pau, near Bordeaux. The Wrights then continued with their European tour and presented the Flyer, first in Rome in March 1909, and then in Germany. Late in 1909 they founded the Wright Company with associated firms in France, Germany and England, for the production of their flying machine. That same year, the Flyer Model A, the first aircraft of the series, was sold to the American army and thus became the world's first military airplane.

Wilbur died in 1912, a victim of typhoid, and Orville gradually reduced his own activities as an aircraft manufacturer. In 1915 he sold his majority holding in the company in order to devote himself to research. He died at Dayton in 1948.

The story of the Wright brothers' success and photos of their airplane gave renewed impetus to aeronautical experimentation in Europe, and a number of French pioneers, including Ferdinand Ferber, Robert Esnault-Pelterie and Gabriel Voisin, began testing crude copies of the Flyer.

40 top left and right Alberto Santos-Dumont, the French pioneer. In 1906, already famous for his experiments on dirigibles, he turned his attention to powered planes. His first aircraft was the 14bis, a canard-type biplane that succeeded in flying 650 feet at a height of 20 feet on 12 November 1906 at the French Aero Club's Grand Prix. This was thus the first official flight in Europe.

40 center right and bottom Two crashes by the Frenchman Louis Blériot with planes of his own construction. His repeated accidents led to him being nicknamed "the man who always falls."

Alberto Santos-Dumont, who was already a pioneer aviator with dirigibles, made the first controlled powered flight in Europe. He built his "14 bis", a bulky canard biplane (an aircraft fitted with lifting surfaces on the nose in front of the main wings) equipped with a 24-hp Levavasseur-Antoinette engine in 1906. On 12 November, after a number of attempts that covered a dozen meters, the Brazilian participated in the Grand Prix de France, organized by the Aéro Club of France, with a prize for the first flight of over 100 meters (327 feet). Santos-Dumont, who had installed a 50-hp engine, covered over 650 feet in 21 seconds, without the aid of any auxiliary launching system. His most famous model was, however, his "19," known as the "Demoiselle," which was fitted with a 20-hp engine. A number of these planes were built and sold. The future French ace Roland Garros was one of those who learnt to fly in this machine.

Louis Blériot was unquestionably one of the leading figures in the pioneering phase of European aviation. Blériot studied engineering and then devoted himself to the construction of headlamps for automobiles, setting up a workshop at Neuilly-sur-Seine. His success in this business provided him with economic security, but in 1900 he began to turn his attention to the world of aviation. All that he initially managed to produce, was an ornithopter with beating wings, which clearly proved to be a total fiasco. Blériot had already abandoned his dreams of flight when he met Gabriel Voisin, a pioneer who had already built a gliding floatplane. Blériot's interest in aviation was rekindled and he decided to found Blériot-Voisin with his new friend in order to build a new floating glider. Results were not all that had been hoped for, and in 1906 the two went their separate ways. Blériot was determined to proceed with his experimentation alone but one of his canard monoplanes crashed in 1907. Perhaps aware of his limitations as a designer, Blériot entrusted the construction of the Blériot VI, a new monoplane with conventional wings and a 25-hp Antoinette engine, to his

41 top In July 1909, Louis Blériot, took part in a competition sponsored by the English newspaper, The Daily Mail, to be the first man to fly across the English Channel. He won with a flight of 32 minutes from Calais to Dover. Left, the pioneer poses by his plane; right, as winner, he is acclaimed by the exultant crowd in Paris.

41 bottom Blériot and his airplane, a Blériot XI with a 28-hp engine, photographed from a boat in the English Channel during his flight from Calais to Dover. In being the first to make a powered crossing, he won the prize of 25,000 gold francs offered by The Daily Mail.

chief engineer, Louis Peyret. This time the design was successful and the follow-up Blériot VI bis was capable of achieving 50 mph and an altitude of over 80 feet, exceptional for the time. However, the Frenchman's pioneering flights all too frequently ended with crashes, earning him in 1907 the nickname of the "man who always falls." These failures were also exhausting Blériot's financial resources and, drawing on all his painfully acquired experience, he invested his remaining fortune in the last of his creations, the Blériot XI. This sleek monoplane was equipped with a front-mounted 28-hp REP engine with a tractor propeller. It first flew in January 1909. In July, after a number of triumphant appearances at various aeronautical events, Blériot registered for a race sponsored by the British newspaper The Daily Mail, which offered a prize of 25,000 francs to the first pilot to cross the English Channel in an airplane. Louis Blériot lifted off from Calais on 25 July in his Model XI and 32 minutes later reached Dover to general acclaim. This success brought the pioneer aviator glory and the financial rewards he needed to save his business. After yet another flying accident a few months later, Blériot hung up his pilot's goggles to devote himself to his activities as an aircraft manufacturer. By 1913 his company had already sold over 800 civil and military aircraft, a quite remarkable success story for that era. Blériot was also the founder in 1914 of the Société pour l'Aviation et ses Détivés, the famous SPAD which went on to produce thousands of fighter, spotter and training aircraft for the Allied air forces. Blériot's fortunes began to decline in the post-war period. He continued to produce aircraft through to 1930s, when he had to close his factories because of a financial crisis, before his death in 1936.

42 top Pioneers of French aviation included the Voisin and Farman brothers. Shown is the Voisin-Farman No.1, designed by the Voisins for Henri Farman, and which flew in 1907. This was a canard biplane with a 50-hp pusher motor.

42 bottom In 1908, Henri Farman founded the Farman Aircraft Company and began to build his own planes. Here he is at Reims on 28 August 1909, the day on which, in one of his own planes, he set a new distance record of 112 miles in 3 hours 40 minutes.

Other pioneers included the brothers Charles and Gabriel Voisin, who early in the century had built box-section biplanes and gliders equipped with floats. In 1906 the two founded the Frères Voisin company and began constructing two fairly similar types of aircraft with tailplanes and pusher propellers. These machines were failures, however. Greater success came the way of the design built for Henri Farman in 1907. The result was the

Voisin-Farman I, equipped with a 50-hp Antoinette engine and structurally similar to the other Voisin planes. Henri Farman, the elder of two brothers famous for their sporting achievements with bicycles and automobiles, was a flying enthusiast. As soon as he had mastered the controls of his new airplane, he decided to participate in the Grand Prix d'Aviation, a trophy instituted in 1904 that was to be awarded to the first aviator to complete a closed-circuit flight of a kilometer (just over half a mile). Despite his lack of aviation experience, Henri made a successful attempt on 13 January 1908 at Issy-les-Moulineaux. Encouraged by this achievement, the Frenchman decided to tackle the flying business seriously and devoted himself to improving his machine. The Farman Aircraft Company was thus born in a hangar at Issy-les-Moulineaux in 1908. Henri Farman then made the first city-to-city flight: between Mourmelon and Reims, at an average speed of 46.6 mph.

In 1991, in light of this success in developing aircraft, Henri joined forces with his brother Maurice and established a new factory at Billancourt. The firm prospered thanks to the outbreak of the First World War, during which it supplied land- and seaplanes to the French air force and those of many of the allied countries. Output at the Billancourt factory, which had been extended to a total of over 950,000 square feet, reached the remarkable rate of 300 aircraft per month. Farman managed to adapt to the post-war economy by converting to the production of passenger planes, beginning with the Goliath (1919), and by founding Lignes Farman as a passenger-carrying airline. In 1936 the company was nationalized and the two brothers retired to private life. Henri died in 1958, while Maurice, who had continued to fly light airplanes until the age of 85, died in 1964.

43 top left Henri Farman at the controls of one of his planes.

43 top right A Voisin biplane without canvas lining and fitted with a forward machine-gun, taken during a display in 1910.

43 bottom On 13 January 1908, Henri Farman was the first person to fly a kilometer (1083 yards). Here he is at the finish line.

Another famous French pioneer of the era was Léon Levavasseur. After having worked on a number of experimental projects, he managed to get his first true airplane into the air on 9 October 1908. In February1909 he introduced an improved design, the Antoinette IV, a monoplane with a large wing area and featuring ailerons, a triangular fuselage with cruciform tailplanes and a front-mounted engine with a tractor propeller. The company produced its own engines which, as noted earlier, were also used by other pioneers. The British pilot Hubert Latham used the Antoinette IV for his attempt to make the first cross-channel flight. In the event, because of an engine failure on 19 July 1909, Louis Blériot preceded him in this endeavor. With the improved 60-hp Antoinette VII, however, Latham collected a number of honors. These included victory in the altitude competition at the Grande Semaine d'Aviation de la Champagne, held at Reims. There Latham reached just over 500 feet, and won second place in the speed event, recording 42.8 mph. Between 1909 and 1911, the Antoinette monoplanes that Latham and others aviators flew consistently finished in the lead in every aviation competition. In 1912, however, the company ceased production and closed down.

Italy too had its pioneers, among whom Gianni Caproni, the founder of a company that was to achieve international status, was of particular note. Born in Massone d'Arco near Trento in 1886, Caproni attended the Munich polytechnic university and graduated in engineering in 1907. He shortly began to take an interest in aviation, designing and constructing his first airplane, the Ca. 1, a biplane that he completed in 1910 in the face of considerable economic difficulties. After settling in the moorland area of Malpensa, Caproni began flight-testing his machine: it proved to be unsuccessful. Displaying great determination, the young engineer founded the Caproni-De Agonstini Company at Vizzola Ticino, the result of a partnership with Gherado Baraggiola and the engineer Agostino De Agostini from Bergamo. The first aircraft produced by the new firm was a monoplane, the Cm.1, which successfully flew on 15 June 1911. Together with his aircraft manufacturing business, Caproni also founded a successful flying school at Vizzola. He then suffered a less felicitous period during which he dissolved the partnership with De Agostini, formed another company with Carlo Comitti and then another with Luigi Faccanoni. Caproni's main problems derived from the Italian army's failure to purchase his aircraft

and its effective removal of one of his sources of income when it founded its own flying schools. Having sold his factories to the Italian state in 1913, Caproni continued to work on his designs. A large three-engined bomber (which initially re-used the Ca.1 name but changed to Ca. 31 in the post-war period) aroused particular interest. This led to the first military orders from Italy and France. In January 1915, Gianni Caproni founded a new company with factories at Vizzola and Taliedo, near Milan. The firm enjoyed prosperity until 1945. During this period, the company developed 175 aircraft, building many of them in series and exporting them throughout the world. Following the disasters of the Second World War, Gianni Caproni unsuccessfully attempted to convert his factories to produce automobiles. He died in Rome in 1957, but his company was then revitalized, if only briefly, thanks to the acquisition of Avimilano, from which it inherited the successful Calif glider series. The Agusta Group finally absorbed the Caproni firm.

44 A striking image of Hubert Hubert Latham in his Antoinette IV during one of his attempts to win The Daily Mail's prize for being the first to fly across the English Channel. Latham twice finished in the sea.

45 top Hubert Latham's Antoinette IV being prepared for flight in July 1909, before one of his attempts to cross the Channel.

45 center Hubert Latham at the controls of his Antoinette IV. Eventually Latham was beaten by Blériot, but only by two days.

45 bottom The Italian pioneer, Gianni Caproni, shown beside his first aircraft, the Ca.1 biplane, built in 1910 and tested over the Malpensa Marshes, in the province of Varese.

Aeronautical progress was by no means a European prerogative, however. In the United States, the Wright brothers were not the only pioneers working in the field. One of the most talented and best known was undoubtedly Glenn Hammond Curtiss. He was born in Hammondsport, New York, in 1878. His father died when Curtiss was just five years old, and he was obliged to work from adolescence in order to help his family. When Curtiss was twenty, he and his partner Tank Waters established their own cycle repair business. As the company developed and began to build its own bicycles, Curtiss was able to give full rein to his passion for engines, and in 1902 he built his first powered bicycle. This was followed by a number of motorcycling victories and records, and Curtiss received an order for an engine that was to be installed in Thomas Scott Baldwin's dirigible, California Arrow. The success of this 5-hp twin-cylinder engine led to the founding of the Curtiss Manufacturing Company in 1905. That same year Graham Bell, the inventor of the telephone, contacted Curtiss to agree upon terms for the development of a joint aircraft-manufacturing project. Curtiss accepted Bell's offer, although not until 1907, and became the chief designer and engine expert for Bell's AEA company. After the construction of five planes, the last of which proved incapable of take-off, Bell lost interest in aeronautical research and the company closed its doors. Curtiss took over the on-going projects and in June 1909, with Augustus Herring he founded the Herring-Curtiss Company, the United States' first aeronautical manufacturing concern. The new company began to prosper, partly because Curtiss threw himself into new feats of aviation. These led to the winning no less than three Scientific American magazine trophies that brought fame and prestige to the company.

Early in 1911, one of Curtiss's pupils and clients, Eugene Fly, made the first deck landing and take-off from a ship, the cruiser USS Pennsylvania. Curtiss thus began to take an interest in the requirements of the armed forces and above all those of the US Navy, devoting himself to the development of seaplanes. In 1912, the company presented its Model F, featuring a central float. The US Navy ordered 150 of the Model F aircraft. At the outbreak of the Great War in 1914, Curtiss was also working for the US Army, developing a biplane trainer, the soon to be legendary JN or "Jenny," of which over 6,000 were eventually produced and used until the end of the 1920s. After the war, Curtiss concentrated primarily on racing planes, producing excellent designs for a number of years. In the late 1920s, exhausted after giving his all to his work, he retired to Florida, where he died in 1930.

In just ten years from the Wright brothers' first powered hop aerodynamics and aircraft engine design had progressed so much that aircraft could now be controlled easily and even to carry considerable loads, frequently passengers. During the Italian-Turkish war fought in Libya in 1911, the airplane was used as an offensive weapon for the first time, with Italian aviators dropping a number of bombs by hand on enemy troops. The psychological effect was more significant than the physical damage caused, but the new practice showed that the time was now ripe for the airplane to become a protagonist on the world stage. It was, however, no longer audacious sporting feats but rather the First World War that catalyzed aeronautical development.

46 One of the most brilliant American pioneers was Glenn Hammond Curtiss, who began as a bicycle repairman in 1898. Here he is at the command of his biplane, Gold Bug.

46-47 and 47 top Two pictures of the Curtiss biplane, June Bug, piloted by Glenn Curtiss himself. The Herring-Curtiss Company – the first aeronautical company in the US – was founded in June 1909. In the years that followed, the company specialized in hydroplanes and became the supplier to the US Navy in 1912.

Chapter 4

The First World War

The uncertain, courageous, faltering and fascinating world of aviation had made considerable progress in the ten years following the Wright brothers' first flight. While giant leaps had yet to be made, the airplanes of 1913, no matter how rudimentary, already boasted fundamental features such as internal combustion engines driving propellers, one or more sets of wings providing lift, stabilizers, pilot-actuated moving control surfaces, and a fuselage capable of accommodating the crew. The factor that imposed a revolution in the design and production of airplanes was the demands of a war such as the world had never seen before. With the First World War, in fact, aviation developed at a truly remarkable rate.

Even prior to 1914, powered aircraft had already been used in conflicts such as the Italian-Turkish War of 1911, the Mexican Revolution led by Pancho Villa in 1912 and the Balkan War of 1913. However, these were isolated operations conducted in the main with civil and sporting aircraft perfunctorily adapted for military use. The armed forces themselves had only vague ideas regarding the use of aircraft in any sphere other than that of observation and reconnaissance.

A number of farsighted precursors had, it is true, already envisioned the potential of air power. An Italian army officer, Giulio Douhet, laid out very clear ideas on the military use of aircraft: "The sky is about to become a new battlefield, one no less important those of land and sea, he wrote in 1909. In 1911, Bertram Dickson, an English officer, also predicted that aircraft would assume important reconnaissance and combat duties in order to secure control of air space. Other trailblazers included the Frenchman Ferdinand Feber, the American Billy Mitchell and the German Helmuth von Moltke. Rather than receive the attention they deserved, they all came up against the traditionalism of their military superiors and were scorned or even punished. Within a few years, however, history was to prove them right.

48 left The use of aircraft in war demanded the development of counter-measures. Here self-transporting launchers belonging to the French army are used to bring down a Zeppelin airship on 21 February 1916.

48 bottom Anti-aircraft defenses came into being with cannon and machine-guns; the photograph shows some German machine-gunners preparing a placement on the Western Front.

48 right Any kind of defense was used to bring down enemy planes. Shown here is a soldier in the British cavalry firing a portable machine-gun against a German airplane in summer 1916.

49 top In the early months of World War I, the planes in service were still fairly primitive and with little equipment, like this French Morane-Saulnier.

49 center At first fighters did not exist. To bring down enemy planes, various solutions were used, like this French two-seater with a machine-gun raised to avoid firing into the propeller.

49 bottom left Logistical and technical mobility began to assume importance: this truck was fitted out for the unusual task of transporting disassembled airplanes.

49 bottom right This photomontage shows the new war in the skies as it was during World War I.

The Great War of 1914-1918 revolutionized the world of aviation, bringing not only sweeping technological changes, but also operational innovations that led to the development of specialist aircraft such as fighters, bombers, and photo-reconnaissance planes. Entirely new military services, the air forces were created to take advantage of the new technology.

Early in the war, most planes in service had been derived from sporting models and were unarmed. The French, whose army air force was composed of over 30 squadrons, flew diverse types such as the Blériot XI, the Morane-Saulnier H, the Maurice Farman MF.7 and MF.11 and the Caudron G.3. The British Army Air Corps and the Royal Naval Air Service also used these aircraft, together with others built in Britain such as the Sopwith Tabloid, the Bristol Boxkite, the Royal Aircraft Factory BE.2 and the Avro 504. The French and British forces boasted a combined total of around 250 front-line aircraft.

For their part, the Germans also had almost 250 aircraft, mostly Aviatik B.I and B.II and Albatros B.II biplanes and Taube LE-3 monoplanes. No army had made plans for the use of aircraft in offensive roles, and the first such actions took the form of isolated individual initiatives: on 13 August 1914, Lieutenant Franz von Hiddeson hand-dropped two grenades on Paris from his Taube.

50 top With the advent of military aircraft, the first terrible bombing raids began. This is the Belgian village of Poperinge in May 1916.

50-51 The crew of a Russian bomber demonstrate the commonest bombing system of the era: a second crew member lines up the target through the sight and releases the bombs by hand.

51 bottom left Industry had to adapt to the new war requirements in a hurry. Here at the Ansaldo factory, static resistance tests are being carried out on a complete wing structure.

50 bottom Spring 1916: workers in a French munitions factory preparing a series of plane bombs.

51 bottom right The French inventor, G. Fabre, shows the bomb-launching device that enabled greater precision in aerial bombardment.

52 top Wide use was made of observation balloons during World War I. This 1916 photograph shows a French army balloon being prepared for launching during the battle of the Marne.

52 bottom Another French army observation balloon is prepared for launching. The observers in the balloon performed a very risky task.

53 top left An observer in the basket of a controllable balloon (known as a "sausage") telephones enemy positions to the ground.

From the first weeks of the war there were encounters between enemy aircraft in the skies over the front, but many pilots never even considered the possibility that they might fight against each other. Neither did the respective chains of command; in fact, pilots were frequently prohibited from carrying and using firearms on aircraft for safety reasons. The first air-to-air conflicts were fought with hunting rifles, shotguns and pistols, almost as if they were duels on horseback.

At first, the pilots acted on their own initiative. There were those who decided to try to ram enemy aircraft at the risk of losing their own planes and their own lives. Some dropped hand-grenades or attached hooks to their plane with which they tried to gash their enemies' wings, while others let ropes dangle which they hoped would become tangled up in their counterparts' propellers. However, most authorities were in agreement that only by mounting a fixed or traversing machine-gun could the problem be solved, and thus the era of true aerial combat came into being. The ideal layout for fighter aircraft was a fixed machine-gun mounted in front of the pilot so that he could shoot forward and aim by pointing his airplane at the target. The bullets would, however, have to pass through the disc created by the spinning propeller and they would inevitably, and sooner rather than later, saw through the blades. Morane-Saulnier devised a solution to this problem: apply armor plating to the back of the propeller blades, allowing those bullets striking them to be deflected. Thanks to this idea the Morane-Saulnier L became the first true fighter aircraft, and allowed Roland Garros to become the first pilot to shoot down an enemy aircraft with a machine-gun mounted in the so-called "fighter" position. This plane, designed in 1913, was powered by an 80-hp Gnome rotary engine, had a maximum take-off weight of 1,500 lb and a top speed of 71.5 mph at sea level. The path had now been traced and future fighters – whether mono-, bi- or triplanes – all to adopt this layout.

53 top right A British poster of 1915 highlights the differences between friendly and enemy airplanes to familiarize the public with aircraft and help them recognize danger.

53 bottom A French engineer installs a reconnaissance camera on a twin-engined Caudron. World War I also led to the development of aerial reconnaissance.

54 left, third picture In 1915 the French Nieuport 11 Bébé fighter appeared, the first of a successful family of planes. It had an 80-hp engine and could reach 97 mph.

54 left, fourth picture World War I obliged the warring parties to produce huge numbers of combat aircraft. Their fragility and vulnerability were one of the reasons for their frequent replacement. Here a French squadron of Nieuports is displayed.

54 bottom right One of the most successful fighter aircraft was the SPAD VII, used by the Italian aviation forces. The photograph shows Major Chiappirone and his airplane at Villaverla airfield in June 1917.

In 1915, the principal aircraft operated by the various forces were the Royal Aircraft Factory FE.2b, the Bristol Scout D, the Fokker E.III, and the Nieuport Bébé. The first true modern fighter, introduced in the summer of 1915, was the Fokker E.III, a monoplane armed with a 7.92-mm Spandau machine-gun and powered by a 100-hp Oberursel rotary engine providing a top speed of 87 mph. The supremacy of the E.III was due above all to the adoption of a revolutionary mechanical device, the synchronizer. Thanks to this mechanism, devised by the aircraft constructor himself, Anthony Fokker, the movement of the propeller governed the firing of the machine-gun, with the firing interrupted as the propeller blades passed in front of the muzzle. Moreover, the E.III was a monoplane, lighter and much more agile than the many heavy biplanes still in service with the Allies. Such was the E.III's superiority during the winter of 1915-1916 that the embattled Allied forces nicknamed it the "Fokker Scourge." Early in 1916, the Royal Flying Corps ordered that each of its spotter plane was to be escorted by three fighters when flying over the enemy lines.

The Allies were not slow to respond to the E.III, the British producing the Royal Aircraft Factory FE.2bn, a two-seat biplane with a pusher propeller that was equipped with two machine-guns and a 160-hp engine that gave it a top speed of 91 mph. The summer of 1915 saw the début of the first of a line of successful fighters, the French Nieuport 11 Bébé, a fast (97 mph), agile biplane propelled by an 80-hp Le Rhône 9C engine and

54 left, first picture This French air force Nieuport 17 of 1916 boasts two armaments. The pilot is an American volunteer: non-commissioned officer Lufberry.

54 left, second picture One of the most famous German fighters was the 180-hp, 117- mph Albatross D.V. It was brought into service in summer 1917 and more than 1800 were built.

armed with a 7.7-mm machine-gun. The aces Guynemer, De Rose and Nungesser inflicted serious losses on the Germans with this machine during the battle of Verdun, fought during February 1916. Shortly afterwards, Nieuport presented the improved version of the Bébé, the Nieuport 17, which was larger, more robust, better armed and much faster than its predecessor. The Nieuport 17 had a 110-hp engine and a top speed of 110 mph at 6,550 feet. It was fitted with twin 7.7-mm machine-guns. This became the favorite plane of many aces, including Guynemer, Nungesser, Fonck, Navarre and the British pilots Ball and Bishop.

By 1916 the war in the air had already changed considerably and the new aircraft that were introduced were ever more sophisticated and deadly. The Germans leapfrogged the Allies once more with the introduction of an aerodynamic new biplane, the Albatros D.I, which reached the front in the autumn and was immediately followed by the D.II. This last model boasted a 160-hp Mercedes D III engine that gave the aircraft a top speed of 108 mph. The armament comprised two machine-guns. Another aircraft with notable specifications was the Hansa-Brandenburg D.I, an Austrian plane equipped with a 160-hp Austro-Daimler engine capable of 116 mph. This fighter, however, proved to be rather unstable, and limited the pilot to a restricted range of vision; it was consequently blighted by a series of accidents earning it the nickname "coffin" from its pilots. Autumn 1916 also saw the Allies' introduction of an excellent new airplane, the French-built SPAD VIII, probably the best fighter of the first half of the Great War. The aircraft was designed around a V8 engine, the Hispano-Suiza 8 producing 150 hp. This was the engine of the future and was soon to replace obsolete rotary-architecture engines. The fuselage was highly aerodynamic with the engine was enclosed in a metal fairing; the armament comprised a Vickers 7.7-mm machine-gun. The fighter made its first flight in April 1916, and its performance immediately proved to be remarkable: a top speed of almost 122 mph and the ability to climb to 10,000 feet in just 15 minutes. France immediately ordered 268 of the SPAD VIIIs, with further orders flooding in from the Allied countries. The plane entered service on 2 September 1916, and was warmly welcomed by pilots, including the Italian ace Francesco Baracca who first flew the new plane in March 1917. The SPAD XIII, the upgraded version of the VIII, equipped with two machine-guns, made its maiden flight on 4 April 1917. This version, powered by a 220-hp engine and capable of almost 140 mph, was heavier and less maneuverable and many pilots, including Baracca, preferred the old VIII. The XIII was nonetheless widely adopted by the French, British, American, Belgian and Italian forces and no less than 8,472 of them were built.

55 top A German Aviatik B.I ready for take-off. This 1914 spotter plane had a 100-hp engine and a maximum speed of 62 mph. It had a two-man crew.

55 bottom left An RAF F.E.2b at Ste. Marie-Chappelle in France, 1916-17. This was a two-seater spotter-fighter with a 160-hp engine and a pusher propeller. It entered service in 1915.

55 bottom right A German Fokker E.III monoplane undergoing maintenance in a wood. It was the first airplane fitted with a propeller synchronizer (invented by Anthony Fokker) that allowed bullets to be fired through the revolving propeller blades.

1917 saw the débuts of a number of other excellent and famous airplanes. The included the British Sopwith F.1 Camel and Triplane, the Royal Aircraft Factory S.E.5, and the French Hanriot D.1. The Germans introduced the Fokker Dr.I, Pfalz D.III and Albatros D.II and D.V. Following its successful Pup, Sopwith designed a highly maneuverable new small fighter equipped with two synchronized machine-guns. The Pup downed no fewer than 1,294 enemy aircraft in about one year's fighting. The F.1, known as the Camel because of the humped fairing over its guns, flew for the first time on 22 December 1916, and began to reach the front-line squadrons in May 1917. Powered by a 130-hp Clerget 9B rotary engine, the Camel was capable of 115 mph, but the great torque its engine generated made it demanding to fly. However, the most skilful pilots could exploit this negative characteristic to make the plane even more agile. Almost 5,500 Camels were built between the end of 1916 and 1918.

For a time in 1917, aeronautical technology seemed to be moving in the direction of the triplane layout. In fact, in February of that year, the Sopwith Triplane made its début at the front, a fast maneuverable fighter that proved its worth and, most significantly, threw the Germans into a panic as they hurriedly set about catching up with what they perceived to be a notable technical advance. In reality, despite the Triplane's undoubted qualities, by July 1917 the more traditional but more effective Camel was replacing it; in all, only 144 Triplanes were built. German research had, in the meantime, involved 14 companies. Of the resulting prototypes, a Fokker triplane was judged to be the best. Following flight-testing, it was immediately put into production as the Dr.I and entered service in August 1917. Despite its tricky handling, the plane was immediately popular with its pilots, thanks to its exceptional maneuverability and climbing speed. Among the aces of particular note who flew it were Werner Voss and Manfred von Richtofen, the famous "Red Baron." The Dr.I remained in production until May 1918, with 318 being built before it made way for the more modern Fokker D.VII.

1917 also saw the début of another famous German fighter, the Albatros D.III/V. (This introduction actually occurred earlier than that of the Dr.I, triplane, and the Albatros D.III/V was built in far greater numbers than the Dr.I.) The D III/V was developed from the D.II. It had sesquiplane wings with V-shaped struts and was equipped with a more powerful engine than its predecessor. This new engine, the in-line 6-cylinder Mercedes D.IIIa, produced 176 hp and provided a top speed of 109 mph. Beginning in January 1917, almost all the fighter squadrons were re-equipped with the D.III/V, which in the hands of aces of the caliber of von Richtofen, Voss and Udet, helped the Germans regain aerial superiority above the battlefields. This period of German predominance culminated in the month of April, known by the British as "Bloody April" owing to the losses they suffered. In the summer of 1917, the

56-57 The best British fighter of the war was the S.E.5a, which entered service in June 1917. It was fitted with a 200-hp engine that gave it a top speed of 139 mph.

57 right, first picture The Sopwith Triplane was the British attempt to build a faster, more agile fighter than its predecessors.

57 right, second picture The Sopwith F1 Camel entered service in May 1917. It had a 130-hp engine and reached 116 mph. The name was given by the hump created by the machine-gun cowling. Over 5500 were built.

57 right, third picture The Fokker Dr.I, in service from August 1917, was the German response to the Sopwith Triplane. Difficult to handle, it passed into history as the favorite airplane of Manfred von Richthofen, the "Red Baron."

57 bottom Shown during combat, the Fokker D.VII entered service in 1918 and was thought by many to be the best fighter in the war. It touched 119 mph and remained powerful up to 19,500 feet.

Albatros D.V was introduced. It had a180-hp engine providing a top speed of 116 mph; it also had stronger wings than its predecessor (which had had an unfortunate habit of losing its lower wings in prolonged dives). More than 1,800 D.Vs were built, and the plane remained in service until 1918.

During the last year of the war, when Germany's defeat became increasingly likely though not yet certain, the nation produced its finest fighter aircraft, the Fokker D.VII and the Siemens-Schuckert D.II/IV. The Fokker D.VII appeared in April 1918, and was perhaps the best fighter either side produced in the First World War. Early in 1918, the aircraft won a competition held by the War Ministry to find a new fighter; it also won von Richtofen's approval. The D.VII, a biplane of traditional lines, was equipped with a 160-hp Mercedes D.III engine, offering a top speed of 118 mph at 3,280 feet. The War Ministry immediately 400 D,VIIs, and the plane was supplied to the leading fighter squadrons, with von Richtofen's personal group being the first to receive them. The pilots were impressed by the D.VII's qualities, in particular its high climbing speed and high-altitude performance. The model F, powered by a 185-hp BMW IIIa engine that maintained an appreciable power output to an altitude of 19,500 feet, was in production by August 1918. While the Fokker D.VII climbed to 16,500 feet in 38 minutes, the D.VII F took just 14 minutes. About 1,000 were built prior to the armistice.

The Siemens-Schukert II/IVs were also notable fighters. The D.III model was equipped with a 160-hp engine driving a four-bladed propeller and could attain a top speed of 112 mph.

By the end of the First World War, the aircraft industry had achieved great technological progress. All fighters were now being fitted with in-line rather than now obsolete rotary engines. Performance levels had increased constantly: the Italian Ansaldo A.1 Balilla boasted a 220-hp engine and a top speed of almost 137 mph, while the British Martinside F.4 Buzzard, powered by a 300-hp Hispano-Suiza 8F engine was capable of 133.6 mph. The same year (1918) also saw the introduction of the Junkers D.I, the first all-metal monoplane fighter.

Giant steps had also been taken in bomber design. In 1915, still early in the conflict, bombing raids were carried out by aircraft such as the Caudron G.4 (twin 80-hp engines, top speed 82 mph, 250-lb bomb load); the Farman F.40 (single 160-hp engine, top speed 84 mph, 110-lb bomb load); and the Caproni Ca.32 (three 100-hp engines, top speed 72 mph, 771-lb bomb load). Over the years, larger and more capacious aircraft were introduced. By the end of the war bombers were carrying highly respectable bomb loads: the Vickers Vimy (twin 360-hp engines, top speed 102.5 mph, 4,800-lb bomb load load) and the Handley Page V/1500 (four 375-hp engines, top speed 90 mph, 7,500-lb bomb load).

While the First World War increased the airplane's romantic appeal and legendary status, it above all favored the diffusion of aeronautical technology. During the five years of the conflict, Germany built 48,537 aircraft, Great Britain 58,144 and France no fewer than 67,987. Italy, the United States, Russia, and Austria contributed a further 45,000 aircraft. The air corps had expanded considerably: while in 1914 France had 138 front-line aircraft, by 1918 this total had risen to 4,511. Great Britain's air strength rose from 113 aircraft to 3,300 and Italy's from 150 to 1,200. The size and importance of the new corps actually led Great Britain to create a third branch of its armed forces, the Royal Air Force, which in 1918 incorporated men and machines from the Army and Navy. Other countries shortly followed suit.

All aircraft had become much more reliable, robust and controllable than four years earlier. Thanks to technological developments and the surplus of airframes, engines and pilots, the post-war years were to see the establishment of the aircraft as a means of transport. At the end of the war aeronautical development was ensured by the thirst for adventure and glory of both the pilots and the manufacturers, who would turn to the fields of record breaking and aerial exploration.

59 center France also produced various types of bombers. Shown is a Caudron G.4 from 1915. It had twin engines of either 80 or 100 hp and could carry a bomb load of 248 lb.

59 bottom The Vickers Vimy, introduced in 1918, was another British success. It was powered by two 360-hp Rolls Royce engines and carried a bomb load of up to 4800 lb. Like the Handley Page V/1500, after the war it was used for civilian transport.

58 The Handley Page V/1500, a British bomber introduced in 1918, was an exceptional plane for its era. It was powered by four 375-hp engines, had a maximum take-off weight of 67,200 lb (including a bomb load of up to 6780 lb), and could remain airborne for 6 hours.

59 top An Italian air force Caproni Ca.3 in flight in 1915. This biplane bomber, powered by three 150-hp Isotta Fraschini engines, a 3100 lb bomb load. This successful plane was also made in France under license.

The Knights of the Air

While the Great War signified blood, suffering and death for millions of men condemned to an anonymous fate in the mud of the trenches, the new aerial warfare in which the individual courage, altruism, honor and heroism of the medieval knights were revived soon caught the imagination of the masses. Fighter pilots became the objects of unprecedented admiration and fame. Frequently exploited by the powers-that-be to boost public morale, the pilots themselves did their best to feed the legends, with audacious, bold and daring behavior on and off the field of combat. They were showered with honors to a degree that had not been seen for centuries, both during their lifetimes and after their deaths. Despite the ferocity of their aerial duels, they fought a different war, one ruled by a gentlemanly code and distinguished by mutual respect.

The greatest pilots became known as "aces," a term initially coined to describe someone who excelled, but it was then attributed only to those who had shot down five or more enemy aircraft.

The most famous of all was without question the German Manfred von Richtofen, the legendary "Red Baron," so named after the color in which he had his personal aircraft painted. Prior to his death, shot down by a Canadian, Captain Roy Brown, on 9 April 1918, von Richtofen set an unmatched record of 80 kills.

A list of the leading First World War Aces follows, subdivided by nation, and giving the number of enemy aircraft each brought down:

Germany: M. von Richtofen (80), E. Udet (62), E. Loewenhardt (53), W. Voss (48), F. Rumey (45), O. Boelcke (40). Austria-Hungary: G. Brumowski (40), J. Arigi (32), F. Linke-Crawford (30). Great Britain: E. Mannock (73), W.A. Bishop (72), R. Collishaw (60), J.T.B. McCudden (57). Italy: F. Baracca (34), S. Scaroni (26), P.R. Piccio (24), F.T. Baracchini (21), F. Ruffo di Calabria (20). France: R. Fonck (75), G. Guynemer (54), C. Nungesser (45), G. Madon (41). United States: E. Rickenbacker (26), W. Lambert (22), F. Gilette (20). Belgium: W. Coppens (37). Russia: A.A. Kazakov (17).

61 top left The German ace, Baron Manfred von Richthofen, better known as the Red Baron because his planes were painted that color. He commanded the 11th Pursuit Squadron (Jasta II) and was brought down behind British lines on 9 April 1918.

60 top Major Francesco Baracca, Italy's most successful war ace, posing in front of a captured enemy airplane. He knocked out 34 enemy planes before being brought down by anti-aircraft fire. His "rearing horse" coat-of-arms was handed on to certain Italian fighter squadrons; his mother also donated it to the car-maker Enzo Ferrari.

60-61 bottom Georges Guynemer, France's second most successful ace, shot down 54 enemy planes before being killed himself.

61 top right General Bongiovanni, Major Gabriele D'Annunzio and other pilots of the Serenissima Squadron's famous flight over Vienna, made on 9 August 1918. The purpose was to drop propaganda leaflets and bombs.

61 center right The funeral of Baron von Richthofen, held in Berlin in 1925. For seven years his remains had rested in a French cemetery.

61 bottom right Albert Ball, one of the most famous British aces. He died in combat after shooting down his forty-third victim.

Chapter 5

The Birth of Commercial Aviation

Up until the First World War, popular imagination had above all seen the airplane as the instrument of sporting bravado and a weapon for intrepid warriors; in both cases a vehicle of the élite, reserved for a few fearless souls.

The Great War radically changed this situation, in terms of numbers at least, if not the underlying concepts: by 1918 there were thousands of trained and experienced pilots, along with thousands of aircraft about to be sold off as military surplus. Flying had begun to spread and its commercial potential had begun to be appreciated.

The birth of civil aviation took place in post-war Europe and was the fruit of necessity rather than chance. The war had, in fact, destroyed many rail links, and routes which involved a sea crossing, such as the popular London-Paris route, were slow and complicated for any land-based means of transport. In contrast, America still benefited from an excellent and undamaged rail network, and the characteristics of contemporary aircraft meant that the railways still enjoyed a measure of superiority.

France, Great Britain and Germany all saw the birth of the first airline companies within a few months, early in 1919. The first scheduled service was actually established in Germany, with Zeppelin dirigibles linking Berlin and Weimar. The Paris-Brussels and Paris-London routes inaugurated by French and English companies operating airplanes soon followed.

62 top left The postwar period marked the beginning of many commercial aviation projects, like the then very sophisticated Caproni Ca.60 Transaereo hydroplane. Despite its eight 400-hp engines, this gigantic flying craft was too heavy to fly and was destroyed during trials.

62 bottom left A crowd gathers round a de Havilland DH.86 biplane during the opening of Gatwick Airport by Britain's Minister of Aviation.

62 right Two posters advertising KLM (Royal Dutch Air Services), the Dutch national airline.

63 left A splendid poster advertising Air France's "Arrow of the East" service, begun in 1933.

63 top right A twin-engined Air France Bloch 220 over France. The airline was formed in 1933 by the merger of Aéropostale, Farman and Air Orient.

63 bottom right The bare interior of the cabin of the Caproni Ca.60 Transaereo, which was designed to carry 100 passengers.

The first of the British companies were Aircraft Transport, Handley Page Transport, Instone Air Line, and Daimler Airway. They were emulated in Holland by KLM, in Belgium by SNETA, in France by Farman, Latécoère and Aéropostale, and in Germany by Junkers and Deutscher Aero Lloyd. As the years passed, many of the pioneering air transport companies disappeared or merged as a response to the demands of both the market and national governments. In 1924, the British companies merged to form Imperial Airways, directly supported by the government and charged with maintaining links with the far-flung corners of the empire. Two years later, all the German companies were brought together at the government's behest to form a single company, Deutsche Luft Hansa Aktiengesellschaft, better known as Lufthansa. The year 1933 also saw the birth of Air France as a result of the merger of Aéropostale, Farman and Air Orient. Other companies sprang up throughout Europe: Det Danske Luftfartselskab in Denmark, Linee Aeree d'Italia in Italy, Sabena in Belgium, and so on.

Various constraints limited the individual companies' commercial initiatives, market capacity, and development potential. In fact, the aircraft themselves were the principal difficulty facing the nascent civil aviation industry: in the immediate post-war period there were no purpose-built passenger planes specifically designed to operate on commercial routes.

64 left The SAS sign in Bodo, Norway, indicates flight times to various destinations. With air travel, the world began to get smaller.

64 top right An 'air porter' of the German airline Luft Hansa helps a passenger board a Fokker F.II. This passenger plane entered service with the newly formed airline in 1926.

64 bottom right The luxurious passenger cabin in the four-engined, long-range Handley Page HP.42, in service with Britain's Imperial Airways from 1930.

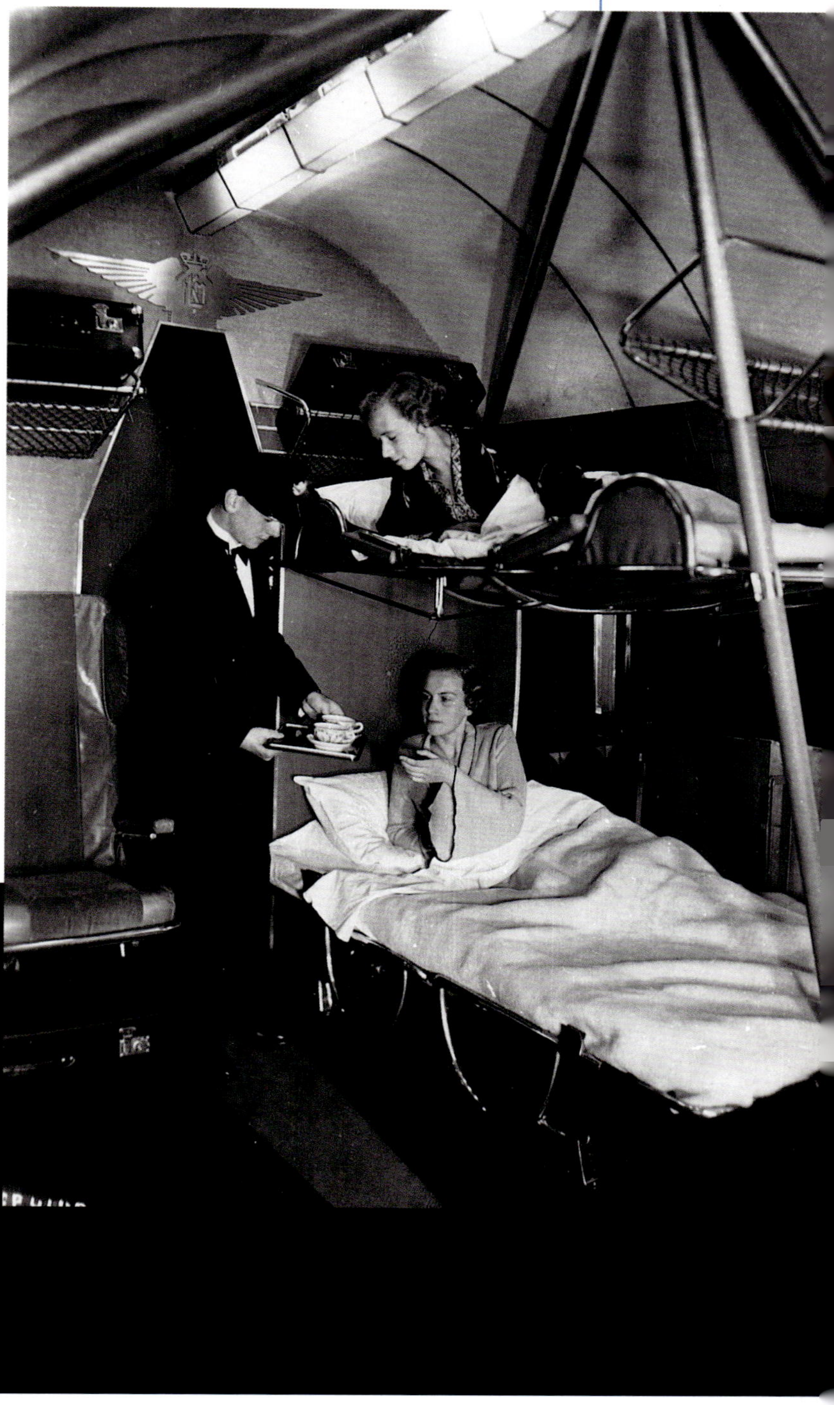

65 top left Passengers disembark from a Luft Hansa Junkers G.24 at Berlin's Tempelhof airport in 1928. A shuttle bus took them right into the city center.

65 center left A poster advertising the 1927 Cologne air race, for civilian planes only.

65 bottom left Luft Hansa was the first airline in the world to serve food and drinks on board. Stewards are shown serving an in-flight meal in the cabin of a three-engined Junkers G.31 in 1928.

65 right With the advent of new and larger airplanes, on-board services increased. This Fokker F.XXXVI has couchettes, like KLM organized in 1935 for its night flights.

70 top The Handley Page HP.42 was the largest land-based aircraft of the 1930s; eight were built for Imperial Airways. Despite its antiquated appearance, it was made entirely of metal and first flew in 1930.

70 center left Two single-engined Hamilton monoplanes in the United States at the end of the 1920s. These transport planes were also made entirely of metal.

70 center right Publicity for an air race held in Miami, probably in the 1930s.

70 bottom A TWA Douglas DC-3 over Manhattan in 1935. For the period, passenger transport was very developed in the United States thanks to the famous twin-engined Douglas.

In the meantime, civil aviation had also begun to establish a foothold in America, thanks above all to the government's issuance of airmail contracts that gave the pioneering companies a much-needed boost. In fact, the Kelly Act of 1925 enabled American air transport to establish itself, and before long the United States outstripped Europe in this sector.

The first American carrier of the post-war period was Aeromarine West Indies Airways, which was founded in 1920 but ceased operations in 1923 owing to financial difficulties. After 1925 numerous companies began to appear, mainly competing for postal routes which they integrated with passenger transport. The period between 1925 and 1926 saw the birth of Stout Air Services, Florida Airways, Western Air Express and Pacific Air Transport, among others.

71 top Two pamphlets – one English and the other American – providing information on the growing air travel sector of the 1920s and 30s.

71 bottom A photograph from 1930 shows the first group of air hostesses; they worked for the American company, United Air Lines.

Henry Ford, the celebrated automobile manufacturer, also sought to establish himself in aviation. He founded Ford Air Transport Services and developed a modern transport aircraft: the Ford Tri-motor. It was soon carrying America into the skies while attaining legendary status.

The Tri-motor's origins lay in a 1922 design for the Stout 2-AT Pullman, an eight-seater high-wing monoplane equipped with a single 406-hp Liberty engine. In 1926, at Ford's request, Stout developed this design into the 3-AT, but the new aircraft proved to be unsuccessful and led to Stout's dismissal. However, that same year saw the appearance of the Ford 4-AT, an all-metal, high-wing, three-engined monoplane skinned in corrugated metal and seating eight passengers. This plane was very similar to the 3-AT, and first flew on 11 June. The first and most important client for the new plane was Maddux Air Lines of Los Angeles, which purchased 16 of them. In 1928, the 5-AT version was introduced; it was larger, more robust and featured a greater wing area. It had a wingspan of 77 feet 6 inches, a length of just over 50 feet, and a maximum take-off weight of 13,503 lb. It was powered by three Pratt & Whitney SC-1 engines producing 425 hp. They provided a top speed of 143 mph and an operational range of about 550 miles. Between 1929 and 1931, 117 of these 17-seater 5-ATs were built, while a total of 198 Ford Tri-Motors in over 30 versions were sold to numerous air lines such as American British Columbia, Curtiss Flying Service, Eastern, Northwest, and Transcontinental. The American armed forces used the Tri-Motor for military purposes, as did Australia, Canada, Colombia and Spain. The aircraft proved to be so robust and reliable that a number were still being used for tourist flights in the 1980s.

72 top One of the first Ford Tri-Motors. The Tri-Motor first flew in June 1926 and was a development of the unsuccessful single-engined 3-AT.

72 center The passenger cabin of a Ford Tri-Motor 4-AT which could seat up to 12 passengers. The plane had two pilots.

72 bottom The Ford/ Stout 2-AT Pullman of 1922, first of a series that led to the definitive Tri-Motor 5-AT of 1928. It could carry up to 17 passengers.

73 top A three-engined Fokker F.VII with French identification just after landing at a European airport, surrounded by a crowd. The F.VII was the Ford Tri-Motor's main competitor.

73 bottom An excellent photograph of a Ford Tri-Motor 5-AT belonging to American Airways.

A new generation of transport planes was emerging in Europe, the most common types were the Handley Page H.P.42, the Junkers Ju-52, and the de Havilland Dragon biplanes.

In Germany, shortly after the introduction of its successful W33/34 series, Junkers began work on a large new single-engined aircraft designed primarily for cargo work. Designated the Ju-52, the aircraft was completely skinned in corrugated metal, and at the wings' trailing edge had Junkers patented movable aerodynamic surfaces that acted as flaps and ailerons. The Ju-52 flew for the first time on 13 October 1930. Thanks to the power output of about 800 hp guaranteed by its engine, the Ju-52 could carry up to 17 passengers. In 1931, however, Junkers' design department decided to transform it into a tri-motor with the installation of 530-hp Pratt & Whitney Hornet engines. The prototype of this particularly fast version (188 mph) known as the Ju-52/3m took off in April 1932, and was an immediate commercial success. The Ju-52/3m had a wingspan of almost 96 feet, a length of 62 feet and a maximum take-off weight of 23,148 lb. The first two Ju-52s were delivered to Lufthansa before the end of 1932; the airline eventually purchased a further 230. Others were exported to Finland, Sweden and Brazil. In all, the Ju-52/2m saw service in over 25 countries in Europe, Africa, Asia, Australia and South America. It enjoyed continuing success after the Second World War when numerous planes were built in France (as the AAC 1 Toucan) and Spain (as the C-352-L) and was used for both military and civil applications for many years. In 1934 the Ju-52/3mg3e military version was introduced, and thousands were built to serve as cargo planes and bombers.

In contrast to the Ju-52, the Handley Page H.P.42 was an aircraft with antiquated

74 bottom A de Havilland DH.89 Dragon Six belonging to Britain's Olley Air Service. A total of 930 Dragons were built in various versions for both civil and military use.

75 top The interior of a Handley Page HP.42 'Heracles' transport plane belonging to Imperial Airways. It was normally used on the London-Paris route and could carry 38 passengers.

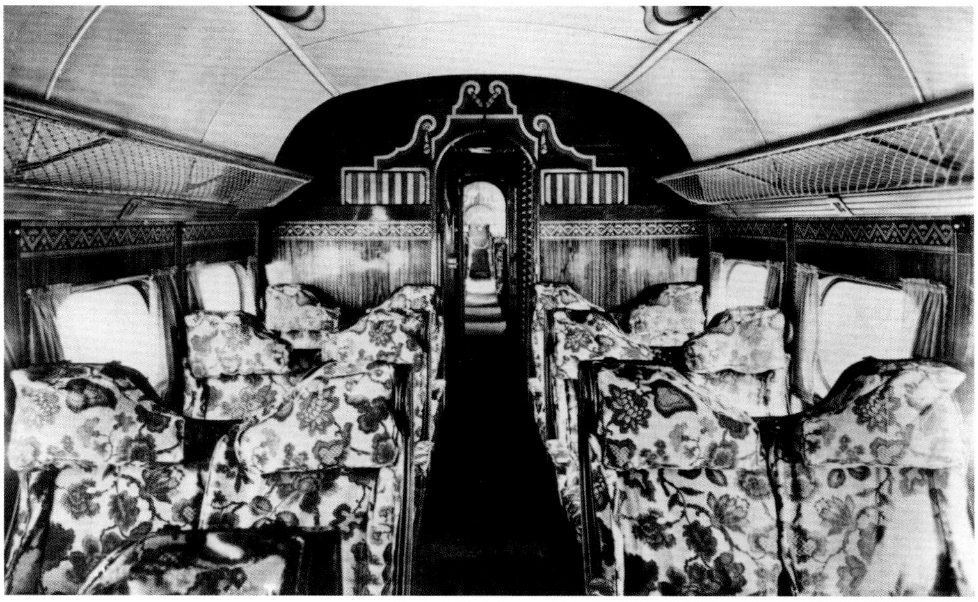

architecture built in response to the specific demands of Imperial Airways and its need to maintain links with the distant British colonies. The aircraft was designed with mechanical reliability and passenger comfort in mind rather than speed and large payloads. The result was a four-engined metal biplane with a biplane tail and triple rudders. It was powered by Bristol Jupiter XIF engines that produced 500 hp and gave the plane a top speed of 127 mph. The wingspan was 130 feet, overall length was 89 feet 6 inches, and the plane had a maximum take-off weight of about 29,500 lb. The first H.P. 42 took to the air in November 1930. Only eight were built; four 24-seater E versions for use on Asian routes and four 38-seater W versions (with 560-hp engines) for use on European routes.

The de Havilland Dragon series of twin-engined biplanes designed for short-haul work was far more successful. The first in the line, the D.H.84, was developed in response to a civilian airline's requirements for an efficient twin-engined passenger aircraft to link England with Paris. The Dragon was an elegant, six-seat airliner with fixed undercarriage and two 132-hp de Havilland Gipsy Major engines. The plane made its maiden flight on 12 November 1932. A total of 115 were built in England, along with a further 87 constructed under license in Australia. The Dragon was the basis for the D.H. 89 Dragon Six, which was introduced in 1934. This model was equipped with two 203-hp Gipsy Six engines. The production version, which could carry up to ten passengers, was known as the Dragon Rapide. In total, 728 D.H.89 Dragon Rapides were built, with many of them seeing military service during the Second World War. The type went on to form the backbone of many post-war airlines such as Jersey Airways, KLM and Iraqi Airways. A number of Rapides were still in use in the 1980s.

75 bottom left Seen from above, a Junkers Ju.52 flying Austrian colors. The Ju.52 first appeared in 1930 as a single-engined transport plane; it became successful when, from the seventh plane on, it was fitted with three engines.

75 bottom right A lovely view of the Handley Page 'Heracles' taking off. This four-engined plane could only just reach 130 mph and manage a distance of 500 miles.

74-75 A Misr Airwork's de Havilland DH.84 Dragon flies above the Pyramids in Egypt. The Dragon was first manufactured in 1932 to carry passengers on the London-Paris route.

In the United States, in the meantime, thanks to the Waters Act of 1930 which revolutionized the airmail contract system, the air transport scene was changing: the major monopolistic companies such as United Air Lines had to give up portions of their market to permit competition. It was within this environment of rationalization that TWA was born out of the merger between Transcontinental and Western Air Express. American Airways then acquired Southwest Air Fast Express, while on the East Coast, Eastern Air Transport took the place of Ludington Air Lines. In 1933, four major companies, American, TWA, Eastern and United Air Lines, dominated the American market, while Pan American monopolized the international routes.

The technical revolution that characterized the period leading up to the Second World War came about in the mid-1930s with the introduction of the first truly modern aircraft. These were to change the face of commercial air transport. The precursor of this new category of aircraft was the Boeing 247, designed toward the end of 1932. This airliner was derived from the Model 215 bomber, but was an all-new and highly innovative design that may rightly be considered as the first of the modern passenger planes. The 247 had a very aerodynamic shape, with a cantilevered low wing into which two engines were faired, and cruciform tailplanes. Moreover, the plane was of all-metal construction, had a retractable undercarriage and was equipped with a pneumatic de-icing system along the length of the wing and tailplane leading edges. The 247 had a wingspan of 74 feet, a length of just over 49 feet and a maximum take-off weight of 13,688 lb. It was powered by two 600-hp Pratt & Whitney Hornet engines and could carry ten passengers over distances of up to 750 miles at a top speed of 200 mph; more importantly, it could maintain a cruising speed of no less than 189 mph. The prototype first flew on 8 February1933, and Boeing immediately received orders for 60 of the 247s from the four United Air Lines companies, while other firms ordered a further 15 planes. In 1934, Boeing introduced the 247D, an improved version of the 247. This variant adopted the new variable pitch propellers, new NACA engine cowlings and an aerodynamic windshield.

In spite of its avant-garde specifications, the initial Boeing 247 was simply too small to be economically viable, and Boeing's production capacity was too low to meet the initial demand for the plane. In 1932, Jack Frye of Trans World Airlines (TWA) was looking for a new plane to replace his Fokker Tri-Motors, but Boeing was obliged to quote a lead-time of two years for the 247. Frye thus invited other constructors to present designs, eventually selecting the one proposed by Donald Douglas. Although TWA's requirements were demanding, the Douglas engineers were determined to produce an aircraft that could out-perform the Boeing 247 while incorporating all the latest technological innovations such as a retractable undercarriage, NACA engine cowlings and flaps (a feature the Boeing lacked) to reduce landing velocity without penalizing cruising speed. The prototype of the new plane, the DC-1 (Douglas Commercial 1) was ready to fly on 1 July 1933. For that era it was an enormous machine, with a wingspan of over 85 feet, a length of 59 feet, and a maximum take-off weight of 17,361 lb. Two 710-hp Wright Cyclone engines powered the DC-1; these were new and were fitted with the new variable-pitch propellers. The plane was capable of flying and climbing on just one engine. TWA was very satisfied and ordered 25 DC-1s. These were modified, however, to allow sufficient space for flexibility. The fuselage was lengthened by 23.5 inches, which allowed two more seats to be added, for a total of 14. The engines were also uprated. The DC-2 (as the modified plane was designated) stunned the aviation world. Anthony Fokker applied

76 top An American Airlines Douglas DC-2. The DC-2 was 25 inches longer than the DC-1 and could carry 14 passengers rather than 12.

76 bottom left TWA was the first company to introduce the revolutionary Douglas DC-1 in 1933. Here it is about to take off from an American airport.

77 top left *The passenger cabin of the revolutionary twin-engined Boeing 247. However, it could carry only 10 passengers and had no wing flaps and was soon outclassed by the more profitable DC-3.*

77 top right *A Boeing 247 ready for flight. This was the first metal monoplane with low, cantilevered wings, a retractable undercarriage and de-icing equipment. It first flew in 1933, the same year as the DC-1.*

77 bottom *A spectacular view of a DC-3 at full power. This airplane is considered to have been the first modern commercial aircraft and formed the basis of air travel for many years. It was also built for military use during World War II.*

for distribution rights in Europe, immediately placing a number of DC-2s with KLM, the Dutch national airline. In 1934, in order to satisfy an American Airlines request for an aircraft with couchettes, the Douglas DST was introduced. This variant, which featured 14 couchettes in a wider fuselage, entered service in 1936. However, the plane could also be fitted with up to 21 seats, and in this configuration was designated the DC-3. It was a plane significantly different to the DC-2 in that it had a hydraulic retractable undercarriage, new propellers, new dampers, a redesigned, thinner wing, and a wingspan extended by almost 10 feet.

The DC-3 was considerably easier to fly than its predecessor had been, but what particularly attracted the airlines was its ability to carry 21 passengers, a fact that made it the first airliner in aviation history able to pay its way on passenger ticket sales alone. A company equipped with the DC-3 was much more likely to make a profit than a loss, and the DC-3 soon became the standard American transport plane. In 1939, 75 percent of all American passenger traffic was carried on DC-3s.

The Second World War saw the mass-production of the DC-3 for the American armed forces (as the C-47 Skytrain for the US Army, as the R4D for the Navy), as well as the Russian air force and the British RAF, where it was known as the Dakota. Many historians have claimed that one of the principal factors in the Allies' victory was, in fact, the C-47, the indefatigable cargo and parachute drop plane. Its widespread use during the war ensured that once peace had been established there were countless C-47s in various forms and with various engines that continued to be flown successfully throughout the world – up to the present day.

76 bottom right *A British European Airways DC-3. This definitive model of the DC family was developed from the DST model. It entered service in 1936 and could carry 21 passengers.*

Chapter 6

Seaplanes and Marathon Flights

During the inter-war period the seaplane contributed more to the development and diffusion of aviation worldwide than any other type of aircraft. The fact that the seaplane required no particular infrastructure for take-off and landing operations, but could use any stretch of open water made it ideal for the development of aviation, for the great explorations of distant lands, and for the commercial linking of the continents.

Henri Fabre, a French engineer, built the world's first seaplane. Fabre, who was born into a Marseille shipping family in 1882, was infected with a passion for flight at a very early age. From 1906, he began devoting himself to his hobby, meeting with French aviation pioneers such as Voisin, Blériot and Farman and beginning to build the tethered float gliders he tested on the lake at Berre. In 1909, after developing working models capable of take-off and landing, Fabre constructed a seaplane with three floats and three 125-hp Anzani engines. However, owing to its excessive weight and the poor hydrodynamic shape of the floats, the design proved incapable of flight. Fabre then built a new, smaller hydro-aeroplane, as it was known, with canard aerodynamic surfaces (i.e., with the main wing and the engine mounted at the rear and the stabilizers at the front). This new model also had three floats, but was powered by a single 50-hp Gnome engine. It took off for the first time on 28 March 1910. The era of the great seaplanes had dawned.

Significant developments soon followed, especially in the military field. In 1911, Italy saw the introduction of two new seaplanes built by the Italian navy officers, Alessandro Guidoni and Mario Calderara, and notable progress was made in France, Belgium and the United States. Glenn Curtiss of America was particularly successful in developing seaplanes, and his Triad first took off on 26 January 1911. In 1912, bolstered by this experience, Curtiss and his Triad participated in a rally in Monaco. In the same year, Curtiss also produced the

78 bottom Mario Calderara, an Italian Navy officer, was one of the first hydroplane pilots. Here he is in Brescia at the start of the Olofredi Prize race in 1909.

79 top The world's first hydroplane was built by the Frenchman, Henri Fabre. Here he is testing the plane on Berre Pond on 28 March 1910.

79 center Low over the water, Henri Fabre at the controls of his first successful hydroplane, following the botched attempt of 1909.

79 bottom left Henri Fabre next to one of his planes. He was born in Marseilles in 1882 and died in 1984, aged 102.

79 bottom right Another picture of Fabre's hydroplane, here seen over the Mediterranean near Monte Carlo in 1914.

Model F, a flying boat with a central hull. The Curtiss Company sold a150 of the Model Fs to the U.S. Navy, and additional ones to other navies, including Italy's.

The Schneider Trophy, instituted in 1912, greatly stimulated progress. The event was an air race reserved from seaplanes, and was the first competition of its kind. The debut race was held at Monaco in 1913, and the winning seaplane was won by a Deperdussin monoplane powered by a 160-hp engine, which managed to complete the course at an average speed of 61 mph.

By the outbreak of the First World War, the seaplane was already a fairly common type of aircraft. During the conflict, naval air services invested heavily in seaplanes, and thousands were manufactured around the world. The planes were particularly useful for spotting and patrol duties, especially in the North Sea and Adriatic regions. The most common types were the Austrian Lohner E and L, the Italian Macchi L.1 (based on a captured Lohner L), and the Anglo-French F.B.A.C. Subsequently, fighter floatplanes such as the Hansa-Brandenburg KDW, W.12 and W.29, the Sopwith Baby, and the Macchi M.5 made their appearance. Other successful seaplanes included the Curtiss H-1 and H-16, the N-9 trainers, the Macchi M.9 and the British Felixstowe F.2A patrol plane.

After the war, conflict no longer provided a stimulus for aircraft development. The armed forces set about promoting and completing great feats of aviation, aiming to demonstrate the aircraft's potential and overall worth. Prior to the war, aircraft capabilities were very limited and aeronautical exploits such as Blériot's crossing of the English Channel in 1909 and Roland Garros' trans-Mediterranean flight (from the Côte d'Azur to Tunisia) in 1913 were relatively tame. The Great War, however, had provided an enormous boost to aeronautical progress and by 1919 there were already land and seaplanes capable of crossing one of the world's traditional barriers, the Atlantic Ocean.

80-81 top Hydroplanes were used extensively during World War I. Here a French plane is being loaded onto a military ship for transportation overseas.

80-81 bottom The Curtiss Hydroaeroplane, first successful American hydroplane, developed in 1911.

81 top The German Arado Ar.196, one of the best-known military hydroplanes, first flew in 1937. More than 600 were built and used in World War II.

81 center After capturing an Austrian Lohner, Italy too was able to build hydroplanes in World War I, such as this Macchi M.5.

81 bottom right The Hansa-Brandenburg W.12, one of the first fighter hydroplanes, was used in World War I. This one belonged to the Austrian Navy.

On the 15 May 1919, the US Navy initiated the first trans-Atlantic flight. Three Curtiss seaplanes took off from Newfoundland to complete a coordinated naval mission with 41 destroyers deployed across the ocean that were to serve as route markers. The aim was to reach the Azores and then Portugal. Owing to a storm only one plane actually reached the Azores intact and was able to fly on to Lisbon, where it touched down on 27 May. The next great feat involving seaplanes was the round-the-world flight, organized by the American army's Air Service in 1923. For this event, the Douglas company built four examples of what it named the World Cruiser (DWC), which was derived from the DT-2 torpedo bomber. These two-seater biplanes were powered by a single 420-hp Liberty engine and were equipped with floats, but could also be fitted with a wheeled undercarriage to allow touch down on dry land. The four aircraft took off from Seattle on 6 April 1924, and followed a route that crossed Alaska, Japan, China, India, France, Great Britain and Newfoundland. Two of them made it through to Washington, D.C. on 9 September, and then on to the West Coast after a flight of over 26,500 miles.

In 1925, Francesco De Pinedo scored an even greater triumph. De Pinedo was an Italian naval officer and a veteran of the Italian-Turkish and Great Wars. With the creation of the Italian Royal Air Force in 1923, De Pinedo left the navy and became a promoter of the seaplane within the new service. It was in order to demonstrate the seaplane's potential that in 1924 he began planning an adventure that would outdo even that of the Americans. He plotted a route of 34,200 miles that took in Rome, Melbourne and Tokyo and included about 15,500 miles of flying close to the coasts, about 5,000 over the open sea, and 4,350 over dry land. Having received authorization to undertake his flight, De Pinedo chose to fly a standard Savoia Marchetti S.16ter seaplane, a biplane with a pusher propeller powered by a 400-hp Lorraine engine, which he baptized Gennariello. Lieutenant-Colonel De Pinedo, assisted by Marshal Ernesto Campanelli (a general mechanic), took off on 20 April 1925 from the waters of Lake Maggiore. The flight was not without difficulties. The major ones were due to problems with the lubrication system, but all the stages, from Baghdad to Java, Melbourne, Sydney and Tokyo were eventually completed and the Gennariello successfully touched down on the waters of the Tiber in Rome on 7 November 1925, to great celebrations.

On 13 February 1927, De Pinedo set out on a new adventure aboard the Savoia Marchetti S.55 flying boat Santa Maria, accompanied by Captain Carlo Del Prete and engineer Vitale Zacchetti. During this second marathon, De Pinedo visited South America and Amazonia and then headed north toward the United States. At Phoenix, Arizona, the plane caught fire while refueling and had to be replaced. The three intrepid aviators eventually reached Ostia on 16 June, having covered almost 28,000 miles. With his promotion to the rank of general, De Pinedo subsequently organized two great formation cruises for military seaplanes, flying over the southern and eastern Mediterranean in 1928 and 1929. However, the Italian air ministry's upper echelons did not appreciate De Pinedo's his success and the popularity he enjoyed, and he was passed over for the most prestigious appointments.

82 top One of the US Navy's three three-engined Curtiss NC-4 planes that attempted to cross the Atlantic in 1919, but without success.

82 bottom Two Italian Navy FBA hydroplanes during World War I. Francesco De Pinedo, seated on the plane on the left, was to become famous in 1925 as an intercontinental flier.

82-83 In 1923, the US Army planned to circle the world with four Douglas DT-2 World Cruisers. Two managed to complete the trip, between April and September 1924.

83 center The Douglas World Cruiser could be fitted with floats for landing on water and with a conventional undercarriage for on land.

83 bottom De Pinedo and Campanelli on 'Gennariello', the Savoia Marchetti S.16ter they used for the Rome-Melbourne-Tokyo-Rome race.

The last great exploit in which seaplanes played the leading role occurred in 1933. Italo Balbo, the Italian Navy's under-secretary for aviation at that time, initiated the event, following up on the success of the1930 aerial cruise from Italy and Brazil. In 1933, on the tenth anniversary of the founding of the Italian Royal Air Force, Balbo set about organizing a huge flight from Italy to the United States and back, involving an entire squadron of war planes. The aim was to demonstrate to the world at large the Italian air force's technical and military resources and its ability to dispatch not just one, but a significant number of aircraft to distant lands. After months of preparation at the Orbetello sailing school, the 25 twin-hulled Savoia Marchetti S.55X monoplane flying- boats, each powered by two 930-hp Isotta-Fraschini Asso engines, were ready. The expedition, composed of over 100 men including pilots and support staff, took off from Orbetello on 1 July 1933. Despite a tragic accident at the first stopover in Amsterdam, where an aircraft and one crewman were lost, the rest of the flight went perfectly. Thanks to the crews' training in close formation flying, they dealt brilliantly with the typically poor Atlantic weather. The 24 aircraft reached Chicago, site of the World's Fair, on 15 July and then proceeded to New York. Balbo and his men were given exceptional welcomes in both cities. The expedition then returned to Rome on 12 August, with 23 aircraft, a second having been lost in the Azores on the way home. The exploit so impressed the English-speaking world that since then the term Balbo has been used in aeronautical jargon as a synonym for a large formation of aircraft.

In the 1920s and 1930s, however, the seaplane was by no means an exclusively military option; it was also used for commercial purposes. The S.55, for example, had made its maiden flight in July 1923 as a military plane but was soon converted to civilian use, given the Italian air ministry's diffidence about what it saw as an overly innovative machine. In 1926, the S.55, converted to carry ten passengers, entered service with Aero Espresso Italiano and was subsequently exported to various countries. It was only later that the air ministry took an interest and purchased about 100 planes, using them until 1939. Prior to this, in 1919, the SIAI S.16 had been developed as a single-engined monoplane flying boat with a single central hull, built specifically for carrying passengers. It enjoyed considerable success with hundreds of planes were manufactured under license in France and Spain.

84 top North America was the Decennial Long-Distance Flight's goal. Here are some of the 24 Savoia S.55Xs that reached America, taking off from the St. Lawrence River, Montreal, Canada.

84 bottom left The formation of S.55Xs flies over the Statue of Liberty in New York in July 1933 during the Decennial Long-Distance Flight.

84 bottom right Italo Balbo, prime developer of the Italian air force and Under-Secretary of the new Armed Forces, was a great supporter of long-distance competitions which increased the prestige of Italy's government, air force and industry around the world.

85 top left. The Savoia X.55 was the Italian Air Force's most important plane in long-distance flights. It was a single-wing seaplane, with a double-hull, powered by two 930-hp Isotta Fraschini Asso engines that gave it a top speed of 175 mph and a flight distance of 2200 miles.

85 top right. The goal of the Decennial Long-Distance Flight was the Chicago World Fair. Here Balbo receives an official celebratory welcome in the city.

85 bottom. Some of the S.55s on the Decennial flight berthed at one of the stops.

86 top The German Dornier Do.X, built in 1929, was the largest plane in the world at the time. This gigantic 12-motor seaplane was built for the Atlantic passenger routes.

86 bottom left The cross-section of the Dornier Do.X shows the plane's interior layout.

Dornier was another company that made a name for itself by building seaplanes, and in 1922 the Dornier J Wal, which many termed the greatest cargo seaplane of the Twenties, made its maiden flight. The J Wal had a twin-stepped hull, two short stabilizing wings, and a parasol main wing on which two engines were mounted in tandem. It carried a crew of four and had a rear compartment that could accommodate passengers or cargo, with various configurations available to suit purchasers' demands. The plane was designed for both civil and military use. As a clause in the Treaty of Versailles prohibited Germany from building heavy aircraft, serial production was entrusted to the Società Costruzioni Meccaniche in Pisa. The Do. J Wal was immediately successful and was also built under license in Switzerland, Spain, Holland and Japan. It was also used for a number of geographical expeditions; in 1925, the explorer Roald Amundsen acquired two to use on a trip to the North Pole while in 1932 a number of Wals were used for a round-the-world flight. In 1934, Lufthansa used the seaplane to inaugurate an air mail service between Europe and South America, a route from Stuttgart to Natal in Brazil that took four days and required the presence of a support ship in mid-ocean.

In 1926, Dornier introduced a successor to the J Wal, the Do. R Super Wal. The new plane featured an elongated fuselage and a greater wingspan, and was powered by two 660-hp Rolls-Royce Condor engines. Like the J Wal, the Super Wal carried a crew of four, but passenger capacity was increased to 19 persons. The series continued with the Do. R4 which was introduced in 1927; it was equipped with four 525-hp Bristol Jupiter engines enabling it to carry a 30% greater payload. The last of the seaplane line developed from the Wal was the Do.X, which at the time of its maiden flight on 25 July 1929, was the world's largest aircraft. The concept behind the Do X's design was that of creating an aircraft capable of carrying 100 passengers across the Atlantic while offering the standards of comfort found on liners and dirigibles. The aircraft had a single central hull, three decks, and was even equipped with cabins with bunks, salons, bathrooms and dining rooms. The fuselage was 131 feet long and the wingspan extended to over 157 feet. The original twelve 500-hp Siemens Jupiter engines proved insufficient to provide the plane, which had a maximum take-off weight of over 11,000 lb, with an adequate safety margin. They were subsequently replaced with twelve 640-hp Curtiss Conqueror units that allowed the Do.X to complete a test flight with no fewer than 170 passengers. The inaugural cruise got underway on 2 November 1930, but was so dogged by accidents and mechanical problems that after visiting Amsterdam, the Canary Islands and Brazil the Do.X reached New York only after ten months – and became the object of much criticism and derision in the press.

While the magnificent but overly ambitious Do X seaplane never entered regular service this did not mean that the era of the great cargo and passenger carrying flying-ats was over.

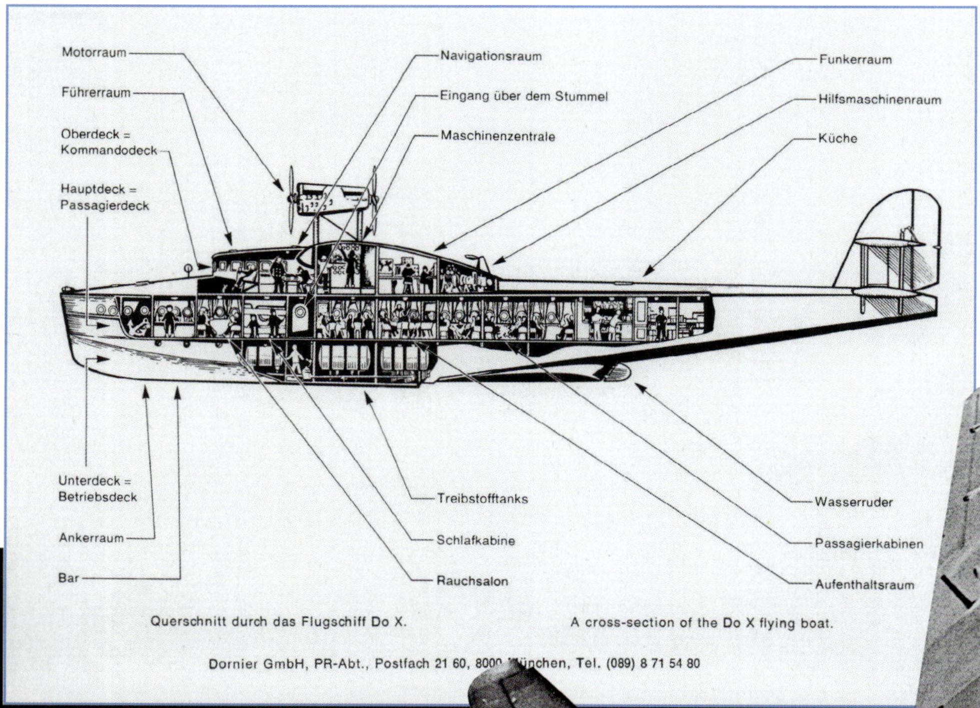

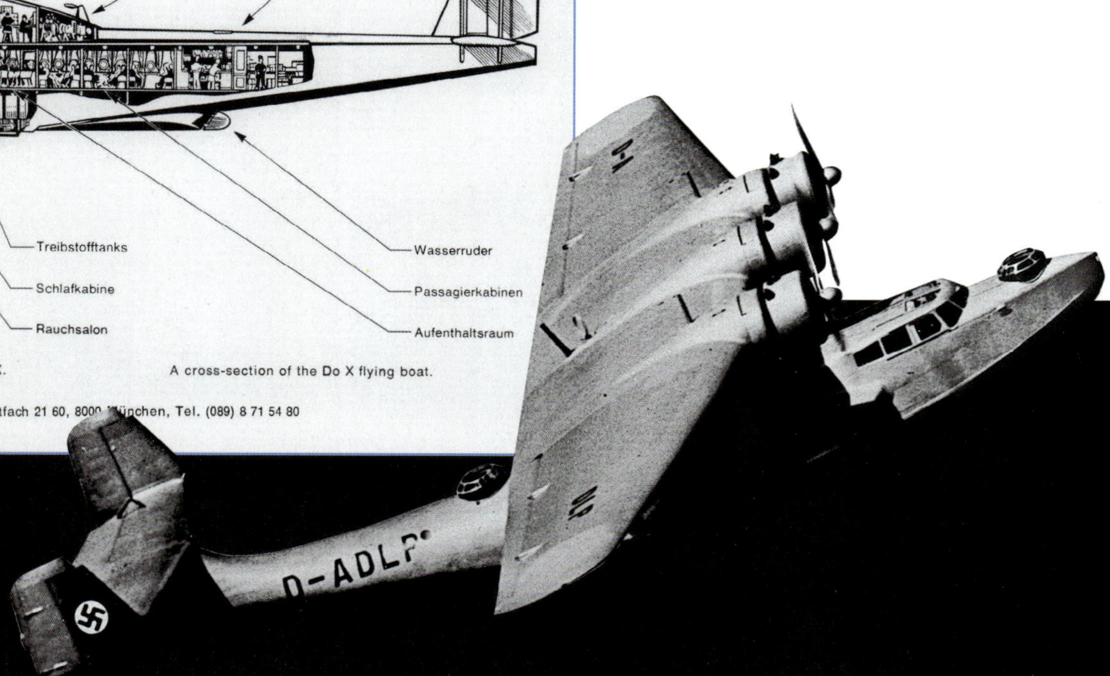

86 bottom right Designed in 1935 as a spotter and patrol seaplane, the Dornier Do.24 was also used during World War II. The German, Dutch, Spanish and French air forces all used it.

87 top left The Do.X's elegant passenger cabin of attempted to offer passengers every comfort of the period.

87 top right The Dornier Do.J Wal was perhaps the best transport hydroplane of the 1920s. It made its maiden flight in 1922 and roughly 500 were built. This one is being loaded on the German ship 'Westphalen.

87 bottom The arrival of the Dornier Do.X in New York on its maiden flight. Owing to various difficulties, the trip lasted from 2 November 1930 until 27 August 1931, giving rise to largely well-founded criticism.

88 top The Boeing 314 Clipper was the largest passenger seaplane in history. It entered the trans-Pacific and Atlantic services for Pan America in February 1939. It could carry up to 74 passengers and had a maximum distance of 3500 miles.

88 center The interior of the Boeing 314 was spacious and comfortable. Shown is part of the crew section where the radio operator and flight engineer were housed.

88 bottom The Sikorsky S.42 was one of the B.314's predecessors. Pan Am put the S.42 into service in August 1934 on Caribbean routes, and later on Pacific routes.

88-89 The Martin M-130, the S.42's competitor, entered service in September 1935; it too was purchased by Pan Am for use on Pacific routes. Here it is taking off in San Francisco Bay. Below, the Golden Gate Bridge was still being built.

In fact, in the United States during the 1930s, a number of new aircraft were developed to satisfy the demands of Pan American, the carrier that thanks to its American government contract enjoyed a monopoly on international airmail. In order to support the expansion of Pan Am's routes, Juan Trippe, the firm's managing director, asked Sikorsky (which had already produced excellent transport amphibious aircraft), for a new larger and more capacious plane. The result of this commission was the S-40, introduced in 1929.

The S-40 was a single-hull amphibious flying boat with twin tail booms powered by four 580-hp Pratt & Whitney Hornet engines. The plane was capable of carrying 40 passengers on long-haul routes of up to 1,000 miles. The first S-40, named American Clipper, entered service in November 1931. The Pan Am consultant, Charles Lindbergh, was already working on proposals for an even better plane. Sikorsky came up with the S-42 while Martin's response was the M-130. Pan Am purchased both. The S-42 was ready in August 1934, and was immediately put to work on the Caribbean routes.

Pan Am, however, also wanted to fly Pacific routes and to this end an S-42 was modified for use on path-finding flights. In 1935, the company began work on the construction of landing fields on the Central Pacific route at Honolulu, Midway, Wake and Guam. In the October of that year, the Martin M-130 also entered service, an enormous, elegant four-engined plane capable of flying for over 3,170 miles at 150 mph. During this period Pan Am acquired the last of the permits required to establish its network of routes, and the airline began to operate services to the Pacific regions and New Zealand. But in 1937 two accidents involving the loss of M-130s and many lives put the company under pressure. Juan Trippe's reaction was decisive: he unveiled a new strategy: the Atlantic route using the new aircraft he had commissioned the previous year, the Boeing 314. This was without question the most beautiful commercial flying boat of its era, and through to the 1960s, the Boeing 314 remained the world's largest transport plane. The B-314 was derived from the XB-15 bomber design, from which it retained the broad wingspan of 152 feet coupled to a new twin-deck fuselage that was 106 feet long. The new plane was equipped with a huge centrally mounted cantilever wing and four 1,520-hp Wright R-2600 Cyclone engines. It too carried the name Clipper, and could carry 40 passengers in particularly comfortable couchettes or 74 in seats on routes up to 3,500 miles long, at cruising speeds approaching 186 mph. Its maximum take-off weight was a huge 82,122 lb. The B-314 entered service on 22 February on both the Pacific and Atlantic routes. Shortly afterwards, however, the outbreak of the Second World War was to put a halt to all commercial transport aircraft development.

90 top Two British amphibian Supermarine Walruses are escorted by other hydroplanes. The Walrus entered service in 1935. Both the RAF and Fleet Air Arm used it throughout World War II as a marine assistance plane.

90 bottom A CRDA Cant Z.501 in low-level flight. Nicknamed 'Seagull,' it first flew in February 1934 and remained in service with the Italian Air Force until 1948.

91 top A British Short S.25 Sunderland in flight. The flexible and effective four-engined patrol plane was used in attacks against submarines and shipping.

91 bottom Perhaps the most famous amphibious plane was the American Consolidated PBY Catalina, which first flew in 1935. More than 4500 were built. At first it was just a hydroplane, but it was improved with various versions until the PBY-5A of 1940.

Seaplanes and amphibious aircraft had continued to find favor in the military sphere, at least with those sectors in which their particular capabilities were of use: maritime reconnaissance, patrol and rescue duties. During the war the Heinkel He-115, Dornier Do.18 and Do.24, Supermarine Walrus, Arado Ar.196, Kawanishi H6K and H8K and the Blohm und Voss Bv.138 all proved their worth. By the most famous examples of the breed were the CANT Z planes, the Short Sunderland, and the Consolidated PBY Catalina.

Thanks to the uncommon talent of the engineer Filippo Zappata, during the 1930s the Cantieri Riuniti dell'Adriatico (CRDA) company produced two successful series of military and civilian seaplanes, with hundreds of examples being constructed. The 7th of February, 1934, saw the maiden flight of the CANT Z.501, a single-hull, monoplane seaplane featuring a parasol wing and powered by a 900 hp Isotta Fraschini Asso XI engine. The aircraft was originally conceived for long-range reconnaissance and bombing duties and had a maximum speed of 171 mph and a range of around 1,500 miles. The Z.501 entered service in 1937 and was used in both the Spanish Civil War and the whole of the Second World War, remaining in service with the Italian air force through to 1948. The Z.501 was shortly followed by the CANT Z.506, which first flew on the 19th of August, 1935. Derived from the Z.505 postal aircraft, this plane was instead destined to carry passengers for the Ala Littoria company. The Z.506 was a low-wing tri-motor equipped with traditional tailplanes and two large floats below the wings. Its 750 hp Alfa Romeo 126 RC 34 radial engines provided it with a maximum speed of 227 mph and a range of around 1,700 miles. It had a maximum take-off weight of over 13 tons and could carry up to 14 passengers. The Italian Royal Air Force ordered the first 32 examples in 1937, adapting them for use as bombers, but with the outbreak of war the Z.506 was relegated to less hazardous duties such as reconnaissance, search and rescue and medical evacuation. The airplane remained in service in Italy through to 1960.

The Short S.25 was instead born in 1937 in response to a British ministry requirement for a new four-engined maritime patrol aircraft. The plane completed its maiden flight on the 16th of October that year while the production version, known as the Sunderland Mk. I, appeared in 1940. This large single-hull monoplane had a high wing in which four 1,024 hp Bristol Pegasus XXII engines were buried. It was crewed by no less than 13 men and its armament of eight 7.7 mm machine-guns earned it the nickname "flying porcupine." As the war proceeded, the Sunderland's general excellence ensured that continually improved versions were developed and it was also used in anti-submarine and anti-shipping roles. The last version, the Mk.V, was equipped with anti-shipping radar and powered by four 1,217 hp Pratt & Whitney engines providing it with a maximum speed of around 211 mph and a range of over 2,670 miles. It could carry a weapons load of up to 5,000 lb of depth charges or mines and had a maximum take-off weight of around 33 tons. The Sunderland remained in production through to 1945, with around 740 examples being built.

Perhaps the most famous flying-boat (and amphibious aircraft) of all, however, was the Consolidated PBY Catalina, born in 1933 in

response to a US Navy brief calling for a new patrol flying-boat.

The first prototype made its maiden flight on 28t March1935; it was a monoplane parasol wing carrying the two radial engines, a tapering fuselage and a raised tailplane. That year the US Navy ordered the first 60 planes of the production version, the PBY-1. Prior to the outbreak of World War II, Consolidated had introduced a further three versions with gradual improvements, most relating to the engines and armament. The PBY-5A was introduced in 1940, and its amphibious capabilities made it particularly flexible. This version was powered by 1,217-hp Twin Wasp engines and flew at a top speed of 179 mph, with a range of about 2,500 miles. Armed

with five machine-guns and carrying a load of up to 4,000 lb, the PBY-58 had a maximum take-off weight of over 35,250 lb. The Catalina was built at a frenetic pace in order to satisfy the demands of America and her allies (of whom the RAF took 700 planes), and was used as a spotter, a patrol plane, a bomber, and for search-and-rescue and anti-submarine duties. A total of over 4,500 Catalinas were manufactured, of which about 1,000 were built under license in the Soviet Union). The Catalina was an exceptionally robust and reliable aircraft, and numerous planes remained in service throughout the world into the 1960s. There are still a number of privately owned Catalinas in air-worthy condition.

The end of the Second World War nevertheless marked the onset of the seaplane's inexorable decline. In the field of post-war air transport, the great flying boats were replaced by more economical four-engined land planes which, thanks to their better aerodynamics and lighter weight, flew better and faster. Moreover, the demands of war had led to the widespread building of airfields, which meant that there was no longer any need for planes that could exploit economical stretches of water for take-off and landing operations.

The post-war period did however see the production of numerous seaplanes for specific duties, above all in regions such as Canada, Alaska and the Pacific Islands. A notable example is the Canadair CL214, a high-wing, twin-engined amphibian designed in the mid-Sixties as a specialist fire-fighting plane. Equipped with 2,129-hp Pratt & Whitney radial engines providing it with a top speed of 190 mph, the CL214 can scoop up no less than 5,450 litres of water, which can then be dumped on the nearest forest fire. The plane was updated in 1989 with the introduction of the CL215T version, which was equipped with turboprop engine in place of the original piston units. Then in 1993 the CL415, a production turboprop version featuring a number of aerodynamic and avionics innovations, made its first flight. About 200 planes of the Canadair amphibian series have been sold around the world, and the aircraft is still in production.

92 top Alcock and Brown make their historic take-off in their Vickers Vimy from St. John's airfield, Newfoundland, at 4.13pm on 14 June 1919.

92 bottom The inglorious end of Alcock and Brown's Vimy. After their successful trans-Atlantic flight, they landed in a water meadow in Ireland, which they mistook for a suitable landing spot.

Land-based planes were also involved in marathon flights and important expeditions that contributed significantly to the widening of man's horizons and the development of aviation. Perhaps the two most famous exploits were John Alcock and Arthur Brown's flight in 1919, and Charles Lindbergh's solo flight in 1927.

Since 1913 the London Daily Mail had been offering the considerable sum of £10,000 for the first aviator to cross the Atlantic Ocean in less than 72 hours, and with at most one refueling stop. The prize lay forgotten during the First World War but the race to win it began immediately after the signing of the armistice, since by 1918 there were both pilots and airplanes capable of the feat. In the spring of 1919, only three weeks apart, two men, both ex-RAF officers, knocked on Vickers' door to propose themselves as candidates for an attempt. One was John Alcock, an expert pilot and instructor; the other was Arthur Brown, an observer who had specialized in the modern techniques of astronomical and blind navigation. Vickers had an appropriate aircraft for the flight, the twin-engined Vimy bomber that had been modified with the installation of Rolls-Royce Eagle VIII power-units and enlarged fuel tanks that could hold over 750 gallons. After a single test flight, the pair, like the other competitors, decided to make their attempt by flying from Newfoundland to Ireland to take advantage of the Atlantic's prevailing winds, which blow from west to east. On 14 June 1919, after completing a sea voyage, reassembling the aircraft in a field near St. John's, and waiting for favorable weather forecasts, Alcock and Brown prepared for departure. After a nail-biting run and last-second lift-off, the Vimy headed through the fog in the direction of Ireland.

The flight was not trouble-free. First, the radio died, then one of the engines shed an exhaust manifold, and finally the crew's heated

The Great Land Plane Exploits

93 top John Alcock and Arthur Brown, the two intrepid aviators who on 14 June 1919 became the first to fly the Atlantic, after a flight of 16 hours, 27 minutes.

93 bottom The Vickers Vimy suffered substantial damage in its landing in Ireland. The two pilots, however, were much luckier.

flying suits stopped working. Ten hours into the flight the pair encountered a storm; they had no choice but to fly straight into it. The Vimy was at the mercy of the air currents within the clouds and fell into a spin; Alcock managed to regain control just fifty feet above the ocean waves. Despite the dark and rain, he and Brown gradually managed to regain altitude, but then they started to meet problems with icing. When they finally emerged into the sunshine Ireland was there in front of them, no more than 125 miles away. The flight had a less than heroic conclusion as the Vimy finished nose-down in a bog near Clifden that, from above, had appeared to be a suitable landing site. However, the crossing had been made – and in just 16 hours and 27 minutes. The two were fêted throughout the country and along with the £10,000 prize, they also received knighthoods.

Charles Lindbergh's Challenge

Eight years later, the American Charles Lindbergh completed another epic flight. In this case too, a monetary prize as well as thoughts of fame and glory stimulated the pilot's ambitions. In fact, in 1919, Raymond Orteig, a French entrepreneur living in New York, had announced a $25,000 prize for the first person who completed a non-stop flight from New York to Paris. Lindbergh, a 25-year-old pilot who had learned the ropes with flying circuses, the army, and the airmail service, decided that the best solution was that of a special plane with a single pilot and no radio. The Ryan

Company of San Diego built Lindbergh's aircraft, and it was named the NYP Spirit of St. Louis in honor of its financiers' hometown. The Spirit of St. Louis was a small high-wing monoplane dominated by an enormous 375-gallon fuel tank set between the engine and the minuscule cockpit. After meticulous preparations, Lindbergh took off from New York's Curtiss Field on 20 May 1927 at 7.52 in the morning; he had passed an almost sleepless night. The flight was relatively uneventful, apart from the continual attacks of drowsiness Lindbergh suffered for almost all of the 33.5 hours that it lasted before he reached Paris Le Bourget at 10.22 in the evening on 21 May. His flight of over 3,500 miles brought Lindbergh universal honors and an indelible place in the history of aviation.

95 top right Charles Lindbergh with his flying helmet. The 25-year-old American was the first to fly solo across the Atlantic, winning himself a place in history.

95 bottom left The cover-page of a famous song of the period illustrated with Lindbergh's Ryan NYP.

95 bottom right The 'Spirit of St. Louis' was an experimental plane. It was fitted with an enormous 450-gallon fuel tank, but had no radio and its single-seat cockpit offered no view directly forward.

94 top After his arrival in Paris, Lindbergh stopped in London: here is the Spirit of St. Louis surrounded by a cheering crowd at Croydon airport.

94 bottom The City of New York gave Lindbergh an exceptional welcome. Here the pilot is receiving a traditional Lower Broadway ticker-tape reception.

95 top left The Ryan NYP in flight. Lindbergh completed the New York-to-Paris trip in 33 hours, 30 minutes.

The Schneider Trophy

The Coupe d'Aviation Maritime Jacques Schneider (the Schneider Trophy race) was instituted by Jacques Schneider, the son of an arms factory owner of the same name. He wanted the trophy to promote development of seaplanes, the aircraft he felt to be most worthy of attention given that seven tenths of the earth's surface is covered by water. The race was to be held annually at a location decided by the previous year's winners, and a nation winning the trophy three times in five years would retain it in perpetuity.

The first race was staged in Monaco on 16 April 1913, over an aerial circuit of 10 kilometers (6.21 miles) that was lapped 28 times. The winner was a Frenchman, Maurice Prévost, who flew a 160-hp Deperdussin floatplane. On 20 April 1914, the second race, organized by France, was also held in Monaco, and was won by the Englishman Howard Pixton, flying a Sopwith Tabloid at an average speed of 86.82 mph. The First World War then interrupted the series, but in the post-war period, the spirit of competition and interest in the event were higher than ever. The third race was eventually staged at Bournemouth, England, on 10 September 1919. However, the weather was bad and only Italy's SIAI S.13 actually completed the full distance. As its performance could not be verified because of the fog, the organizers made the decision not to declare a winner but to entrust Italy with the organization of the 1920 race. This was held at Venice, with four aircraft competing, all Italian. Lieutenant Luigi Bologna, flying a SIAI S.12, won the contest. On 11 August 1921, again at Venice, three Italians and a Frenchman contested the trophy, which was won by De Briganti who, flying a Macchi M.7, averaged 117.91 mph. The sixth race was held at Naples, but the event did not go quite as expected: the Briton Henry Baird defeated the Italian and French teams. He was flying a Supermarine Sea Lion II powered to an average speed of 145.79 mph by a 456-hp engine. The 1923 race, held off the Isle of Wight, saw the first American victory thanks to XX Rittenhouse whose Curtiss CR-3 had

the better of the British and French teams, finishing at an average speed of 177.1 mph. Owing to a lack of competitors, the race was not staged in 1924. In 1925 the British and the Italians were beaten by Lieutenant Doolittle, whose Curtiss R3C-2, powered by a 600-hp engine, completed the circuit at the stunning average speed of 232.67 mph. In 1926, both the French and the British chose not to participate. They had had only one year in which to develop new planes and believed that it would be impossible to compete with the Americans. The Italians instead pulled out all stops, with the full support of the Fascist régime, which was eager for a prestigious international victory. Macchi developed the M.39 while Fiat created the 800-hp A.S.2 engine. The race was held at Norfolk, Virginia, and was an Italian-American challenge. The Macchi proved to be superior to the Curtiss, and the Italians finished first and third, with De Bernardi winning at an average speed of 246.6 mph.

The 1927 race was held at Venice, and this time the British were out in force with an excellent seaplane, the Supermarine S.5. This plane had been developed from the S.4, which had been unable to compete in 1926 because of an accident. Despite the Italian expectations placed on the new Macchi M.52, Captain S.N. Webster emerged victorious, flying an S-5 at an average speed of 281.79 mph. The Italians' defeat led to the founding of the Italian Royal Air Force's High Speed School at Desenzano, where one aim was to reclaim the trophy. After 1927, in recognition that the design of new aircraft was an increasingly complex affair, the race was staged every two years rather than annually. In 1928, however, in a non-Schneider trophy event, Italy did set a world air speed record at Venice, where De Bernardi reached a top speed of 297.96 mph with his Macchi M.52R.

In 1929 the Schneider trophy race was held at Calshot in England. The home team, with the new Supermarine S.6, won at a canter after the two Italian Macchi M.67s had to retire on the second lap.

Italy then plotted revenge with a fabulous new seaplane, the Macchi MC.72, but unfortunately it was not ready in time for the 12th race, held on 13 September 1931. Great Britain completed the event alone with Captain J.N Boothman flying the S.6B, powered by a 2,332-hp Rolls-Royce R engine. By winning this race, Bootham ensured that the Schneider Trophy would remain in Great Britain permanently.

Having lost her chance for victory in the Schneider Trophy race, Italy could take some consolation in the Macchi MC.72's victory in the 1933 Blériot Cup contest. Italy also set the world air speed record in 1934 when Sergeant-Major Francesco Agello achieved 440.90 mph, a record that remained unbeaten until 1939 and still stands as a seaplane record to this day.

The Winners

1913	France
1914	France
1919	No race
1920	Italy
1921	Italy
1922	Great Britain
1923	United States
1924	No race
1925	United States
1926	Italy
1927	Great Britain
1929	Great Britain
1931	Great Britain

96 top The first Schneider Trophy was held in Monte Carlo. It was won by Maurice Prévost piloting this 160-hp Deperdussin single-wing hydroplane.

96 bottom The most important plane missing from the 1931 Schneider Trophy competition was the Macchi MC.72, an exceptional plane for the period but not yet perfected. In 1934, it won a world speed record for hydroplanes of 443 mph, still unsurpassed.

97 top An advertising poster for the Gloster announces the Model VI's victory in the King's Cup Race of 1929.

97 bottom The French businessman Jacques Schneider started the race that bears his name with the aim of developing hydroplanes.

The Hurricane's decline as a front-line fighter was not a major problem for the RAF as its successor had been available for some time. This was the famed Supermarine Spitfire, an aircraft so beautiful and effective as to gain the respect and envy of even enemy pilots.

The Spitfire, one of the greatest fighters in the history of aviation, was designed shortly after the Hurricane and completed its maiden flight on 5 March 1936. The aircraft was the triumph of Reginald Mitchell, a designer who had gained a solid reputation by creating the world's fastest planes, the winners of the Schneider Trophy. In 1931 he began work on a fighter, the Supermarine Type 224, but probably underestimated the complexity of the project. The resulting design was a failure. Mitchell then decided to shorten the wingspan and the fuselage and to adopt a retractable undercarriage. The result was the Type 300, which was still too slow when climbing. The solution to the problem arrived in 1934, when Supermarine reached an agreement with Rolls-Royce for the supply of a new engine, the Merlin; with this power unit the Type 300, eventually to become known as the Spitfire, was transformed. After its maiden flight, the test pilot Joseph Summers exclaimed, "Don't touch anything."

The Spitfire was a sleek, elegant machine, but it was also innovative and highly efficient, so much so that a number of its features such as the elliptical wing and the semi-stressed skin initially caused production problems. In terms of both the structure and the engine, the Type 300 had grown out of work done on aircraft developed for the Schneider Trophy contest. The plane was equipped with the same armament as the Hurricane but boasted a more powerful engine and handled magnificently. The first production Spitfires were delivered in 1938, and by 1939 the Air Ministry had ordered 4,000 aircraft to be constructed by a number of companies. At

100 top left The Supermarine Spitfire was Britain's most famous fighter in World War II; it owed its success to its manufacturer's experience in the Schneider Trophy.

100 top right A Spitfire Mk. XII with trimmed wings during an exercise over Britain in 1944. The Spitfire remained in production until 1948; in total, almost 23,000 were produced.

the start of the war only nine squadrons were equipped with Spitfire Mk.Is, but their numbers grew rapidly. The Mk.V appeared in the March 1941 with heavier armament and a more powerful engine (the 1,440-hp Merlin). This version, which was capable of over 373 mph at 19,500 feet, was also used as a fighter-bomber.

In July 1942 the Mk.IX was introduced to meet the threat of Germany's new Focke-Wulf 190. In 1943, the Spitfire began to be equipped with the 2,050-hp Griffon engine that, together with a five-bladed propeller, made the Mk.XIV one of the fastest aircraft in the world, capable of matching the best the Germans could offer.

Production of the Spitfire continued through to 1948 with over 22,800 planes being built in some 30 versions. More than 20 air forces, including those of the Soviet Union and the United States, used the plane. Numerous aces flew the Spitfire. From the RAF's standpoint, the most important of them was James "Johnny" Johnson, who shot down 38 enemy fighters. In no fewer than 515 missions, he was only hit once and survived the conflict, continuing to serve with the RAF in the post-war period. It is sad to record that Reginald Mitchell, designer of the Spitfire, was never able to enjoy his creation's success. He died of cancer in June 1937, a year before the Spitfire entered service.

100 bottom The Spitfire production line, Southampton, 941. Some of the plane's most advanced technical characteristics initially gave problems when mass-produced.

101 top Duxford airdrome, 1939: a squadron of Mk. I Spitfires taking off in an emergency exercise.

101 bottom A formation of Mk. XII Spitfires in flight in May 1944 before the D Day landings in Normandy. This model had the more powerful 2050-hp Rolls-Royce Griffon engine.

Two other important Second World War British fighters were the Hawker Typhoon and Tempest. The Typhoon was developed in the late 1930s in response to an Air Ministry requirement for an eventual successor to the Hurricane fighter. Hawker came up with two prototypes, each of which was fitted with one of two new engines specified in the brief, the Rolls-Royce Vulture and the Napier Sabre. The design using the first of these engines, the Tornado, was abandoned, while the second, the Typhoon, went ahead, despite myriad difficulties encountered during the final development of the engine. The prototype first flew on 24 February 1940. However, problems with the engine were then compounded by deadly adversary for enemy vehicles. Assigned to the 2nd Tactical Air Force, the Typhoons could operate from advanced bases close to the front lines, frequently little more than fields with improvised runways, stands and maintenance hangers. Used in this way, the Typhoon could be kept in the air continuously and could easily support the land forces, among which there would have been an air force officer (the so-called FAC, Forward Air Controller). When required, the FAC would call up the planes and guide them to enemy groups that were obstructing the advance of the Allies' armor. It was above all with the Typhoon that the Close Air Support strategy was born and developed.

102 The Gloster Meteor, Britain's first jet aircraft, first flew in March 1943. In 1945 it was sent to the Belgian and German fronts but was not up to the class of the Messerschmitt Me.262.

103 top left A formation of Mk. IA Typhoons on a training flight over England. Initially, the plane was supposed to be an aerial-defense fighter, but proved to be very dangerous to pilot.

103 top right A Hawker Typhoon fighter-bomber with four 20-mm cannons and eight air-to-ground underwing rockets, which were deadly against armored vehicles.

103 center Close-up of an RAF Gloster Meteor; it shows the central position of the wings, the two jet engines and the four 20-mm cannons in the nose.

certain structural failures, and after a number of fatal accidents shortly after the Typhoon entered service in 1941, it was threatened with withdrawal. Remedial work was doe on the design, and in 1942 the Air Ministry decided to continue using the airplane, not as an interceptor but as a fighter-bomber. In all, more than 3,000 of these fighter-bombers were manufactured.

The Typhoon's new career began in August 1942, and from 1943 on the plane proved to be an excellent tactical support weapon, known to be at its best in low-level flight. The Typhoon was never an easy plane to fly, but it built its reputation in the summer of 1944 when it participated in the Normandy landings and the Allied troops' advance across northern Europe into the heart of Germany. It was armed with four 20-mm cannons and 2,000 lb of bombs or rockets, making it a

The Typhoon's successor was the Tempest, a variant design (initially known as the Typhoon Mk.II) intended to be operated as an interceptor. The Tempest Mk.I made its maiden flight on 24 February 1943, but continuing problems with the Sabre IV engine resulted in it being abandoned in favor of the Mk.V, which was equipped with the Sabre II engine. This version, which first flew on 21 June1943, was approved for serial production. It proved to be powerful and extremely fast at all altitudes. The Tempest Mk.II was then introduced with the Centaurus engine and improved aerodynamics, thanks to adoption of an annular radiator that made the nose more streamlined. More than 1,500 Tempests were built and the plane was retained in service in the post-war period until 1951.

The last of the RAF's famous Second World War fighters was the Gloster Meteor. Its fame derived largely from the fact that it was the first and only Allied jet fighter that actually took part in the war. The origins of the Meteor dated back to 1940 when the Air Ministry decided to press on with research on jet propulsion. The experimental Gloster E.28/39, propelled by a Whittle W.1 engine, was first British jet aircraft; it made its maiden flight on 15 May 1941. Gloster then built a Meteor equipped with a later version of this power unit, the Whittle W.2B/23, producing 1,697 lb of thrust, with the twin engines buried in the mid-mounted wing. Eight such prototypes were constructed. The first new Meteor to fly, on 5 March 1943, was the fifth in the series. An initial batch of 20 planes of the production version was ordered, and they entered service in June 1944. They were immediately employed over England to intercept German V1 flying bombs. About

103 bottom The Typhoon's successor was the Tempest Mk.V, which could be used as a fighter or a fighter-bomber. It had the same armament as the Typhoon but its Centaurus 2550-hp engine gave it a maximum speed of 442 mph.

200 planes of the subsequent Meteor F Mk.III version were built and assigned to number of squadrons. They saw action in 1945 in Belgium and Germany. The Meteor was not, however, a particularly fast aircraft and was generally inferior to the Messerschmitt 262, which it never actually met in combat. The Meteor had a highly successful post-war career. More than 25 different versions, including a trainer, were developed; a total of over 3,800 Meteors were built and were purchased by 14 different countries. The RAF employed the Meteor until 1961; it saw combat action in Korea and in the Arab-Israeli wars.

104 top The four-engined Halifax bomber entered RAF service in 1940. More than 25 versions were developed and over 6000 planes built. Shown here is a Handley Page HP.57 Halifax Mk.III.

104 center The Short S.29 Stirling was a contemporary of the Halifax. It first flew in May 1939 and entered service in August 1940, but was quite soon surpassed and in 1943 was relegated to secondary duties.

104 bottom A daytime bombing raid on Cologne, 18 August 1941. The RAF did not take long to take the war in the air directly to the heart of enemy territory.

105 Cold, uncomfortable and noisy: inside a Halifax bomber during a raid on Germany in October 1941. The machine-gunners search the sky for enemy fighters.

During the Second World War, the RAF successfully employed various types of bombers. Heavy bombers included the Short Stirling, Handley Page Halifax and Avro Lancaster; light bombers included the Bristol Blenheim and Beaufighter and the de Havilland Mosquito. The Stirling and the Halifax had almost parallel careers, both being designed in 1936 in response to the Air Ministry's requirement for a new medium bomber. Both aircraft were equipped with four engines and both flew for the first time in 1939, the Stirling on 14 May and the Halifax on 25 October. The two planes' shared characteristics ended there, however. The Halifax was by far the more modern of the two and could reach a considerably higher altitude than the Stirling. The Halifax Mk.III version that went into service in 1943 was fitted with Hercules radial engines that improved its performance and ensured that together with the Lancaster, the Halifax remained in front line service throughout the war. In addition to use as a bomber, the Halifax was used for transport and glider-towing duties and for dropping parachutists. In total, 6,176 planes were built. From mid-1943 on, the RAF diverted the Stirling from bombing to transport and glider towing.

The most important British bomber of the war was in fact the Lancaster, an aircraft that evolved curiously out of an unsuccessful project. In response to an Air Ministry brief, Avro had built a heavy twin-engined bomber, the Manchester, powered by Rolls-Royce Vulture engines that proved to be insufficiently powerful and somewhat unreliable. In mid-1940, in light of this unsatisfactory performance, Avro decided on a complete and urgent redesign, focusing on a four-engined layout using Rolls-Royce Merlins, redesigned tailplanes and a greater wingspan. The result was the best British bomber of the war: nine factories built 23 different versions, for a total of more than 7,300 Lancasters.

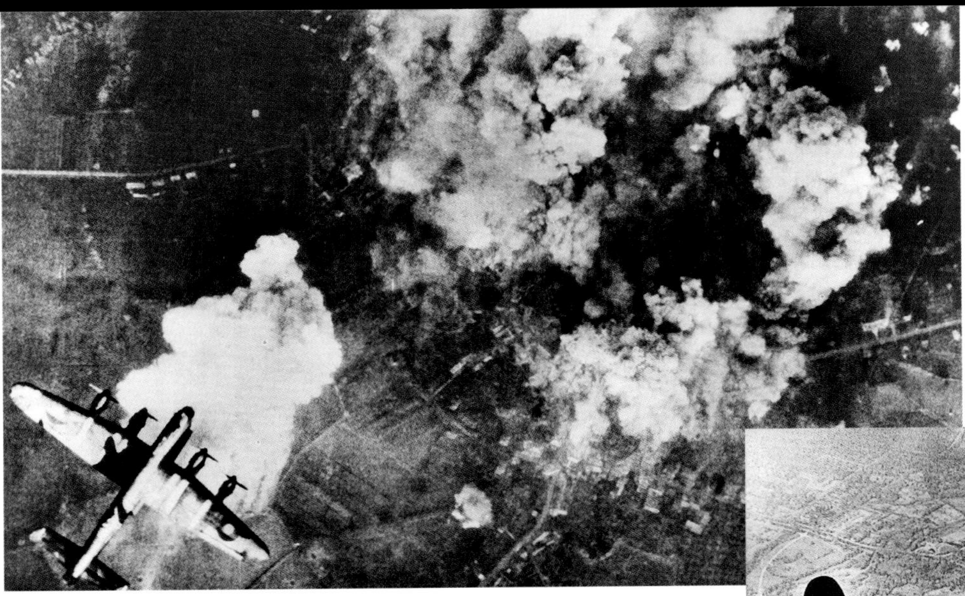

106 bottom The aggressive take-off of a light Bristol Beaufighter Mk. VIC, an anti-shipping plane equipped with a torpedo. The Beaufighter entered service in 1940; 6000 were produced.

107 top left A fighter-bomber de Havilland Mosquito Mk.IV in 1942. Made from wood and offering exceptional performance, it was quickly nicknamed "the Wooden Wonder."

106 top left A four-engined RAF Avro Lancaster on a bombing raid on a German city.

106 top right The Lancaster, possibly the best British bomber in the war, entered service in January 1942. It was produced in 23 versions and could carry up to 14,000 lb of bombs.

The first Lancaster flew on 9 January 1941, and the type entered service in January 1942.

The Lancaster was also used for radar countermeasures, reconnaissance and maritime patrol duties, and remained in service until the 1950s. It was the RAF's principal bomber; it could carry the heaviest bombs and it shouldered most of the weight of the British night bombing offensive over Germany and occupied Europe. The Lancaster dropped several special-purpose bombs. They included the "Tallboy," an armor-piercing 12,000-pounder used, for example, to sink the Tirpitz; the 22,000-lb "Grand Slam" that achieved supersonic speed during descent; and a remarkable Barnes Wallis-designed bouncing bomb – a 9,250-lb cylinder bomb used to destroy the dams servicing the Ruhr's hydroelectric power stations. This delicate mission, perhaps the most famous of all those flown by the RAF during the war, was entrusted to 617 Squadron, the "Dambusters" as they became known, specially formed for the task and led by the veteran Lieutenant-Commander Guy Gibson. The mission was difficult because of the extreme precision required. The dam-busting bomb was set rotating at 500 rpm by a hydraulic motor before the drop; it had to be released at 250 mph from a height of 60 feet above the water and between 1,200 and 1,345 feet from the dam. Only within these parameters would the bomb strike the face of the dam and sink to a depth of 40 feet before exploding. Squadron 617 had only six weeks in which to prepare for the raid. The squadron took off for Germany on the night of 16 May 1943. Results met expectations: two of the three dams were breached. Success had its price, however: eight of the 18 crews failed to return. Guy Gibson himself died in September 1944, shot down while returning from a raid over Germany. It was to have been his last mission.

The first of the famous British light bombers was the Bristol Blenheim, originally designed as a fast, twin-engined passenger plane. The Blenheim made an immediate impression during its maiden flight on 12 April 1935: it was faster than any fighter then in service. A military version of the plane was developed and was flown for the first time on 25 June 1936. Bristol built almost 6,400 Blenheims, and the aircraft remained in service throughout the war, even though its performance advantage became eroded. The Bristol Beaufighter, a heavily armed twin-engined aircraft, made its first flight on 17 July 1939, and entered service in September 1940 as a night-fighter. Such was the quality of the design that day-fighter and Coastal Command strike versions were subsequently developed. The Beaufighter achieved its full potential in the Mk.VI and Mk.X versions, and the plane eventually carried rockets and torpedoes used against shipping and submarines. Almost 6,000 Beaufighters were built in 22 versions, and it remained in service for the duration of the war.

A plane almost as famous as the Spitfire, the de Havilland Mosquito, together with the Supermarine fighter and the Avro Lancaster, became a RAF symbol in the Second World War. The Mosquito came into existence in 1938; it was a product of de Havilland's private development program. Initially, the experts considered the plane to be too advanced: for this reason the Air Ministry

107 top right A two-engined Bristol Blenheim Mk.IV making a sharp turn early in the war. When introduced in 1935, it was the RAF's fastest fighter.

107 bottom Loading bombs into an RAF Lancaster. The air offensive against German cities and factories was one of the keys to Allied success in World War II.

rejected it no less than three times. Only the tenacity of Patrick Hennessy, a Ford manager seconded to the development of British aeronautical production, enabled the Mosquito to go into production. Once the Mosquito had demonstrated its merits in flight, there was no longer any question as to its suitability. In fact, following the outbreak of war, the authorities realized that an aircraft of this type, with a frame made of wood (a non-strategic material), would be extremely useful. Developed as a light bomber, the Mosquito first flew on 25 November 1940 and entered service in 1941. Thanks to its outstanding performance (a top speed of 400 mph, great maneuverability, a rapid rate of climb, a 1,000-lb bomb load, a range of about 1,500 miles) and economical construction that earned it the nickname "Wooden Wonder," the Mosquito was soon developed for reconnaissance and night-fighter use. On 17 September 1941, the Mosquito undertook its first mission. Given its great speed, this was a reconnaissance flight extended as far into France as Bordeaux. More than 7,700 Mosquitoes were built in more than 40 versions, and the plane was used by the air forces of 18 countries. The Mosquito remained in service with the RAF through to the early 1960s.

29629
48
48

108 top left The Bell P-39 Aircobra entered service in January 1941. It had several defects and was never appreciated by American pilots. The Russians, however, did like it and used about 4800 of the fighter-bombers, which they received from the United States.

108 The Curtiss P-40 Warhawk entered service in 1940. This fighter enabled the American air force to take on the Japanese in the first few months of the war; however, it was not up to scratch and was soon relegated to minor duties.

109 top The Japanese attack on Pearl Harbor on 7 December 1941 forced the United States to enter World War II. Here the NAS base on Ford Island is under enemy fire; in the background, the USS Shaw and USS Nevada are in flames.

During the 1930s and 1940s, the United States' aeronautical industry, like that of Great Britain, had to accelerate its output if it hoped to keep up with and eventually overtake the technical standards set by its adversaries. On the outbreak of the war in Europe, the USAAC's most modern aircraft were the Bell P-39 Airacobra and the Curtiss P-40 Warhawk, neither of which were top-drawer fighters. The former had aroused great enthusiasm when it first appeared in 1937, equipped as it was with a 37-mm cannon, a turbocharged engine located centrally behind the cockpit, and the first example of a retractable tricycle undercarriage with a nose-wheel. However, after military modifications that included elimination of the turbocharged engine in favor of a more traditional unit, the P-39 proved to be a heavy, clumsy aircraft unsuited for use as a fighter. Nonetheless, the airplane was approved for serial production and entered service in January 1941. The Americans never loved it and the British rejected it, with only Russia appreciating its qualities as a robust fighter-bomber. In fact, the Soviet air force received almost 4,800 of the 9,588 P-39s that Bell manufactured.

The Curtiss P-40 Warhawk fared better than the Bell P-39. It completed its maiden flight on 14 October 1938, and was immediately ordered into serial production, with the first Warhawks coming into service as early as 1940. The P-40 offered the advantage of immediate availability as well as respectable performance, and this allowed for high-volume production in a period of emergency. The P-40 was immediately adopted by the RAF as the Tomahawk, and was also used by the American volunteer "Flying Tigers" in China between December 1941 and July1942. Until the appearance of the new Bell P-38s and Curtiss P-47s, the P-40 was the best American fighter of the war, and it remained in production to satisfy the Allies' requirements. By December 1944, more than 13,700 Warhawks had been built.

109 bottom Innovative and unmistakable, the Lockheed P-38 Lightning fighter had exceptional performance and long-distance capability. It was used in the most important American assaults; more than 10,000 were made.

After the United States entered the war following the Japanese attack on Pearl Harbor on 7 December 1941, two new and significantly better fighters were introduced. The first was the Lockheed P-38 Lightning, designed in response to a brief issued by the USAAC in 1937 that called for a high-altitude interceptor capable of 528 mph at 19,500 feet and of climbing to its optimum operating altitude in 6 minutes. Many companies felt that these demands were impossible to meet, but Lockheed decided to make an attempt. The resulting design incorporated two turbocharged engines mounted in twin fuselages connected by the wing and twin tailplanes; the cockpit and the armament were set in a pod at the center of the wing. This was a decidedly innovative layout, but when the prototype made its first flight on 27 January 1939, its performance impressed the army and it was approved for serial production. The P-38, of which more than 9,900 were built, began to enter service at the end of 1941. The plane was at its best in the Pacific theater, where its great range was particularly advantageous.

The P-38 was flown by great American aces, such as Bong and McGuire. Richard Bong was sent to the front in November 1942, assigned to the 35th Fighter Group, then fighting in the Pacific. He completed a number of duty tours in this theater, attaining the rank of major and scoring a USAAF record total of 40 victories. He received the Congressional Medal of Honor, the highest American military award. Bong continued his career as a test pilot, but on 6 August 1945, while testing a Lockheed P-80, his aircraft crashed in flames and he was killed instantly.

In the second half of 1942, the P-38 began operating in Europe and Africa, but in these situations the Republic P-47 Thunderbolt proved to be the better of the two. This fighter developed out of a number of reworkings of a basic project headed by the aircraft designer Alexander Kartveli and the air force's own experts. Born as the AP-10, a light fighter, the plane was eventually enlarged, strengthened and equipped with the most powerful engine available at the time, the 2,000-hp Pratt & Whitney Double Wasp. In June 1940 it received approval as the XP-47B. It was to be the heaviest single-engined fighter of the Second World War, with a maximum take-off weight of almost 20,000 lb. The P-47 completed its maiden flight on 6 May 1941, and went into production in February 1942. The 56th Fighter Group, stationed in England, was the first operational unit to use the p-47 in combat, and employed it as a bomber escort. In spite of its tubby, clumsy appearance, the P-47 was a fast, well-armed and robust fighter that proved its worth in dogfights as well as in fighter-bomber duties. By December 1945 more than 15,600 P-47s had had been built, and in the post-war years the plane remained long in service with numerous Allied air forces.

Of all the fighters flown by the US Army Air Force (as it was known from 1942 on), the North America P-51 Mustang was the greatest and most famous – a true thoroughbred of the air. This plane was the fruit of close collaboration between the United States and Great Britain. In April 1940, Britain was desperately seeking fresh arms with which to thwart the German offensive, and discussions with North American led to the definition of a project that was to use the same Allison V12 engine as the P-40, but mated to a better airframe. The only condition imposed by the British was that the aircraft should be ready in 120 days. North American actually took three days less than that, producing an extremely advanced plane equipped with a laminar wing. The RAF named the new fighter the Mustang Mk.I. It first flew on 26 October 1940, and its performance was excellent at low level, but dropped off notably at higher altitudes. The British thus put it into service in April 1942 as a reconnaissance and ground-attack plane. Both the British and the Americans felt that the Mustang's performance could be improved by replacing the Allison engine with the more powerful Rolls-Royce Merlin, produced under license in America by Packard. The new version, designated as the P-51B, was decisively superior to any other fighter then in service and was immediately put into production: it first saw action in Europe in 1943. The definitive version, the P-51D, was introduced early in 1944 and adopted a new bubble cockpit canopy that reduced the dimensions of the fuselage. The Americans used this plane to escort their bombers deep into the heart of Germany, and it proved to be a supreme interceptor. In 1944, the Mustang was the world's fastest fighter, with the exclusion of the first, very rare, jet aircraft. It was nonetheless a adversary to be reckoned with even by the Messerschmitt Me.262: on 7 October 1944 Lieutenant Urban Drew of the 361st Fighter Group swooped on two Me-262s as they were taking off, and shot them both down. On 25 February a formation of P-51s from the 55th Fighter Group came upon a German airfield just as an entire squadron of Me-262s was taking off and managed to shoot down six: the American pilots, confident of their P-51s' capabilities, were not easily intimidated by the new German jets. The Mustang's success is easily demonstrated by the statistics: over 15,400 were built and over 50 nations' air forces flew them until the 1960s.

110 top A North American P-51 Mustang is directed by a USAAF specialist before take-off on a mission escorting American bombers over Germany.

110 bottom left The Republic P-47 Thunderbolt looked heavy but was one of the fastest aircraft of the war, capable of 431 mph. Seen here is a reconditioned P-47D in the colors of the 366th Fighter Squadron, 358th Fighter Group, in France in 1944.

110 bottom right P-47D first series aircraft still featured the old-style cockpit canopies. The plane shown here has been restored and has the colors of the aircraft flown by Cameron Hart, 63rd Fighter Squadron, 56th Fighter Group, at Boxted in December 1944.

111 left A P-51D Mustang performing acrobatics. With two underwing, releasable fuel tanks, the Mustang proved an excellent fighter-escort plane able to accompany bombers into the heart of Germany.

111 right The North American P-51 Mustang was probably the best fighter of the entire Second World War and stayed in service with the USAF from 1943 till 1957. More than 15,400 Mustangs were built and more than 50 air forces round the world used the plane.

In fact, the planes saw action in a number of conflicts after the end of the Second World War. The American aviation industry also had to satisfy the navy's demands for carrier-borne fighters. In this field, Grumman seized the lion's share of orders, with the company turning out a series of increasingly advanced aircraft. The US Navy's first modern fighter was the Grumman F4F Wildcat, the firm's first all-metal monoplane. The F4F was developed in 1936 and completed its maiden flight on 2September 1937. After a number of modifications had been made, it entered service in 1940, first with Britain's Royal Navy (where it was known as the Martlet), and then from December of that year with the US Navy. The Wildcat, with production totaling 7,800 planes, was no better than the Japanese Zero, but thanks to the tactics developed by the American pilots, it managed to fend off the Zero until 1943, when it was replaced by the F6F Hellcat. Th F6F resulted from wartime experience that had shown the need for a more powerful engine, cockpit armor, more ammunition, and more accommodating flying characteristics. The first Hellcat flew on 26 June 1942. Given the climate of emergency, the testing and development processes were extremely rapid; the plane was in fact pressed into service as early as January 1943. In total more than 12,200 Hellcats were built. The US Navy employed the F6F throughout the Pacific theater, where it proved to be superior to the various Japanese fighters, including the Zero. The Royal Navy also used the Hellcat.

At much the same time as it adopted the Hellcat, the US Navy introduced another carrier-borne fighter, the even faster Vought F4U Corsair, albeit after a more difficult period of gestation. The F4U's design dated from 1938, and the prototype made its maiden flight on 29t May 1940. During the course of the year the 4FU's stunning performance became apparent; it was the first plane to break the 400 mph barrier in horizontal flight. However, modifications were required. These included the adoption of heavier armament and the relocation of the fuel tanks, with the consequent setting back of the cockpit to the detriment of forward vision. This last factor caused considerable problems in carrier-based use of the planes, and the first ones to enter service were actually assigned to the Marines, who from February 1943 flew them from land bases. It was only in April 1944 that the F4U was approved for carrier operations. The Corsair nonetheless proved to be an excellent fighter and fighter-bomber and remained in production until 1952, with over 12,500 being built.

112 bottom The Grumman F4F Wildcat, which entered service in 1941, enabled the US Navy to compete during the first few months of the war. Though later surpassed in performance, it remained in service until 1943. Britain's Royal Navy also used the Wildcat, which it called the Martlet.

113 top The Grumman F6F Hellcat (another 'cat') was the Wildcat's successor on the American aircraft carriers, entering service in January 1943. Shown here is a Hellcat taking off from the USS Yorktown. The propeller-tip aerodynamic effects indicate the humidity of the air.

113 bottom A Hellcat fighter photographed on the deck of the USS Lexington on 13 July 1944, during the battle of Saipan in the Mariana Islands.

112 top The Vought F4U Corsair, an excellent fighter/fighter-bomber, was in service with the US Navy and Marines from 1943. Shown here is an F4U-1A in 1944, flown by Navy Lieutenant Ira "Ike" Kepford of Squadron VF-17.

112 center The Grumman F8F Bearcat, the US Navy's most advanced fighter during World War II, entered service in 1944. It had a 2129-hp engine providing a speed of 430 mph.

The US Navy's best fighter, and indeed the best carrier-borne piston-engined fighter of all time, however, was the F8F Bearcat. A result of Grumman's policy of continual refinement of its successive models, the Bearcat was designed in 1943 and made its first flight on 21 August 1944. Compact, fast, maneuverable, and with an outstanding rate of climb, the Bearcat was immediately put into production, though the Allied victory in 1945 meant that many of the ordered planes were cancelled. About 1,260 Bearcats were built, and the plane remained in service with the US Navy until 1952 and throughout the 1950s with the navies of other countries, including France, South Vietnam and Thailand.

The most advanced Allied bomber was, however, the Boeing B-29 Superfortress, which completed its maiden flight on 21 September 1942. The army required a bomber with extreme range that could be employed for strategic operations in the Asian theater; Boeing responded with an avant-garde airplane that even before that maiden flight had attracted 1,500 orders. The B-29 had a completely pressurized crew compartment, a retractable tricycle undercarriage, and four huge turbocharged engines, each producing 2,231 hp. However, the complexity of these engines caused major reliability problems and drastically reduced the B-29 squadrons' initial effectiveness. Nonetheless, the B-29 was the key to the rapid end of the war against Japan. The final thrust was marked by an incendiary bombing campaign started in March 1944 and culminated in the B-29s dropping atomic bombs on Hiroshima and Nagasaki on 6 August and 9 August 1945.

The USAAF's strike forces were completed by a number of twin-engined planes for medium-light bombing and ground-support duties. The most famous models were the North American B-25 Mitchell, the Douglas A-20 Havoc, the Martin B-26 Marauder and the Douglas A-26 Invader. The best was unquestionably the Invader, which entered service in April 1944. This plane was extremely fast (over 350 mph), powerful (two 2,028-hp engines), and heavily armed (up to ten 12.7-mm machine guns and 4,000-lb bomb load). The fact that it remained in service with the USAF until the Vietnam War and was used by a number of other countries is testimony to its worth. More than 2,400 Invaders were built.

A number of non-combat models featured among the most important Allied aircraft. For example, transport planes were crucial to the war effort, not only to maintain supply lines to troops scattered around the world, but also to play a key role in a various air transport operations – first and foremost the D-Day landings in Normandy. The leading transport plane was the twin-engined Douglas C-47 Skytrain, the military version of the famous DC-3. The civil airliner had completed its maiden flight on 1 July 1933, and in 1940, when it was selected as the standard military transport for the American armed forces, it was the leading airplane on the civil market and already in service with the US Army and Navy. During the war, some 12,700 Skytrains were constructed, along with a further 2,000 built under license in the Soviet Union as the Lisunov Li-2. The Skytrain was a simple, rugged aircraft, but with features that made it an avant-garde design for its time. The RAF designated it the Dakota, and it has continued to fly with civil and military operators throughout the world, until the present day.

117 center The Douglas A-26 Invader was another medium bomber and assault aircraft was. It entered service in 1944 and was armed with ten 12.7-mm machine-guns and a bomb load of to 4,000 lb.

117 bottom B-25 bombers from the 345th Bomber Group attack Japanese-held Dagua airfield in New Guinea in 1943.

116 top A Martin B-26 Marauder medium bomber during Operation Overlord, the assault on the beaches of Normandy in June 1944. The black and white stripes worn on the occasion are easily seen.

116 center A formation of USAAF Douglas A-20 Havoc bombers over the North Sea, looking for German ships. The RAF also used this aircraft, designating it the Boston.

116 bottom The Douglas C-47 Skytrain transport plane, the military version of the famous DC-3, was one of the main reasons for the Allied victory in World War II. This airplane was used with success in many roles, including medical transportation, as shown here.

117 top The Boeing B-29 Superfortress used against Japan was undoubtedly most powerful bomber of the war. The two 468th Bomb Group Superfortresses shown here are dropping incendiary bombs on targets in Burma in 1945.

118 top Mustered in front of their MiG-3s, Soviet pilots take an oath before combat. The Mig-3 was hurriedly brought into service in 1941.

118 center Soviet workers building aircraft during 1943. Despite dreadful work conditions and inferior technologies, the Russians managed to produce over 80,000 airplanes during the war.

118 bottom left The Polikarpov I-16, the standard Russian fighter at the start of the war, was an excellent aircraft in 1935 but heavily outclassed by 1941.

118 bottom right The Ilyushin Il-2M3 was the Soviets' best attack plane. Armed with 37-mm cannon and 2,200 lb of bombs and rockets, it earned the name 'flying tank.' Including several postwar versions, a record number of 58,192 were built.

Mention should be made of the most important aircraft that the Soviet Union fielded during the Second World War. Stalin, who had signed a non-aggression pact with Hitler, was caught completely by surprise by Operation Barbarossa, the attack the Germans launched along the length of the Russian border on 22 June 1941. The Soviet armed forces were unprepared for conflict: their training was inadequate; their equipment too often obsolete and virtually useless. The situation was no better for the Soviet air force; in the first days of the war, it was almost annihilated on the ground and in the air.

At the outbreak of war, the Soviet Union's primary fighter was the Polikarpov I-16, a plane from 1933. In its day, it had been the first monoplane fighter with a retractable undercarriage, but by 1941 by the German Bf-109 outclassed it. However, in 1939, the Soviet Union introduced the MiG-1, a fast, modern aircraft equipped with an in-line engine, but which suffered from problems of instability. After just 100 MiG-1s had been built, the plane was replaced by the MiG-3, which incorporated a more powerful engine, greater range, and other improvements. The MiG-3 entered service in 1941, and it remained in use until 1943. During this period about 3,300 were built. However, the aircraft that enabled the Russians to compete on even terms with the Germans were those produced by the design departments of the Lavochkin and Yakovlev enterprises. Lavochkin, after having launched the LaGG-1 and 3 models, two relatively unsuccessful machines that were only lightly armed and dangerous to fly, in 1941 introduced a new model using the powerful 1,600-hp Shvetsov M.82 radial engine. The result was the La-5, which entered service in 1942. The design was soon uprated to La-5FN specifications (more powerful engine, composite wood-metal structure), and then to La-7 specifications (an even more powerful engine, heavier armament and a new all-

119 top Yakovlev Yak-9 fighter, an excellent aircraft that remained in production from 1942 to 1946. It was used in the Korean War in the 1950s.

119 center Yak-9 fighters on a training flight. In total, 36,000 were built.

119 bottom The Lavochkin La-5FN was another successful Soviet fighter. It was equipped with a 1700-hp engine and It could reach 400 mph.

metal wing), introduced in 1943 and 1944 respectively. The La series was produced until 1946, with a total of more than 22,000 units being built.

Yakolev introduced its first design, the Yak-1 in 1940. This was a small, elegant fighter that entered service in 1942. Such was its success that its designer was honored with the Order of Lenin. This first model was gradually improved and modified, giving rise to the Yak-3 (an excellent low-altitude fighter-bomber), the Yak-7, and lastly the Yak-9, which entered service in 1942 and was to remain in production until 1946. Fast, well-armed and agile, over 36,000 piston-engined Yak fighters were built. Yak-9s fought in Korea between 1950 and 1953.

The most famous Soviet aircraft of the Second World War was, however, the Ilyushin Il-2 Shturmovik, the aircraft produced in the greatest numbers in the history of aviation. The USSR had been looking for a tactical support aircraft since 1938. A requrement was issued to which the Sukhoi and Ilyushin design departments responded. According to the Soviets the air force's duty was to provide tactical support for ground forces. The first Ilyushin prototype flew on 2 October 1939. In 1940, a second prototype, with an improved engine, armament, armor and maneuverability, was approved for production entered service in summer 1941. The Shturmovik, a single-engined, two-seat fighter-bomber with a low wing and a heavily armored engine and cockpit, became the scourge of the German armored divisions. There was no escaping the Shturmoviks: they were armed with 37-mm cannons, fragmentation bombs, bazookas and 132-mm anti-tank rockets. After having dropped their fragmentation bombs from about 2,500 feet, these aircraft would usually press home their attack by skimming over the ground, aiming directly at the armored vehicles. On the morning of 7 July 1943, during the famous Battle of Kursk, the greatest tank battle of the Second World War, the fighter-bomber provided a dramatic illustration of its offensive capabilities. Swarms of Shturmoviks attacked at dawn. Within 20 minutes, the 9th Panzer Division had lost 240 of its 300 armored vehicles, while 270 the 3rd Panzer Division's 300 vehicles had been hit.

The Il-10 Shturmovik, an improved model, first flew in 1944. It proved to be a truly excellent machine, equipped as it was with a 2,000-hp engine, four 23-mm cannons and up to 2,200 lb of bombs and rockets. The Shturmovik series remained in production until the 1950s, being adopted by all the Communist countries. It was manufactured in the USSR and Czechoslovakia, and over 58,000 were built.

122 top left Detail of a Heinkel He.111 medium bomber, taken from the fore-station in another plane in the formation.

122 top right The glass nose of a Heinkel 111. Note the placement of the machine-gunner that half-lay in the fore-station of the plane. The pilot sat behind him to the left.

At the start of the war the Luftwaffe was also operating three well-known medium bombers that were responsible for most of the attacks on Poland, France and, above all, on the United Kingdom during the Battle of Britain. These bombers were the Dornier Do-17, the Heinkel He-111 and the Junkers Ju-88. As already mentioned, Germany never built or used four-engined strategic bombers. This colossal error was largely due to Hitler's impatient insistence on the immediate creation of a large air force. Thus, in 1936 Goering cancelled two large bombers projects that would have required greater resources and longer lead times in favor of full scale mass-production of the proven Dornier Do-17 and the Heinkel He-111. At the time Goering stated, "The Führer will not ask me how big the bombers are, but how many of them are there."

The twin-engined Dornier Do-17, that first saw the light of day in 1934, was designed to meet Lufthansa's demand for a fast airmail plane that could also carry six passengers. The plane was not particularly well suited to these duties but, like all German aircraft of the era, its design was valid from a military point of view and it was soon transformed into a bomber. The first production versions (the E-1 bomber, the F-1 reconnaissance plane) were tested under operational conditions in 1937 during the Spanish Civil War. The experience gained there led to the development of the Do-17Z, with its more spacious cockpit and improved armament. This version remained in production until 1940 and in service until 1942. Over 600 Dornier Do-17s were produced, and the basic design was also used to develop the Do-217, conceived specifically for military use in 1938. By 1944, more than 1,900 Do-217s had been built; the Luftwaffe used them for bombing, reconnaissance, anti-shipping and night-fighter duties.

The Heinkel He-111 had also been designed officially as a fast commercial plane, but it was soon – from the third prototype – adapted for its true vocation as a bomber. The He-111 completed its maiden flight on 26 February 1935. It too saw action in Spain. The He-111H version was then developed, fitted with the more powerful Jumo 211 engines. Early in the war, in the Blitzkrieg in Poland, over Scandinavia, and on the western front, the He-111 was used in massive formations and proved to be a fearsome instrument, completing the first carpet bombings of enemy cities. During the Battle of Britain, however, when the HE-111 came up against a determined defense, it began to reveal its shortcomings. From 1941 on, it was relegated to night bombing, glider- towing, and finally transport duties. If fulfilled this last role until the end of the war. In all, more than 7,000 Heinkel He-111s were built.

By 1941, the high-performance, twin-engined Junkers Ju-88 was recognized to be the Luftwaffe's best bomber. The Ju-88 was designed in 1935 as a fast bomber; the first prototype flew on 21 December 1936. The production version, the Ju-88A-1, entered

122 bottom Heinkel 111s over France in May 1940. The plane began to show its limitations during the Battle of Britain in May 1940 and was relegated to secondary duties.

123 top left Though its design dated from 1934, the Dornier Do.17 was one of the Luftwaffe's strongest bombers at the start of the war and it remained in service until 1943. This is the Z version.

123 top right The Junkers Ju.88 was perhaps the best German bomber. It entered service in 1939 and proved flexible enough to be used as a night-fighter and even a glider puller. Almost 16,000 planes of the 88, 188, and 388 series were built.

service in 1939, on the very eve of the war. It was initially used only as a pure bomber, but reconnaissance, night-fighter, anti-shipping, anti-tank, dive-bomber, ground attack, transport and glider-towing versions of the aircraft were subsequently developed, confirming the validity and versatility of the basic design. The Ju-88 was also built in France and Czechoslovakia; ultimately it was even transformed into the Mistel guided flying bomb. This program, developed in 1944, involved the installation of a Bf-109 fighter on the back of a Ju-88 with a modified nose containing an 8,350-lb explosive charge. The pilot flew the paired aircraft from the Messerschmitt as far as the target; he then uncoupled the Ju-88 and returned to base in the fighter. Having seen that the format worked, the Germans also created the Mistel 2, using a Fw-190 fighter. In all, about 250 Mistels were constructed in four different versions. By the end of the war, nearly 16,000 Ju-88s had been constructed in over 100 different versions. This production total includes 1,076 Ju-188 aircraft, a version with an extended fuselage and larger wing. The Ju-88 was used not only by the Luftwaffe but also by another five Axis air forces.

123 bottom left A Heinkel 111 during a raid on Warsaw during the Polish campaign in September 1939.

123 bottom right This Heinkel 111 was photographed over London during the Battle of Britain. The 111H could fly 1220 miles but carry only 1,250 lb of bombs.

124 top The Messerschmitt Bf.108, undoubtedly the most famous German fighter; seen here is the B-1 version. As this and other models were improved, the Bf.108 remained the backbone of the German fighter force from 1939 to 1945.

124 bottom Derived from the civil transport plane of the same name, the Junkers Ju.52/3mg was adopted by the Luftwaffe as a bomber but most of all as a transport plane. It remained in production until 1944 and was used in many support roles.

125 left A formation of heavy Messerschmitt Bf.110 fighters during the Battle of Britain in 1941. Unsuited to the job it was designed for, the Bf.110 was developed and put to other uses like night-fighting and reconnaissance.

125 top right Willy Messerschmitt, the brilliant German engineer who designed several successful military aircraft, including the Bf.108, Bf.109, Bf.110 and Me.262.

Another very famous aircraft from the early years of the war was the Junker Ju-52 transport plane, a solid, reliable machine, affectionately known as "Tante Ju" (Auntie Ju) by its crews. The aircraft was originally designed in 1930 as a single-engined cargo plane, but its fate was transformed in 1932 when the seventh model (the Ju-52/3m) was built in what was to become the definitive configuration, with three radial engines, a fixed undercarriage, and characteristic ribbed metal skinning. International civilian sales success was repeated in 1934 in the military sphere when the Luftwaffe ordered a version for use as a bomber and transport plane. Following the Spanish Civil War, the Ju-52/3mg3 went into mass production, and by the outbreak of the Second World War the Luftwaffe was already using 1,000 Ju-52/3mg2s to transport troops and materials and for dropping paratroopers. These planes were fundamental to the success of the invasions of Denmark, Norway, Holland and Greece. The Ju-52/3mg2 was also used in the airborne invasion of Crete, during which it suffered extremely heavy losses, and in the airlift during the tragic battle of Stalingrad. The plane, which remained in production until mid-1944, was also used for towing gliders and as an air-ambulance, operating with wheels, skis or floats. More than 5,400 Ju-52s were constructed, some of them under license in Spain and France.

Germany's fortunes waned after three years of war, and after 1942, the Nazis were obliged to fight a defensive battle. In spite of Hitler's wishes to the contrary, the Luftwaffe's most important aircraft began to be the fighters needed to combat the Allies' devastating bombing raids on the Fatherland and the occupied territories. The Germans had begun the war with two principal fighters, the singe-engined Messerschmitt Bf.109 and the twin-engined Bf.110.

The first is still the aircraft most closely associated with the Luftwaffe and, with almost 35,000 built, it was the most numerous plane of the Second World War; it was also one of the most famous fighters in the history of aviation. In 1934, the German Air Ministry had issued a requirement calling for a new monoplane, single-seat interceptor to replace the biplanes currently in service. Willy Messerschmitt responded. In just 15 months he designed his first great aircraft, the Bf.109. This was a small, light, all-metal plane with extremely advanced features and the most powerful engine then available, the 695-hp Rolls-Royce Kestrel V. The Bf.109 was not a particularly innovative fighter, but for the first time a fighter combined a number of advanced features. They had a retractable undercarriage, a closed cockpit, a metal fuselage and wings and Handley Page-type slats on the wings' leading edges. The prototype first flew in September 1935. It was soon put into limited production at the expense of its three competitors: the Arado Ar-80, the Focke-Wulf Fw-159 and the Heinkel He-112. This last was in practice the equal of the Bf-109, but lost out because it was more complicated to mass-produce. The Luftwaffe tested the aircraft in the Bf.109G form, powered by the German-built Jumo 210 engine, in air operations in the Spanish Civil War, where it proved to be the best fighter in the world at that time. The 109 design achieved full maturity with the E (Emil) version. It was the first Messerschmitt to be mass-produced (from 1939 on) and introduced significant improvements, including the powerful and reliable new Daimler-Benz DB-601 engine. The Bf.109E participated in the opening phases of the war, above all during the Battle of Britain, when many of its pilots distinguished themselves and became celebrated aces. However, in 1940 the Bf.109 met its eternal rival, the British Spitfire, the aircraft against which it was to be matched throughout the war as designers on both sides searched for ever greater performance.

125 center right Professor Krauss, director of Messerschmitt's aerostatics department. Here he is illustrating the characteristics of a wing with a spar.

125 bottom right Adolf Galland, left, and Werner Moelders, right, were two of Germany's most famous German flying aces. They are shown here in 1941 after being decorated with the Knight's Cross. Galland commanded a squadron of JG.26s, Moelders a squadron of JG.51s.

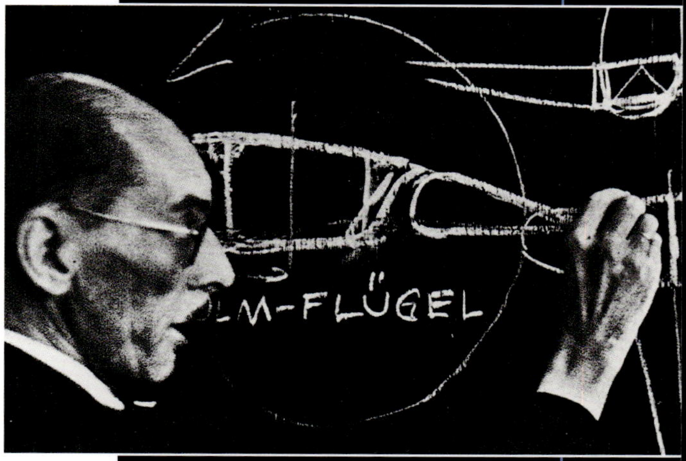

The most important version of the 109 was the G (Gustav), which boasted significant improvements to its 1,800 hp DB-605 engine and armament. It entered service at the end of 1942. Eleven versions of the 109G were developed, and almost 20,000 planes were constructed built. The Bf-109 was the favorite aircraft of Germany's greatest aces, including Gerhard Barkhorn, who accumulated 301 kills, and Eric Hartmann, the most prolific ace of all time, with 352 victories. Hartmann arrived on the Russian front as a sergeant and undertook his first combat mission in October 1942. After his first victory on 5t November 1942, Hartmann developed a refined, ruthless combat technique in which his great success was based on the initial decision to accept or refuse the fight according to the situation at that moment. A year later he had already accumulated 148 victories. In March 1944 Hartmann was promoted to lieutenant; by the end of the war, after receiving the foremost honors (such as the Knight's Cross with Oak Leaves, Swords and Diamonds), he had reached the rank of major and was in command of the 1st wing of the JG.52. After some time, Hartmann's plane began to sport a black tulip as a personal insignia, its petals covering the whole nose of his Bf-109. For this reason he was nicknamed by the Russians the "Black Devil of the Ukraine" and was soon avoided by all enemy pilots. In the later stages of the war he only managed to reach his eventual total of victories by flying an unmarked aircraft.

Toward the end of the war, production difficulties imposed various modifications designed to simplify and accelerate assembly of the Bf-109. The final model was the K-4, introduced in 1944. It had a 2,000-hp engine, and its maximum take-off weight was almost double that of the first series, a testimony to the basic design's underlying excellence and versatility.

The career of the other pre-war Messerschmitt fighter, the Bf.110, was less successful. This aircraft was designed in 1934 to meet the requirement for a long-range, heavy fighter. Despite early successes in the Polish campaign, the Bf.110C was outclassed during the Battle of Britain, unable to match the British fighters in terms of maneuverability and acceleration. In 1941, the Luftwaffe began to use it in other roles. It performed best as a night-fighter, especially in Bf.110G form, equipped with Lichtenstein radar from 1942 on. The 110s were also used as reconnaissance planes and fighter-bombers. By 1945, German factories had built more than 6,000 of the Bf.110s.

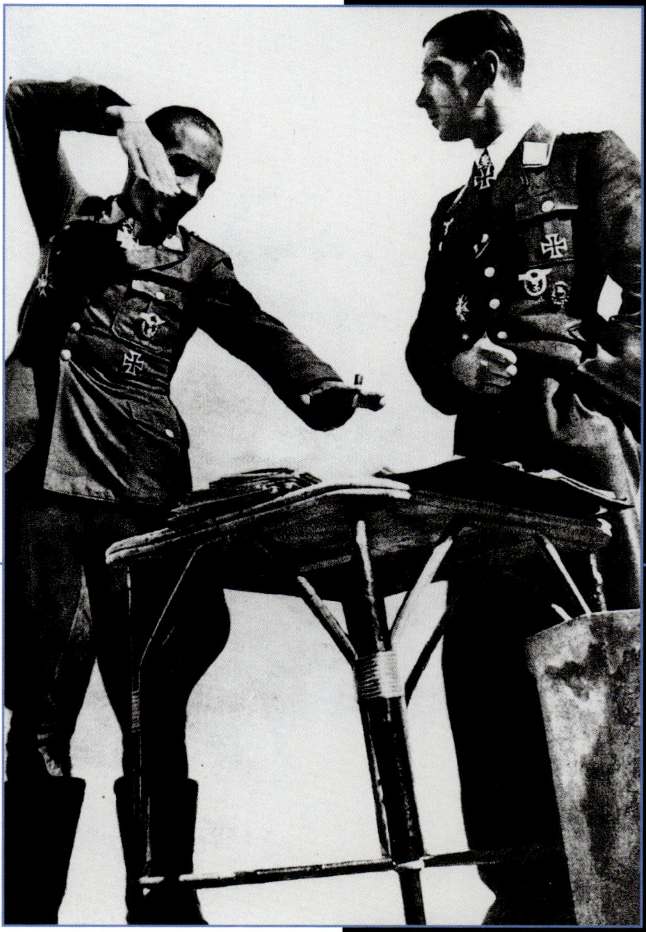

Together with the Bf.109, the greatest German fighter of the war was unquestionably the Focke-Wulf 190; it was in fact superior in many respects to its Messerschmitt stable-mate. In 1937 the German Air Ministry asked Focke-Wulf to create a new interceptor to join the Bf.109. At that time Kurt Tank led the Focke-Wulf design department. He was an engineer who had ventured into aviation in 1924 to design a seaplane. Tank, who was also a fine test pilot, wanted a less extreme airplane than the Bf-109 and the Spitfire; a fast fighter but one that would be larger, more robust and heavier, a true war machine. He succeeded in his intent thanks to BMW's 139 14-cylinder radial engine that produced an impressive 1,550 hp. The prototype first flew on 1 June 1939, immediately demonstrating great promise. After 40 pre-series planes had been built for operational testing, the FW-190A went into production in 1941. From its very first encounters with the Spitfire Mk.V, it proved to be superior, offering attractive flying characteristics and great speed and maneuverability. In 1942, the FW-190A fighter version was followed by the F and then by the G versions. Both were dedicated to the fighter-bomber role. In that same year, General Douglas of the RAF stated, "I have no doubt, and nor do my pilots, that the FW-190 is today the world's best fighter." The last mass-produced version, the FW-190D appeared early in 1944, characterized by a new, longer and tapering nose that housed a1,800-hp in-line Jumo engine. With this model, the German pilots were able to fight on even terms with the best Allies' best fighter aircraft of 1944-45, even though training possibilities and fuel supplies were no longer adequate to allow the Luftwaffe to conduct effective and continuous operations. The last of the FW-109 line was the Ta-152, an extremely fast and heavily armed high-altitude interceptor, of which only 150 were produced. A total of about 20,000 planes in the FW-190 series were built before the war ended.

The German munitions industry's desperate efforts toward the end of the war gave rise to revolutionary aircraft that might have been able to change the conflict's eventual outcome. These were the first operational aircraft to be powered by jet and rocket engines, designs that impressed the Allies but which, for various reasons, were ultimately unable to alter the course of history. The first jet aircraft in the history of aviation to see active service was the Messerschmitt Me.262, the origins of which actually dated back to before the war. The delays in its introduction were largely due to the developmental problems encountered with the BMW engines. In fact, the Messerschmitt Me.262 made its first flight on 18 April 1941, with a piston engine mounted in the nose.

126 top The German Arado Ar.234, the first jet bomber in history, is shown here during trials without an undercarriage. Powered by two Jumo 004 motors producing 890-hp of thrust, it flew at 462 mph and carried 1,250 lb of bombs. It was also used for reconnaissance.

The third prototype, the first to be fitted with the final version of Junkers Jumo 004 engines, flew on 18 July 1942, and immediately showed exceptional qualities in terms of performance and handling.

The Me-262 nonetheless suffered from a serious drawback: the conventional layout of its undercarriage and tail-wheel meant that during take off the jet exhaust flow interfered with the elevators, preventing the plane from gaining altitude. In order to get round the problem, the test pilots developed a procedure as simple as it was dangerous: a short squeeze on the brakes during the take-off run that dipped the nose, modified the gas flow at the tail, and allowed the elevators to function efficiently. Clearly, this was not an option in normal conditions and so the airplane had to be modified with the nose-wheel adapted to align the plane with its aerodynamic requirements. Flight testing also revealed another problem: the engines. They had to be treated very gently because they tended to flood and on occasion catch fire. Moreover, at certain engine speeds anomalous vibrations set in that could cause cracking in the turbine blades.

In order to resolve this problem, the Messerschmitt company turned to a professional violinist who "played" each blade with his bow. He thus identified their natural frequency and the fact that they began resonating at the plane's cruising speed. A decision was made to reduce the blades' thickness and to vary the engine speed at cruising velocity, two measures that effected a miracle cure. Nonetheless, Hitler's insistence on making a bomber of the Me.262 meant that development and production were delayed; the first pre-series planes did not actually fly until spring 1944. In total, 1,433 Me.262 s were actually constructed, but in too many versions and sub-versions (17 in all): including single- and two-seat reconaissance, ground attack, day-fighter and night-fighter variants. Efforts were thus diluted and dispersed, and the Me.262 had little operational effect despite the fact that it in technical terms it was unrivalled.

The Messerschmitt Me.163 Komet was a small rocket-powered interceptor that came out of the engineer Lippisch's pre-war research into tailless gliders. The aircraft's period of gestation was nonetheless lengthy because the technical difficulties associated with using highly unstable dual-component fuels were considerable. Early in the war the Me.163 became a low-priority project. The rocket-powered plane made its first test flight 13 August 1941, and the Me.163B pre-series model first flew on 23 June 1943. Production began in February 1944 and the plane went into service in the summer of that year. The Komet, which was armed with two 30mm cannons, was powered by a Walter HWK rocket engine that gave it a top speed of almost 600 mph, but for a mere 7.5 minutes. After the rocket engine had burned out the plane had to glide back to base and make a belly-skid landing. At the end of 1944, however, Germany was facing defeat and the 400 or so Komets that had been produced could do little to help the aerial defense of the Fatherland.

Another German aircraft, the Arado Ar.234 Blitz, was the first jet bomber in history. This plane, which used the same Jumo engines as the Me.262, also went through a troubled development period; two years elapsed between design and the maiden flight made on 15 June1943. The Blitz went into service with an experimental squadron in summer 1944, but saw only limited action in 1945 as a bomber and reconnaissance plane. A total of 200 Blitzes were produced in 22 different versions. The jet had a maximum take-off weight approaching 22,000 lb, and was flown by a single pilot. It could fly at 460 mph for 1,000 miles, with a 4,400-lb bomb load.

126 center The Focke-Wulf Fw-190, a fast, manageable and sturdy airplane, was the German fighter forces other arm. It created serious difficulties for the Allies when introduced in August 1941.

126 bottom The Messerschmitt Me.262, the first jet fighter in history to be used in combat. Its troubled development resulted principally from Hitler's decision to use it as a fighter-bomber.

127 top The Messerschmitt Me.163 Komet jet interceptor was an unusual aircraft developed to counter Allied bombing raids on Germany. It was, however, difficult and dangerous to fly.

127 bottom A row of Me.262-B Messerschmitts, the two-seater training version. The U-1 version of this model was equipped with a Liechtenstein radar unit in the nose and used for night fighting.

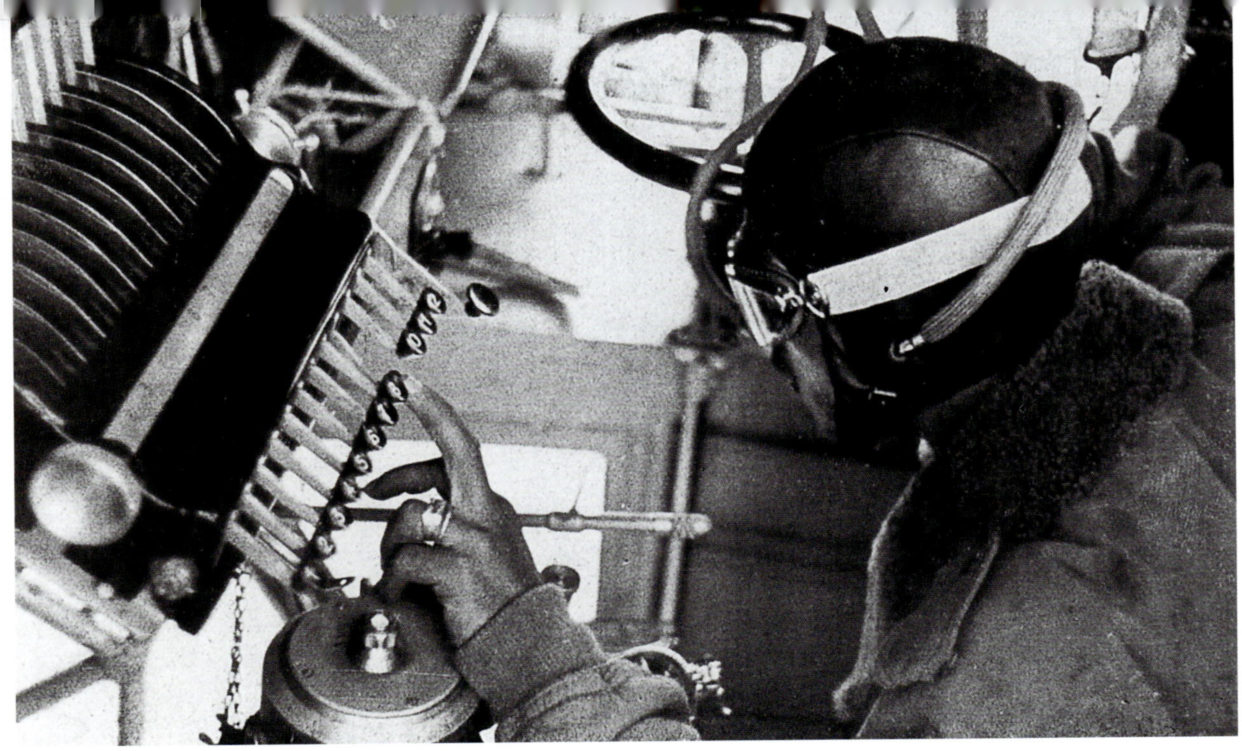

128 top On board an Italian bomber: note the early bomb-release system.

128 bottom left A formation of FIAT CR.42 biplane fighters over the Mediterranean. Already old in 1938, this plane was produced until 1944.

128 bottom right Though more modern-looking than the CR.42, the FIAT G.50 of 1937 offered no better performance than the biplane; it had only two machine-guns and a top speed of only 294 mph.

Germany's principal European ally was Italy, a country that in the 1920s and 1930s had been at the forefront of aeronautical technology, thanks to the exploits of various record-breaking planes and the great ocean-crossing flights organized by the Italian Royal Air Force. Italy had, however, been notably weakened by its involvement in colonial wars and the Spanish Civil War, and in 1940 its aging air force could count on fewer than 1,800 serviceable aircraft. During the course of the war this situation worsened: given the excessive variety of types in service, the national aeronautical industry was unable to produce either a sufficient number of new aircraft or the modern engines needed to power them.

Early in the conflict, the Italian Royal Air Force's main fighter was the Fiat CR.42 Falco biplane. This had a fixed undercarriage already obsolete upon introduction on 23 May 1938. There was, however, still conviction in Italy that a light, maneuverable biplane was preferable in aerial combat. As a result, while Germany initiated production of the Bf.109, Italy began to turn out the Falco, which was powered by an 840-hp radial engine, had a top speed of 267 mph, and was armed with two 12.7-mm machine guns. Thus Italy's CAI (Corpo Aereo Italiano), in accordance with Mussolini's wishes but against German advice, participated in the attack on Great Britain with planes such as the CR.42. Their pilots had to compete with superior airplanes in a fighter that was conceptually outdated and even had an open cockpit that resulted in a number of frostbite cases. Up through 1943, a total of 1,782 CR.42s were built – a significant number for Italy. They were built despite all evidence to the contrary and the availability of better aircraft, on which production should have been concentrated. From 1941, however, the plane was relegated to less demanding duties such as ground attack, reconnaissance and night-fighting.

On 26 February 1937 (a year earlier than the Falco's introduction), another Fiat fighter had made its first flight, the G.50 Freccia, an all-metal monoplane with retractable undercarriage. In 1938, the G.50 lost out to the Macchi MC.200 in the competition for the air force's new interceptor, but it was nonetheless approved for serial production. This plane, equipped with the same engine as the CR.42, was not however, a success; its performance, with a top speed of 292 mph, was only slightly better than that of the Falco. A total of 780 G.50 Freccias were built.

The Macchi MC.200 Saetta was a far superior machine. It had made its maiden flight on 24 December 1937, and entered service in the summer of 1939. The Macchi fighter, despite being penalized by a not particularly powerful radial engine (870 hp), boasted excellent performance and flight characteristics. A total of 1,151 Saettas were constructed up to July 1942, when the model was replace by the MC.202 Folgore. This last was none other than the Saetta re-equipped with the German Daimler-Benz DB.601 engine (built under license by Alfa Romeo), an in-line unit producing 1,100 hp that gave the Folgore a top speed of 373 mph. After its maiden flight on 10 August 1940, the MC.202 went into production: a total of 1,070s were built. The planes operated successfully on all fronts.

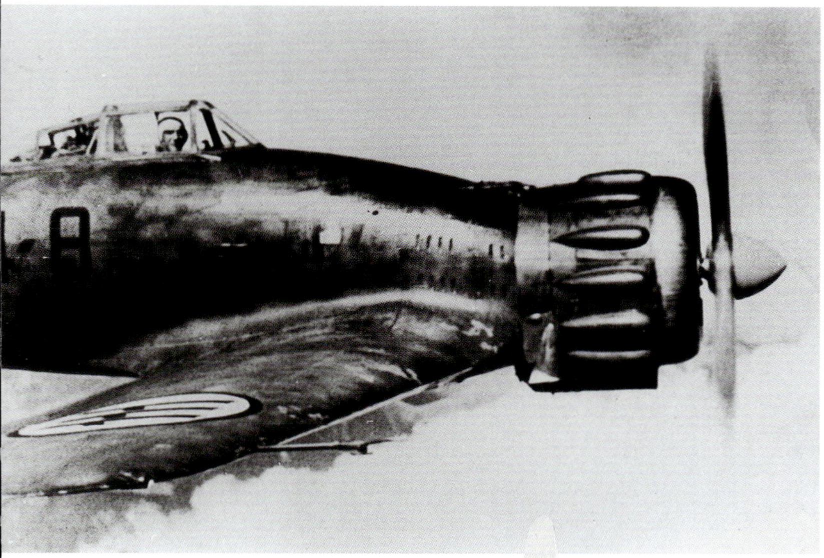

129 center right
A machine-gunner in an Italian bomber in his aft defensive placement during a mission over the Mediterranean.

129 bottom Despite its defects, the CR.42 was agile and reliable and continued to be used by the Regia Aeronautica even after 1941 as a light attack craft and night-fighter.

129 top In the foreground, the Macchi MC.202 Folgore fighters of the 97th Squadron on an Italian airfield. The MC.202 was the best Italian fighter before the advent of the 'Serie 5' fighters.

129 center left A Macchi MC.200 Saetta in the 74th Squadron. The MC.200 was an excellent plane but suffered from a radial motor of little power and just two machine-guns.

130 top Two armorers loading bombs into a three-engined SIAI S.79 Sparviero, the most famous Italian bomber of World War II.

130 bottom Two S.79s over the Mediterranean. The Sparviero also proved competent as a torpedo craft and was the main threat to British shipping in summer 1942.

131 top With the advent of the Daimler Benz DB.605 engine, the best Italian fighter planes of the war were produced. The Macchi MC.205 Veltro was basically a direct development of the 202 model but armed with two 20-mm cannons as well as two 12.7-mm machine-guns.

131 bottom The FIAT G.55 Centauro was perhaps the best Italian fighter of World War II. A sturdy and manageable plane, it carried three 20-mm cannons and two 12.7-mm machine-guns and could touch 394 mph.

The final version of this Macchi MC family was the MC.205 Veltro, which completed its maiden flight on 19 April 1942. This model was designed to overcome the main weaknesses of the 202: it was more powerful and was more heavily armed. The Veltro thus used the 1,475-hp DB.605 engine and carried two 20-mm cannons along with the traditional two 12.7-mm machine guns. It was an excellent aircraft, capable of matching its Allied counterparts, but only just over 200 were built and the plane did not enter service until April 1943. Following the armistice between Italy and the Allies that was signed on 8 September 1943, the National Republican Air Force (ANR) used a further 112 in the north of the country.

Perhaps the best Italian fighter of the war was the Fiat G.55 Centauro, which flew for the first time on 30 April 1942. The Fiat fighter was powered by the DB.605 engine (again built under license by Alfa Romeo), and armed with three 22mm cannons and two 12.7-mm machine-guns. It was robust and maneuverable and could reach a top speed of 391 mph. The ANR used it primarily in attempts to fend off the Allies' bombing raids on northern

Italy. Only just over 250 were manufactured before the already reduced production capacity collapsed entirely. The last aircraft of the famous "5 Series" (fighters equipped with the DB.605 engine) was the Reggiane Re.2005 Sagittario that flew for the first time on 9tMay1942. Though it was an excellent plane, only 29 Sagittarios were built before the armistice.

Among other aircraft Italy used during the war, special mention should be made of the SIAI S.79 Sparviero, a traditional profile trimotor that remained in service for various duties for the duration of the conflict. The aircraft had been conceived as an eight-seat civilian transport plane and was equipped with three 750- hp radial engines, a low wooden wing and retractable undercarriage. The S.79 first flew on 8 October 1934. It immediately distinguished itself for its speed, setting six world records. This performance alerted the military authorities to the potential for a bomber version. Production began in October 1936; subsequently, the aircraft performed well during the Spanish Civil War. At the outbreak of the Second World War, almost 600 S.79s were in service. Torpedo-launching duties were soon added to the traditional bombing missions, and enhanced the S.79's reputation thanks to a number of heroic operations in the Mediterranean. Some of the Italian Royal Air Force's most glorious Second World War achievements featured the 132nd Group's S.79s, first based in Sicily and then at Pantelleria, under the command of Major Carlo Emanuele Buscaglia. Other famous pilots such as Giulio Cesare Graziani and Carlo Faggioni flew with this autonomous unit. It earned great distinction in maritime operations in summer 1942, in particular in the battles of "Mezzogiugno" and "Mezzagosto" (Mid-June and Mid-August), when its pilots succeeded in sinking or damaging numerous Royal Navy warships and cargo vessels.

The setting for these two battles was the struggle for Malta. The S.79 crews displayed great courage in the face of withering anti-aircraft fire. Detetermined British fighters slipped in a few feet below the S.79s' wings and above the waves as they lined up to drop their torpedoes. During the battle of Mezzagosto, the Italian aircraft, in collaboration with the Germans, managed to sink a Royal navy destroyer and five cargo vessels, and hit two aircraft carriers and a further five cargo ships. This was the Italian torpedo planes' last successful mission: the war in the Mediterranean was by now going against the Axis forces. Major Buscaglia was shot down on 12 November and was presumed dead. In fact, he was alive and a prisoner of the Allies. In early in 1944 he returned to command the 132nd Group, now equipped with British twin-engined Baltimores, in the Aeronautica Cobelligerante (the Italian post-armistice airforce assisting the Allies). Shortly thereafter, on 23 August 1944, Buscaglia was killed in a flying accident at Camp Vesuvius. The S.79, nicknamed the "hopeless hunchback" because of the shape of the fuselage behind the cockpit, was also used by the ANR after 1943 as a torpedo plane; more than 1,300 were produced. Exported and constructed under license abroad, in the post-war period the S.79 was used as a cargo plane until the 1960s.

On the other side of the world, the third Axis ally, Japan – a country that had no great aeronautical tradition – had gone into the war with an excellent fighter. This was the Mitsubishi A6M Zero which, thanks to gradual improvements and modifications, was to remain in service throughout the war. The Zero was designed in the late 1930s as a new carrier-borne fighter for the Imperial fleet, and had been developed in response to a very demanding brief that Mitsubishi's rival, Nakajima, had judged to be impossible to meet. The A6M1, which first flew on 1 April 1939 responded perfectly to all requirements save that of top speed, which was lower than hoped-for 310 mph. After the engine had been replaced with a more powerful unit, the A6M2 went into serial production.

The Zero entered service in July 1940, and up to mid-1942 it was probably the best fighter in the Pacific theater. It was fast and maneuverable; its weaknesses were lack of armor plating and self-sealing fuel tanks, making it vulnerable to enemy fire. The best version of the Zero was the A6M5, which was deployed in 1943 to counter the threat of the American Hellcat and Corsair fighters. This model, armed with two 20-mm cannons and two 13.2-mm machine guns, had a 1,130-hp engine and could reach a top speed of 351 mph, but it was still no match for the American designs. A total of 10,499 Zeroes, code-named "Zeke" by the Americans, had been produced by August 1945, but its decline had already become apparent two years earlier, in parallel with the loss of Japan's best pilots. The introduction of the last version, the 16M8 with a 1,560-hp engine, did nothing to improve the situation.

Another Japanese fighter that remained in service throughout the war was the less well-known Nakajima Ki-43, the army's front-line combat plane. This aircraft was commissioned directly from Nakajima without an open competition being held. The Ki-43 made its first flight in January 1939. It was an all-metal monoplane with retractable undercarriage, a low wing and a radial engine that initially proved to be heavy and clumsy. Like the Zero, it subsequently became light and agile; early in the war it actually gave the impression of being invincible. Constantly improved, especially with regard to its engine, the Ki-43 ("Oscar" to the Americans) was nonetheless overtaken by the Allied fighters, above all because the Japanese aeronautical industry proved incapable of producing sufficiently powerful engines. A total of 5,919 Ki-43s were built.

A third famous aircraft employed by the Japanese armed forces was the Mitsubishi G4M. This was a twin-engined bomber that completed its maiden flight on 23 October 1939, and went into production in 1940. Though designed for the navy, the G4M was in fact a land-based long-range bomber capable of

132 top A Japanese machine-gunner in 1942. Japan did not produce any particularly advanced planes after 1942 but their crews' courage was never doubted.

132 center right The Mitsubishi A6M Zero was unquestionably the best-known Japanese plane. An excellent fighter, its only weak point was lack of armor-plating and of self-sealing fuel tanks.

carrying a 2,200-lb bomb load or a torpedo for up to 2,500 miles. It had a top speed of 292 mph and was armed with four 20-mm cannons and two 7.7-mm machine guns. Nonetheless, the G4M (designated "Betty" by the Americans) was only lightly armored and also lacked self-sealing fuel tanks. Because of this, the Americans nicknamed the G4M "The One-shot Lighter." This was the most numerous of the Japanese bombers; 2,466 were built and the plane remained in service until August 1945.

132 bottom Japan had more Mitsubishi G4Ms, known as 'Betty' by the Allies, than any other plane but it too was poorly armored. The Americans nicknamed it "the flying lighter."

133 top A Japanese torpedo plane crashes into the sea after being hit by the 127-mm anti-aircraft guns of the USS Yorktown aircraft-carrier on 25 October 1944.

133 center Japan had a powerful navy equipped with naval aviation. These Zeros are preparing for take-off from an aircraft carrier.

133 bottom Toward the end of the war, the Japanese could find no better tactic to stop the American fleet than kamikaze attacks. In this photograph of 4 May 1945, a Japanese fighter is hit as it tried to crash into the aircraft-carrier escort, USS Sangmon.

Chapter 9

The First Jets and the Korean

By the early 1940s, traditional aircraft characterized by piston engines driving one or more propellers had reached the limits of their potential. During the Second World War, the fastest fighters came up against the 500-mph barrier beyond which propellers begin to lose efficiency. In order to go faster, far more powerful engines would be required, but they would be so large and heavy as to be unacceptable. In addition, their weight would eventually offset any theoretical increase in their performance, bringing the practical speed and efficiency values back down to the levels already achieved.

For some time, however, two men, Frank Whittle in Great Britain and Hans von Ohain in Germany, had been working independently on the development of an engine that to most observers appeared as revolutionary as it would be impossible to construct: the gas turbine engine.

The concept underlying gas turbines was very simple; in fact, these turbines were already being used in industrial applications. Air entered from the front and was compressed and forced into a combustion chamber where it mixed with the fuel; the mixture was then ignited, and the gases exited at high speed from the rear jet, driving as they did so a turbine which actuated the front compressor. The difficulties lay in creating a turbine sufficiently compact, powerful and light to be mounted successfully in an airframe.

Von Ohain, a Heinkel aeronautical company an engineer conducted an initial demonstration of his turbine engine in February 1939. The test was successful, and Ernst Heinkel became a supporter of the new power unit, with his firm building the world's first jet aircraft, the He-178. This small, simple all-metal monoplane had a feature that in a few years became very common: an air intake in the nose where a propeller would normally be found. The He-178 flew for the first time on 27 August 1939, inaugurating aviation's modern era.

In Great Britain, Whittle, a RAF engineer, first had the idea of a gas turbine engine in 1929, but he could only begin to work seriously on the project from April 1937. In June 1939, he was ready to stage a demonstration of his WU engine for the Air Ministry, whose representative received the new invention enthusiastically. The Gloster Company was immediately commissioned to build an experimental plane using the W.1 engine. The result was the Gloster E.28/39, a low-wing, experimental model featuring a tricycle undercarriage and a nose air intake. It flew for the first time on 15 May 1941.

In Great Britain, this line of research led to

134 top The Heinkel He-178, which flew in August 1939, was the first jet plane. Its development was made possible by perfection of the jet engine by the German engineer, von Ohain.

134 bottom The de Havilland Vampire was the first successful jet aircraft. A total of 19 versions and more than 3700 planes were built. In addition to Britain, 25 countries used the Vampire. Shown here is a two-seat training model.

135 top The Italian Caproni Campini N.1 was, together with the German and British attempts, another early jet-engined plane. It made its first flight on 28 August 1940 but its top speed was just 220 mph, and only two prototypes were built.

135 bottom With the advent of jet engines, aerodynamic tests became increasingly important. In this 1945 photograph, two men stand inside a wind tunnel.

the construction of the Gloster Meteor fighter powered by Rolls-Royce engines (see Chapter 7), while in Germany, the development of airframes and engines took the form of a race between two industry groups; Heinkel and Messerschmitt, and BMW, Bramo and Junkers. The German Air Ministry ordered Heinkel and Messerschmitt to produce a jet fighter design to be powered by one of the three other firms involved in the project. Heinkel was the first to complete its design (in September 1940), but as no jets were yet available, its twin-engined He-280 first flew in March 1941 with engines built by Heinkel itself. The He-280 was an attractive aircraft that featured a number of interesting innovations such an ejector seat and a tricycle undercarriage. Even with its provisional engines, the He-280 was capable of 485 mph. However, for technical and political motives, the German authorities preferred the Messerschmitt Me.262, despite the fact that it was not ready until later and did not fly until July1942. Its finalized Jumo 004 engines produced 1,984 lb of thrust. However, the Me.262It had a longer range and greater firepower than the He-280 (see Chapter 8).

In Great Britain, Gloster continued to improve and build the Meteor (before its withdrawal in 1954, about 3,900 had been built, with many exported). The second British jet was the de Havilland Vampire (initially named the Spider Crab). The plane was commissioned in 1941, and made its maiden flight on 26 September 1943. The vampire was small, equipped with mid-mounted metal wings with air intakes at the roots, and a short wooden fuselage; the undercarriage had a tricycle layout with a nose wheel. The tailplanes, similar to those of the Lockheed P-38, were carried on twin booms extending backwards from the wings. A single de Havilland Goblin engine producing 2,700 lb of thrust powered the new aircraft, and gave it a top speed of around 500 mph. Though the Vampire proved to be an excellent fighter, it was not ready in time to play an active part in the war; in fact, the first production model, the Vampire F Mk.I, reached the RAF squadrons in April 1946. The F Mk.1, which was armed with four 20-mm cannons, had a Goblin 2 engine, producing around 3,100 lb of thrust and guaranteeing a top speed of 540 mph. The Vampire was continuously upgraded, and specialized versions were introduced. These included the new fighter (F), fighter-bomber (FB), night-fighter (NF), trainer (T), and naval (Sea Vampire) versions. A new wing and new tailplanes gave rise to the Venom and the Sea Venom. More than 3,700 planes were built, with purchases made by 26 different countries around the world. The last Vampires, those used by the Swiss air force, were only withdrawn in the 1980s.

136 Two photographs of a Lockheed T-33A Silver Star. This jet plane was derived from the P-80 Shooting Star fighter as a training plane for the new type of engine. It first flew on 22 March 1948.

The United States was also involved in the development of jet fighters. Thanks to the exchange of technological information with Great Britain during the war, General Electric (which already had considerable experience with turbines) was chosen to develop a new engine based on Whittle's research. On 5 September 1941, Bell was commissioned to construct the prototype of a plane to be propelled by the new engine, the General Electric I-A turbine, which produced 1,250 lb of thrust. Bell came up with the Model 27, a twin-engined plane of traditional lines, with a straight, mid-mounted wing, a tricycle nose wheel undercarriage, and engines buried in the wing roots. This aircraft, designated the XP-59A Airacomet, first flew on 1 October 1942, from Rogers Dry Lake to Muroc Field in California. In 1944, the three prototypes were followed by 13 pre-production examples of the YP-59A, powered by I-16 engines producing 1,650 lb of thrust. These new engines nonetheless failed to provide an answer to the plane's chronic slowness (a top speed of only about 400 mph) and its instability. A total of 20 of the P-59As were then ordered along with 30 of the P-59Bs, the latter powered by the J31-GE-5 engine, which produced 2,000 lb of thrust. All these planes were, however, used for a vast series of test and evaluation flights and for training by the US Army and Navy. In spite of its relative lack of success, the Airacomet made a significant contribution to the development of jet aircraft technology, and the second American jet plane was an immediate winner.

While Bell was struggling with the troublesome P-59, Lockheed was assigned the task of producing an operational jet fighter. The firm prepared the design in just six months and the XP-80 prototype, equipped with a de Havilland Goblin gas turbine engine, flew for the first time on 8 January 1944. However, the decision was made not to proceed with production under license of the British engine. June 1944 thus saw the maiden flight of the XP-80A powered by a General Electric J33 unit. Operational testing began in October 1944, once the first 13 pre-production YP-80As had been completed, and orders were then placed for manufacture of the large total of 5,000 P-80A jet fighters, but the end of the war drastically reduced the USAAF's need for new planes, and the order was cut back to just 917. The P-80A Shooting Star had a wingspan of 39 feet 6 inches and a maximum take-off weight of 14,500 lb. The GE J33 engine produced 3,850 lb of thrust and a top speed of 560 mph. The Shooting Star carried the same armament as the propeller-driven fighters; six machine guns and up to 2,000 lb of bombs. In 1948, the improved F-80C was introduced, and a total of 798 were built; they saw action in the Korean War. (It should be noted that in 1947, with the constitution of the United States Air Force as

137 top left The USAF's "Acrojets" acrobatic team used F-80 fighters. The plane entered service at the start of 1945 but not in time to be used during World War II.

137 top right This formation of Bell fighters was photographed in 1944. From top down: an XP-77, a P-39Q, a P-63A, and the first American jet airplane, the P-59 Airacomet, here in the pre-production version, the YP-59.

137 center In 1947 McDonnell introduced a new jet fighter, the Banshee, derived from the Phantom. Shown here is the XF2H-1 prototype.

137 bottom left The prototype of the Ryan FR Fireball, 1944. A hybrid between jet and propeller propulsion, it was the first jet aircraft used by the US Navy.

a separate armed service, the designation of fighter aircraft was changed from P = Pursuit to F = Fighter.)

The excellent F-80C also formed the basis for an exceptional two-seat trainer that flew for the first time on 22 March 1948. Initially designated the TP-80C, it then became the T-33A Silver Star, with over 6,500 being built, serving in over 40 countries around the world. A number of T-33s are still being flown today by various air forces, a testimony to the validity and robustness of the design.

In the meantime, the US Navy had also taken cautious steps in the direction of jet propulsion, but due to the greater reliability required of carrier-borne aircraft, its first plane of this type was actually a hybrid.

The Ryan FR Fireball, which completed its maiden flight on 25 June 1944, was in effect an experimental airplane, being equipped with a 1,350-hp Wright Cyclone radial engine and a General Electric I-16 turbojet producing about 1,600 lb of thrust. The end of the war again led to production being restricted; in this case to just 66 planes, with deliveries being made in 1945. By this time the Navy was already looking at its first true carrier-borne jet, the McDonnell FH Phantom, which had been designed in 1943. The Phantom flew for the first time on 26 January 1945, powered by two Westinghouse 19XB-2B engines producing about 1,150 lb of thrust (in the production version, these were replaced by J30 units producing 1,600 lb of thrust). On 21 July 1946, the Phantom became the first American jet aircraft to land on an aircraft carrier. Production was gain restricted, and only 60 of the FH-1s were produced; they were delivered between 1947 and 1948.

Progress in the field was continuous, and McDonnell was already working of the FH-1's successor the XF2H-1 Banshee, which took to the air in January 1947. This plane closely resembled the Phantom, but was significantly better in all respects, including its two Westinghouse J34 engines that produced 3,250 lb of thrust and guaranteed a top speed of about 580 mph. The plane was also more heavily armed, with four 20-mm cannons and two 500-lb bombs. The F2H Banshee went into service in March 1949.

137 bottom right The McDonnell FH-1 Phantom was the US Navy's first full jet, adopted in 1945. It was the first jet aircraft to land on an aircraft carrier.

It should be noted that only in very late 1940s were the Americans able to field an operational fighter that was an improvement on Germany's Second World War Messerschmitt Me.262. However, a number of planes were already in the flight-testing stage, and a number of excellent jet fighters were ready in time to participate in a new international war. Though it was geographically restricted to Korea, this new war was in effect a confrontation between the great Western and Eastern powers, and risked plunging the world into another disastrous global conflict.

At 0400 hours on Sunday, 25 June 1950, the North Korean army made a mass attack on the border with South Korea. The opposing forces were mismatched, and the North Koreans easily broke through the defensive lines and penetrated deep into South Korea. A few days later, the UN Security Council approved the use of force to repel the invaders and re-establish the border on the 38th parallel. The United States assumed responsibility for this military action and, together with 16 other countries, intervened drastically. For its part, North Korea could count on the initially indirect and then open support of the People's Republic of China, and then that of the Soviet Union. The result was a long, hard and bloody war that lasted for three years, into 1953.

The United States went into the war with the planes of the 5th Air Force, stationed in Japan. These were mainly F80 jet fighters and F-82 Twin Mustangs, the USAF's last propeller-driven fighters. The North Koreans opposed these with inferior planes, the Soviet Yak-9 piston-engined fighter and the Ilyushin Il-10 Shturmovik fighter-bomber. Early in the war, the Americans introduced new combat jets, the USAF's Republic F-84 Thunderjet and the Navy's Grumman F9F Panther. Work began on the Thunderjet project

138 top Two Grumman F9F Panther fighters fly over the USS Princeton on 28 May 1951 during the Korean War. They are discharging excess fuel from their wing tips before landing.

138 center A Grumman Panther prototype photographed during one of its first flights. The Panther made its maiden flight 24 November 1947 and the fighter entered service with the VF-51 in May 1950, shortly before the outbreak of the Korean War.

138 bottom The two Republic reconnaissance prototypes from the 84 series with swept wings. The aircraft behind had the definitive RF-84F Thunderflash nose design.

in 1944 when the Republic designer Alexander Kartveli started thinking about a replacement for this P-47 Thunderbolt fighter-bomber. The prototype appeared in 1945 and was designed to use the General Electric J35 engine. The XP-84 first flew on 28 February 1946, and testing demonstrated that the design was valid. The fuselage was large and robust, with an air intake in the nose, and a mid-mounted wing was employed, along with a tricycle undercarriage and traditional cruciform tail. In summer 1947, the operational squadrons began to receive the first production F-84Bs. The plane was subject to development, with subsequent versions featuring improved engines and airframe characteristics. The definitive straight-wing version, the F-84G appeared in November 1950, and 3,025 were built. Numerous allied countries purchased and flew the F-84B. The G version (which had a nuclear capability) had a wingspan of 38 feet, and a maximum take-off weight of 23,500 lb. It was powered by an Allison J35-A-29 engine, which developed 5,600 lb of thrust and provided a top speed of 620 mph at sea level. The G version was armed with six 12.7-mm machine guns and 4,000 lb of bombs or rockets. The last version of this aircraft was the F-84F Thunderstreak, which never saw action in Korea. The Thunderstreak was very different to the original Thunderjet, primarily because it had adopted a 45° swept wing and tailplane and had a redesigned cockpit canopy. The prototype flew on 5 June 1950, and production began to reach the squadrons in 1954. Over 2,700 Thunderstreaks were built, and the plane was supplied to a number of allied countries. The RF-84F Thunderflash reconnaissance version was distinguished by twin air intakes at the wing roots given that the nose was occupied by the photographic gear.

The US Navy's leading fighter was the Grumman Panther, which had made its first flight on 24 November 1947, designated the XF9F-2. The plane was originally commissioned during the war as a four-engined night-fighter. However, its development was delayed by various problems and, in accordance with the navy's mutating requirements, it was redesigned as a day-fighter. The F9F prototypes were powered by a Pratt & Whitney J45 engine producing about 5,000 lb of thrust (the J45 engine was a version of the Rolls-Royce Nene built under license) or the Allison J33 of 4,600 lb of thrust. The best version was however the one fitted with the J42 engine and produced in series as the F9F-2 and later as the improved F9F-5. This fighter entered operational service in May 1949 as the Panther, and more than 1,300 were built. The F9F-5 had a Pratt & Whitney J48 engine developing 6,250 lb of thrust and a top speed of 602 mph. The plane had a maximum take-off weight of 18,700 lb, and was armed with four 20-mm cannons and 2,000 lb of bombs or rockets. In Korea, the Panther was used by both the Navy and the Marines, and was generally employed as a ground attack fighter, but also distinguished itself in aerial combat, scoring a number of victories.

Thanks to the availability of information regarding swept wings, Grumman decided to modify its Panther, just as Republic had uprated the F-84G. On 20 September 1951, the XF9F-6 Cougar made its maiden flight. It was powered by a J48 engine that produced 7,250 lb of thrust; it boasted a 35° swept wing with flaps and slats. The Cougar entered service in November 1952. It provided no clear improvement over the Panther given that it carried the same armament and had a top speed of 646 mph. Nonetheless, more than 1,000 Cougars were produced, and the plane remained in service up until the end of the 1960s.

139 top Two Republic F-84G fighters during the early 1950s. The Thunderjet was the first USAF fighter-bomber designed to carry atomic bombs.

139 bottom The F-84 was widely used by the USAF during the Korean war as a fighter-bomber due to its 4000 lb bomb-load and its rockets. Here a F-84G launches a salvo of rockets against North Korean gun emplacements.

With these aircraft, the Americans succeeded in halting the North Korean advance and moving into the offensive. By 29 September 1950, the South Korean and ONU forces had once again regained the 38th parallel. However, the General MacArthur, the Allied commander, decided to pursue the enemy beyond the border in order to inflict a definitive defeat. The Allies' northwards advance finally provoked the direct intervention of Chinese forces and on 25 October an army of 180,000 men launched a counterattack on the Americans, forcing them to retreat. As if this was not enough, on 1 November a deadly new threat appeared in the skies of Korea, catching the western world by surprise. It was an agile, extremely fast swept-wing jet: the MiG-15.

The benefits of extremely advanced German technological research in the Second World War had not been the exclusive prerogative of the Americans. The Russians too, as they advanced from the East, had managed to capture and take back to the Soviet Union a notable quantity of research material and designs, as well as prototype and production examples of jet aircraft, together with a number of German engineers. Then, in 1946, Great Britain's labor Government gave the USSR the production plans and a number of examples of the Rolls-Royce Nene turbojet, at that time perhaps the most advanced jet engine in the world. These factors, combined with the great progress the Soviet aeronautical industry had made by the end of the war, soon enabled the USSR to bridge the technological gap that separated it from the Americans and to produce, in great secrecy, excellent aircraft.

The first Russian jet fighters were the MiG-9 and the Yak-15, both equipped with German-designed engines. The prototype of the MiG-9, derived from 1944 F project, was a fighter with a mid-mounted straight wing and a nose air intake powered by two RD-20 engines (the BMW 003A unit delivering around 1,750 lb of thrust). This fighter flew for the first time on 24 April 1946. The plane had a wingspan of 32 feet 8 inches, a length of 32 feet, and a maximum take-off weight of 11,150 lb. It was armed with one 37-mm and two 23-mm cannons. After certain structural problems had been resolved, more than 1,000 MiG-9 fighters were produced in various versions including a trainer (MiG-9 UTI).

The Yak-15 also flew for the first time on 24 April 1946. It was a hurriedly designed machine, the result of combining a Junkers Jumo 004 engine (designated the RD-10 by the Russians) complete with nacelle and as many components as possible from the propeller-driven Yak-3 fighter. The Yak-15 was armed with two 23-mm cannons and could attain a top speed of 488 mph. Fewer than 300 were built over a period of two years. The Yak-15 was then used as the basis for the Yak-17, which was powered by the RD-10A engine, a unit that produced 2,200 lb of thrust and was capable of providing a top speed of 516 mph.

These initial Soviet planes were relatively modest and overall their capabilities were inferior to the German aircraft on which they were based. The turning point came with the arrival of the Nene engine that was adopted for the MiG project for a swept-wing aircraft initiated in 1945 as Project S but never developed because of lack of a suitable power unit. Designated the I-310, the prototype made its maiden flight on 30 December 1947. The MiG-15 (known as the "Faggot" by NATO) had a fairly short fuselage with a nose air intake and 35° swept wings and tailplanes. It was powered by the RD-45F, a copy of the Nene engine, which produced 5,000 lb of thrust and provided a top speed of 640 mph. The 32-foot long plane had an identical wingspan (to what), and a maximum take-off weight of 10,595 lb. It was armed with one 37-mm and two 23-mm cannons. The MiG-15 entered service with the Soviet air force in late 1948; it was continually improved and over 9,000 were produced, including trainer and all-weather interceptor versions. Thousands of MiG-15s were built under license in Poland (as the Lim-1 then the –2 and –3), in Czechoslovakia (as the S-102 and 103) and in China (J-2), and the fighter was exported to over 30 countries. It remained in service until the 1980s.

On 8 November 1950, the first dogfight between jet aircraft in the history of aviation took place. A MiG-15 dueled with the F-80

140 top A MiG-15 displaying the Soviet Union's colors. This airplane popped up out of nowhere in the Korean War, giving UN forces serious difficulties.

140 bottom The Soviet Yakovlev Yak-15 fighter, was the outcome of a rushed Russian program to equip its air force with a jet aircraft. The plane combined the German Jumo 004 engine with the Yak-3 fuselage. It first flew in 1946.

141 top The North American F-86D interceptor version, with radar from the daytime fighter. It proved itself during the Korean War. Shown here is the F-86K export model, flying French colors.

141 bottom The F-86A Sabre entered service with the USAF in February 1949 and was used in Korea with the 4th Fighter Interceptor Wing from December 1950. It became the NATO fighter plane of the 1950s.

flown by Lt. Russell Brown, who had the better of the encounter. Nonetheless, neither the F-80 nor the F-84 were a match for the MiG-15, and the USAF decided to send the North American F-86 Sabre, its most advanced fighter, to Korea. The plane (the NE134), from which the F-86 Sabre derived dated back to 1944 and was an experimental fighter for the US Navy, subsequently designated as the XFJ-1 Fury. It featured a straight wing, an engine mounted in the center of the fuselage and a nose air intake. North American also offered the airplane to the army, which designated it the XP-86. The prototype's performance was not particularly impressive, however, and in June 1945, after having received technical information regarding swept wing technology, the company suggested delaying the project so that the new type of wing could be adopted. The swept-wing prototype (powered by an Allison J53 engine producing 3,750 lb of thrust) first flew on 1 October 1947; such was the improvement that the Air Force immediately ordered 221 of the P-86A version, soon redesignated as the F-86A, They planes began to enter service in February 1949. The F-86A version was equipped with the General Electric J47-GE-13 engine, producing 5,200 lb of thrust, sufficient to guarantee a top speed of 671 mph.

On 13 December 1950, the 4th Fighter Interceptor Wing arrived in Kimpo, Korea, with its new Sabre fighters. Things began to change. The MiG-15 retained an advantage in terms of rate of climb, could reach a higher ceiling, and was more maneuverable, but it also had major defects: at high speeds it became very unstable and when turning to starboard its pilots could on occasion lose control. Moreover, its gun-sight was inferior to those of the Sabres, and its pilots were not issued with G-suits. Lastly, the tactics the American pilots employed were frequently superior to those of their adversaries. By late December 1950, the Sabres had achieved a success ratio of 8 to 1 and dominated "MiG Alley" a corridor stretching 100 miles along the Yalu River (marking the China-N. Korea border), that the F-86As patrolled to prevent China-based MiGs from attacking. The Sabres flown by the Americans were gradually uprated; the E version, fitted with fully mobile tailplanes, first flew in September 1950, while the F-86F introduced the new "6-3" type wing with no slats, which made the plane more maneuverable at high speeds. The F-86F had a wingspan of 37 feet and a length of 37 feet 5 inches. It was armed with six 12.7-mm machine-guns and could carry an external load of up to 2,000 lb and had a maximum take-off weight was 20,350 lb. It was powered by the J47-GE-27 engine, which developed 5,900 lb of thrust and provided a top speed of 634 mph. Other versions of the Sabre included the F-86D, an all-weather interceptor equipped with radar mounted in the nose, the computerized Hughes fire control system for its 24 unguided rockets and a J47-GE-17B engine with an afterburner that produced 7,650 lb of thrust. The D proved to be troublesome to develop and entered service only in March 1951; however over 2,000 were eventually built. The F-86K was a simplified version for use by NATO countries, and had a simplified armament composed of four 20-mm cannons. This plane first flew in 1954; subsequently 341 were built for use by the French, West German, Dutch, Norwegian, and Italian airforces. The last versions of the Sabre were the F-86H and L, both equipped with improved avionics, engines and airframes. The F-86 was an exceptional fighter for its time, and about 9,800 were constructed including all the various versions and those built under license in Canada, Australia, Italy and Japan. It was used by the air forces of over 30 countries around the world, and in some case remained in service until the 1980s. For its part, the US Navy, after the brief aside of the FJ-1, also flew a plane derived from the F-86; this was the FJ-2, similar to the F-86E apart from modifications required for carrier-borne use and an armament composed of four 20-mm cannons. The FJ-2 entered service in 1954, and was followed by the FJ-3 and 4 versions; about 1,000 Furies were built, and they remained in service until the early Sixties.

Chapter 10

The Era of the Great Transport Aircraft

The Second World War brusquely interrupted the development of air transport, but the hiatus was not wholly detrimental. The 1930s saw the continued rivalry of land and seaplanes, the latter being preferred primarily for transcontinental routes. The war led to considerable changes thanks both to the construction of numerous airfield facilities around the world and the technical developments that allowed for the introduction of multi-engined aircraft of exceptional range, reliability and payload capacity.

At the end of the war, it was only natural that civil air transport should be revived using the machines and airfields (and in certain cases the routes) already developed by military transportation during the war. The lion's share of the new business was naturally conducted by the Americans, who had had to develop the best cargo planes in order to link their homeland with the distant war theaters in Europe and the Pacific. Even before the war, the main major American airlines had actually been searching for new and more capacious aircraft; the design of the Lockheed L-049 Constellation dated from 1939, while that of the Douglas DC-4 went back to 1936, at least in embryonic form. Both aircraft entered service with the US Army Air Force, the first designated the C-69 and the second the C-54 Skymaster.

In the immediate post-war period, America and its enterprising airlines began to look to the international markets once again and in particular to the transoceanic route linking the USA and Europe. In 1947, Boeing, Lockheed and Douglas were all producing long-range aircraft with pressurized cabins. Thus was born the era of the great piston- engine airliners, the Douglas DC-6 and DC-7, the Lockheed Constellation and Super Constellation, and the Boeing 377 Stratocruiser.

The Douglas DC-6 resulted from a USAAF requirement for an aircraft with an even larger capacity than its successful C-54. The prototype of the new airplane (designated the XC-112A) completed its maiden flight on 15 February 1946, too late to contribute to the war effort, but just in time to be sold to the civilian airlines. The first of the 50 ordered by American Airlines flew on 29 June 1946, and entered service on the New York-Chicago route in April 1947. In design, the plane retained the wing of the DC-4 but introduced new and more powerful engines and a stretched fuselage that could accommodate up to 86 passengers in its high-density configuration. The DC-6B (a version with more power and more interior space that appeared in 1948) was equipped with four Pratt & Whitney R-2800-CB17 Double Wasp radial engines, each producing 2,535 hp, and the plane had a maximum take-off weight of about 107,000 lb. It could sustain a cruising speed of 315 mph at 25,000 feet and had a maximum range of 4,725 miles. Like the C-118A Liftmaster (or R6D for the US Navy), the DC-6 also served with the US armed forces, with over 160 being ordered. A total of 704 DC-6s were built in various configurations. The development of its successor, the DC-7, was a result of American Airlines' desire for an aircraft with performance superior to that of the Super Constellation used by its rival TWA. The prototype of the new plane first flew on

142 top A four-engined Breguet Provence belonging to Air France is guided to a halt by ground staff at Marignane Airport, Marseilles, February 1954.

142 bottom A KLM Lockheed Constellation landing at Schiphol Airport shortly after the end of the Second World War. Note the temporary, makeshift buildings in the background.

143 top The control room at Northolt Airport in 1948. Postwar development of air transport required the development and expansion of ground services.

143 center This TWA four-engined Douglas DC-4 was photographed in 1954. The model was developed from the military C-54 version, which first flew in 1942.

143 bottom A Douglas DC-6, British Eagle Airways. This aircraft was a development of the C-54/DC-4; it kept the same wings but had a new, larger fuselage and new engines.

18 May 1953, and was in effect a DC-6B with a stretched fuselage and the same 3,295-hp Wright R-3350 Turbo-Compound engines that powered the Super Constellation. As the first DC-7s lacked sufficient range to cross the Atlantic safely when fully laden and facing head winds, the DC-7C was thus developed with an increased wingspan, more powerful engines, and a fuselage stretched by 3 feet, which allowed 105 passengers to be carried. The same Wright engines uprated to produce 3,447 hp provided a top speed of 406 mph at 21,500 feet and fully laden range of 4,600 miles. Maximum take-off weight was about 143,000 pounds.

144 top The Lockheed Constellation was originally a military aircraft introduced in 1943, but after the war it was transformed into a civilian plane. Shown here is one belonging to Eastern Airlines.

Boeing's large civil transport aircraft derived from a military design. The Model 377, which completed its maiden flight on 8 July 1947, was closely related to the C-97 transport plane of 1944; this in turn had been derived from the B-29 bomber design. The 227 Stratocruiser could be ordered in various configurations, and was first used by Pan Am and BOAC, but only 55 were ever built. The Super Stratocruiser model with fuel tanks modified for operating on the Atlantic routes was equipped with four 3,549-hp Pratt & Whitney R-4360 Wasp Major engines and cruised at 340 mph at 25,000 feet. The maximum take-off weight was 145,800 lb, and the aircraft boasted a maximum range of 4,200 miles.

After a brief military career during the war (just 22 entered service between 1944 and 1945), the Lockheed Constellation was developed essentially as a civil aircraft from December 1945, when it received the appropriate certification. The airplane immediately entered service with Pan Am and TWA who in February 1946 began operating a scheduled service between the United States and Paris. The first true civilian Constellation was, however, the L-649 that had a more luxurious cabin and could carry up to 81 passengers (48-64 in the normal configuration). In 1949, as air transport became established and spread, increasing numbers of carriers required a larger aircraft. Lockheed responded with the Super Constellation (L-1049) that had a fuselage stretched by 18 feet 4 inches, and could carry up to 109 passengers. A total of 856 Constellations were built, including the military versions of the post-war period. The last model in the line was the L-1649 Starliner, equipped with four 3,447-hp Wright 988TC-18EA-2 engines that provided a top speed of 377 mph at 18,700 feet, with a range of 4,950 miles. This model was not a great commercial success owing to the emergence of jet aircraft.

In the second half of the 1940s, jet propulsion began to gain a foothold in the military sphere, but in the civil sector it was seen as little more than a pipe-dream. The jet engines of the era were extremely thirsty and were thus considered to be unsuitable for commercial applications. However, in Great Britain another factor came into play: the jet engine was the only aeronautical technology in which the British had an advantage over the Americans. In order to regain a leading position in the world of aviation, the British decided to build a commercial jet airliner that would outclass all its propeller-driven rivals. This was recommended by the Brabazon Committee, a government study group charged with the task of reviving the British aeronautical industry in terms of commercial transport. There was only one company in the United Kingdom with the capacity undertake such a project, de Havilland, the firm already having experience in the design and production of an airframe and related jet engine, specifically with the Vampire fighter.

Work began in 1947 and was guided by a key concept: the jet engines had to return acceptable fuel consumption figures at altitudes of 32,000 feet and above, and the reduction in thrust due to the altitude had to be compensated by lower drag. The problem of pressurization also had to be resolved, given that at 32,000 feet a pressure would be exerted on the airframe the like of which had never been experienced before. The chosen architecture was traditional, with a 20° sweep to the wings and tailplanes. The four engines

144 bottom The L-1049 Super Constellation was a development of the basic L-649 model. It was given a fuselage 18.5 feet longer, which allowed it to seat 28 more passengers.

145 top left The de Havilland DH.106 Comet, the first passenger jet, was presented as the plane of the future, but it suffered several accidents which heavily dented its reputation.

were to be buried in pairs in the wing roots. The project was designated the D.H.106 Comet. Recognizing the design's potential and driven by national pride, British Overseas Airways Corporation (BOAC) immediately ordered eight planes, with an option on the planes constructed after the fourteenth to be built.

The Comet, built with ample safety margins and subjected to unprecedented endurance testing, made its maiden flight on 27 July 1949, and immediately proved to be fast and highly promising. Further orders then arrived from British Commonwealth Pacific, Air France, Union Aéromaritime de Transport, Canadian Pacific, Japan Air Lines, the Royal Canadian Air Force, and even Pan Am. On 2 May 1952, Comet service was inaugurated, with a flight from London to Johannesburg, South Africa.

The jet airliner unquestionably represented the future; piston engines had by now reached the limit of their potential and were frighteningly complex. Moreover, the noise and vibration they generated in the large four-engined planes made the flights uncomfortable for passengers who were now pleasantly surprised by the quietness and smoothness of the new jet engines.

The Comet was immediately successful and in the first year BOAC's planes carried 28,000 passengers with 88% of seats being filled. In 1953, de Havilland had orders for 50 more planes and were negotiating a further 100 orders. The Comet was capable of 485 mph, had a ceiling of 36,000 feet, and a range of over 3,100 miles. Despite certain accidents considered to be within the norms, the plane's career began to decline on 10 January 1954, when a Comet flying out of Rome's Ciampino airport mysteriously crashed into the sea near the island of Elba, with 35 people aboard. BOAC immediately grounded its fleet, and the British aviation authorities set up investigations. Despite their efforts, however, they found no clues as to the cause of the disaster. It was thus decided to make 50 modifications of various kinds; all were intended to improve safety and would, it was thought, resolve any problems. The Comets resumed operation in the March of that year, while work continued on the recovery of the crashed Comet's wreckage, which was taken to Farnborough for the examination at the Royal Aircraft Establishment.

Then, on 8 April, another Comet unaccountably crashed while flying from Rome to Cairo. The fleet was grounded once again, and the RAE was put under intense pressure to find a solution to the mystery quickly. After conducting tests of various kinds, the RAE put forward the hypothesis of structural failure. At the end of a simulation carried out in a large tank of water with a complete fuselage, the sought-after proof was revealed. Following a certain number of pressurization cycles, the fuselage of the Comet would crack at the square corners of the windows, a feature that had already been criticized by the American aviation authorities. It was not only the angular design of the windows that caused the problems, but its combination with a fuselage that had been lightened as much as possible to allow the plane to fly higher and thus extend its range. The accidents undermined the passengers' faith in the aircraft, and de Havilland hastily made significant modifications. In the third series, oval windows were introduced along with other changes that made the Comet ever better, but its fate had already been decided. American manufacturers had taken full advantage of the time lost by the British and had begun to turn out their own jet airliners. The Comet continued to carry passengers up to 1980, but only about 80 Comets were built. Great Britain had lost forever its supremacy in the field of commercial aviation.

145 top right The large four-engined Boeing 377 derived mainly from the C-97 military vehicle of 1944. Though able to carry 144 passengers, it never met with great commercial success.

145 bottom The Boeing 377 could also be fitted out with 28 couchettes for night journeys. This 1949 photograph shows a mother putting her children to bed during a flight.

The American firms had certainly not been standing idle, but it was Boeing, the producer of the least successful piston-engined airliner, that was most committed to the creation of a jet plane. Boeing was, in fact, the only major firm to have experience with large jets, having built the B-47 Stratojet, a six-engined bomber with swept wings that first flew in December 1947.

From the outset, the Boeing engineers were convinced that one of the errors made with the Comet was its size: it was too small to be economical. The new Boeing was designed to carry twice the number of passengers. Initially the company considered developing the new jet on the basis of the Stratocruiser, but it soon became clear that this was not a viable option. The design was completely revised even though, for reasons of "disinformation" with regard to its rivals, Boeing maintained the original serial designation 367. In reality, within the restricted circle of those working on the project, the aircraft already had another name: 707, because it was Boeing's 707th project. The aircraft had an accentuated 35° sweep to its wings, which carried four separate pods for the four engines. The fuselage was slim and streamlined and the tailplanes were also swept.

Irrespective of its final form, William Allen, Boeing's president, wanted the new design to be appropriate for both commercial transport and military tanker applications. He knew, in fact, that the Air Force would soon need new jet tankers to replace the old propeller-driven Boeing KC-97s, whose performance was inadequate to service the B-47 and the future B-52.

The American airlines also started to take an active interest in jet airliners. Unfortunately, however, at that time they were carrying heavy debts resulting from expansion programs that had led to the acquisition of hundreds of propeller-driven planes. The USAF also lacked the funds to replace its KC-97s. Boeing would have to start up and finance the 707 program on its own. In April 1952, the board of directors authorized the investment of $15 million in the project. At this point Boeing was running a truly enormous risk; the 707 project meant success or outright failure for the company. The design and development work was carried out with the utmost thoroughness, and the resulting aircraft was above all safe and extremely robust. The fuselage, for example, had stood up to a grueling sequence of 50,000 pressurization cycles.

The first Boeing 707 was a truly large airplane for its time: it was 128 feet long and weighed around 160,000 lb. It completed its maiden flight on 15 July 1954, powered by four Pratt & Whitney JT3P engines each producing 9,500 lb of thrust. The plane immediately displayed great handling characteristics. Boeing initially aimed for a military clientele, and the company demonstrated the prototype (equipped with a rigid probe for in-flight refueling) to the USAF in order to underline its potential as a jet tanker. In March 1955, the USAF ordered a preliminary batch of 29 of the KC-135As, the military tanker version of the 707, eventually the Air Force acquired a total of more than 800. Having received military approval, Boeing then modified prototype and fitted it out for civilian use. In this configuration, the 707 immediately made all other transport aircraft of the era seem obsolete. On 13 October 1955, Pan Am ordered the first six planes of the production version, the 707-120.

In reality, the 707 did not have an entirely trouble-free birth; early on there were problems with the brakes, the thrust provided by the engines was insufficient when the aircraft was fully laden, and the range was inadequate for fully laden transatlantic flights. The 707-120 was therefore immediately equipped with JT3C-6 engines producing 13,500 lb of thrust. At this time, Douglas was the largest company involved in building civil aircraft, and it had by no means been resting on its laurels. Its confidential DC-8 project for a 560-mph, four-jet airliner had not, however, attracted much interest from the airlines to which it had been shown secretly in 1952. They had preferred to order the most recent propeller-driven transport aircraft, the DC-7. Douglas suffered a severe shock in 1955 when Pan Am's management declared that with the appearance of the Boeing 707, Douglas risked missing the boat. Pan Am wanted the competition between the two giant manufacturers to result in better, more economical aircraft, and a Boeing monopoly of the four-engined jet market would not have been at all welcome. The task facing Douglas was difficult, the DC-8 would have to be launched fully developed if it was going to better its rival. The company naturally took advantage of the experience accumulated and the defects identified with the 707. The design process was painstaking; given the costs and delay Boeing had experienced, the Douglas plane had to be right first time. The passenger cabin was larger and the aircraft had a longer range than the 707. Douglas's audacity was rewarded, and less than five months after the announcement of the launch of the program, Pan Am announced that it would be purchasing 25 DC-8s and 20 707s. Two weeks later, United also made public its intention to acquire 30 DC-8s. Boeing was stunned: Douglas was selling more jet liners than it was without even having built a single plane, and the airlines appeared to be more attracted to the DC-8 than to the 707. The DC-8 prototype first flew on 30 May 1958: the aces in hand with respect to the 707 were more powerful engines, a greater range, and a larger cabin. All that Boeing could do was redesign the 707.

Thus was born the 707-320 Intercontinental, which first flew on 10 October 1959. This was the first transoceanic version; it was followed by the 707-320C version, which had JT3D-7 engines producing 19,000 lb of thrust. These engines provided a top cruising speed of over 600 mph, a ceiling of 39,000 feet, a maximum range of 5,750 miles with 147 passengers, and a maximum take-off weight of about 333,500 lb. With this aircraft, and thanks to the design flexibility of the 707 fuselage, Boeing regained supremacy over Douglas. In the meantime, the international airlines had come to accept the need to operate jets. American Airlines

purchased 30 707s, with an option for a further 20. Delta, Eastern, Trans Canada, Western, Japan Air Lines, KLM, SAS, Alitalia and Swissair all went for the DC-8, while the 707 was ordered by Lufthansa, Air France, Sabena, Continental, Western, TWA and Air India. More than 1,000 707s and more than 550 DC-8s were built.

The airlines appreciated the new jets' low running costs, while passengers were more than happy with the increased convenience: greater comfort and speed and, with the increase in the number of flights, even cheaper tickets. The world truly had become a smaller place, and even distant lands and the VIP life-style now seemed to be available to all.

While the struggle for supremacy in the long-range sector was going on in America, in Europe there were those who had recognized the potential market for a medium-short range jet. On the old continent, in fact, large jets were still superfluous and overly expensive. In 1951, the French government announced a competition among the nation's aeronautical firms for the design of an aircraft with a range of less than 1,250 miles and seating about 60 passengers. Six companies responded, and in 1952 the design submitted by Sud-Est (SNCASE) of Toulouse was announced as the winner. The Caravelle as it was named, had an innovative layout, characterized by the positioning of the engines at the tail, either side of the fin.

The Caravelle was equipped with two powerful Rolls-Royce Avon RA.29 Mk.552 engines producing 10,500 lb of thrust. This allowed the plane to fly on a single engine in cases of emergency. The fact that the engines were at the rear provided additional advantages such as a cleaner, more efficient wing and a quieter cabin. The first client to purchase the new aircraft was naturally Air France.

The Caravelle completed its maiden flight on 27 May 1955 and soon it proved to be a successful design. Continuously developed and improved (with seating capacities arriving up to 140 passengers), 281 Caravelles were constructed, and it was used by more than 35 airlines.

146 A Sud-Est Caravelle belonging to the Swiss airline Air City about to land. The Caravelle project was prompted by the French government; it became the first successful European jet, eventually sold to 35 airlines.

147 top The DC-8 was Douglas's answer to the Boeing 707 and was initially an improvement on its rival. Note the DC-8-63 of 1967, which had various improvements in terms of streamlining, engines, flight range and flaps and slats.

147 center A three-engined British Airways Hawker Siddeley HS.121 Trident in descent. This airplane was designed for short journeys but was not popular with airlines; only 117 were built.

147 bottom A Lufthansa Boeing 707 in flight. The 707 originated as a military aircraft, covering its development costs with US government orders.

Having seen the success of the Caravelle and given market demands, the British also focused on the medium-small jet sector. The result was the introduction in 1956 of the de Havilland DH.121, which with Hawker Siddeley's acquisition of de Havilland, was soon to become known as the H.S. 121 Trident. It first flew on 9 January 1962. However, despite a number of interesting features (it was the world's first aircraft with automatic landing gear), the Trident was never particularly successful, by 1975, only 117 had been built. On the other hand, the American firms were now also looking at the medium- and short-haul sector, and a number of the most successful and famous airliners in the history of aviation were beginning to take shape.

The third of the Century Series jets in chronological order of introduction was the Lockheed F-104 Starfighter. It completed its first flight as the XF-104 on 7 February 1954. The Starfighter, whose design drew upon the experience and suggestions of pilots fighting in the Korean War, was the first operational fighter capable of reaching Mach 2, and breaking the sound barrier while climbing. In effect, what the USAF demanded of this plane was acceleration and outright speed, and in order to achieve these objectives Lockheed's chief designer, Clarence "Kelly" Johnson (the "father" of the P-38 and F-80), was uncompromising. The F-104 had a stubby laminar-profile wing and a long, aerodynamic fuselage little bigger than the engine it housed. The pilot and the armament, radar and avionics systems were all crammed into the restricted space of the nose.

The F-104A entered service with the USAF on 26 January 1958; it was 54 feet 5 inches long, with a wingspan of 21 feet 7 inches, and a maximum take-off weight of 22,000 lb. A single General Electric J79-GE-3B producing 14,800 lb of thrust powered the aircraft to a maximum speed of 1367 mph, or Mach 2.2. The F-104A was typically armed with two AIM-9B Sidewinder air-to-air missiles and an M61 Vulcan 20-mm cannon. While its performance was exceptional, the plane as a whole did not satisfy the armed forces because it was too difficult to fly, had a restricted range, and was only lightly armed: it was soon passed on to the reserve squadrons. The successor F-104C, fitted with a more powerful engine and optimized as a fighter-bomber, was no more successful; both versions saw limited action in Vietnam but were in practice rejected, although the Starfighter continued to fly with the National Guard up until the 1970s.

The export version of the aircraft, the

150 left The McDonnell F-101A Voodoo fighter's development was long and difficult, and when the plane entered service, it was assigned the role of interceptor. Canada purchased this aircraft, calling it the CF-101, and kept it in service until the start of the 1980s.

150 top right As an interceptor, the F-101A was able to fire Genie nuclear air-to-air missiles, as seen in the photograph of the National Guard launching the aircraft's weapons in Maine.

F-104G Super Starfighter (and the two-seater TF-104G) enjoyed better fortunes, thanks above all to the fact that America's allies were offered the opportunity to construct it themselves under license, an offer taken up by Germany, Holland, Belgium, Italy, Canada and Japan. The F-104G completed its maiden flight on 7 June1960. The plane featured a strengthened airframe (the maximum weight had risen to around 28,750 lb), improved avionics and a new engine, the J79-GE-11A producing 15,800 lb of thrust on afterburn and providing a top speed in excess of 1,350 mph at 36,000 feet. The armament remained the same as that of the F-104C with the Vulcan cannon and the external load of up to 4,000 lb, which could include supplementary fuel tanks, bombs and rockets or a nuclear bomb. Sixteen countries adopted the Starfighter, and a total of 2,576 (including all variants) were built.

In 1966, Italy decided to develop an advanced version of the Starfighter designated the F-104S. This variant entered service in 1969; it had improved avionics, an even more robust airframe (which featured two new underwing pylons) allowing a maximum take-off weight of 30,850 lb, and a heavier weapons load. The plane's maximum speed was about 1,500 mph and the principal armament consisted of the semi-active radar-guided AIM-7 Sparrow air-to-air missile. Modernized on a number of occasions, with models ASA and ASA-M, in 2005 the F-104S was the last Starfighter to be retired from service by the Italian Air Force.

The fourth of the Century Series was the McDonnell F-101 Voodoo, which made its first flight in definitive form on 29t September 1954. In reality, the original design for the aircraft dated back to 1946, when the USAAF issued a requirement for a jet-powered escort fighter for long-range operations. MacDonnell built the XF-88 prototype in 1947, but owing to a change of policy by the armed forces and the strategic command (as well as problems associated with the engines and the financing) the program was suspended and only revived in the face of the developing war in Korea. The new project was designated the F-101 and was to be a large, swept-wing aircraft with twin engines and a T-tail. However, it proved ill suited to the bomber escort role owing to its insufficient range. It plane was therefore adapted as a two-seater all-weather interceptor for the tactical command. It entered service early in 1957 as the F-101B. The B model was 67 feet 3 inches long, had a wingspan of 39 feet 4 inches, and a maximum take-off weight of 52,400 lb. It was powered by two Pratt & Whitney J57-P-55 engines producing 14,880 lb of thrust on afterburn, which gave it a maximum speed of 1,222 mph, or Mach 1.85. The armament consisted of two GAR-2 Genie rockets with nuclear warheads or up to six AIM-4 air-to-air missiles. More effective planes soon replaced the Voodoo for interceptor duties, while the reconnaissance versions (RF-101A and C) remained in service longer. Just over 800 Voodoos were produced, of which 132 were sold to Canada, where they were kept in service as CF-101 interceptors until 1984.

The Republic company built the fifth aircraft in the Century Series on its own initiative, in the hope that it would be selected to replace its own F-84s, which were already in service. Getting underway in 1951, the project involved the creation of a large, single-seat fighter-bomber capable of supersonic performance while carrying a large, potentially nuclear, weapons load. The USAF approved the project in 1954, and the first YF-105A prototype flew on 22 October 1955. As was traditional with Republic, the plane had a large fuselage housing a single Pratt & Whitney J57 turbojet, a mid-mounted wing with air intakes at the roots, and swept tailplanes. The new and more powerful J75 engine was soon available, however, and the plane was re-engined and designated as the YF-105B, making its maiden flight on 22 May 1956; the production version, the F-105B Thunderchief, began to enter service in May 1958.

After 75 F-105Bs had been built, production shifted to the F-105D, which had a length of 64 feet 4 inches, and a wingspan of 34 feet 5 inches, an uprated engine and new avionics that gave the plane all-weather capability. The engine was the J75-P-19W, producing 24,500 lb of thrust and offering reach 1,250 mph at the plane's operating ceiling. Maximum take-off weight was 52,850 lb, of which external loads – auxiliary tanks, bombs, rockets and missiles – accounted for 14,000 lb. A 20-mm Vulcan cannon completed the armament. The model D was followed by the two-seater F-105F, which featured a stretched fuselage. This model proved to be of great use in combating enemy anti-aircraft defenses and led to the development of the EF-1O5F and F-105G versions, both used in the "Wild Weasel" role with special electronic equipment and Shrike and Standard anti-radiation missiles. All these models were widely used in Vietnam and were withdrawn from service only in the early 1980s.

151 top The Lockheed F-104 Starfighter was the first standard fighter to fly at speeds in excess of Mach 2. Everything in this no-compromise project was designed to boost acceleration and speed. However, the USAF never liked the plane.

151 bottom The Republic F-105 Thunderchief was designed as a supersonic fighter-bomber able to carry 14,000 lb of bombs and rockets. Shown here are two refueling over Vietnam.

152 top The Convair F-106 Delta Dart was for many years the standard American interceptor. It was a direct development of the F-102 and had a maximum speed of 1,525 mph and an armament of four air-to-air missiles.

152 bottom The Vought F8U Crusader was adopted in 1952 to give the US Navy a superior supersonic fighter. The last model to fly for the Navy was the RF-8G reconnaissance plane, shown here as it folds its wings after landing.

153 top The Grumman F11F Tiger was the first supersonic fighter used by the US Navy. It entered service in 1957 and was also used by the Blue Angels acrobatic team.

The last of the Century Series was the Convair F-106 Delta Dart, basically the direct descendent of the earlier F-102. This last had never fully satisfied the USAF, which required an all-weather supersonic interceptor with computerized fire control. The F-012B was intended to be the definitive version, but in June 1956, the in-development plane was redesignated the F-106. This aircraft was equipped with a new and far more powerful engine (the J75) and a Hughes MA-1 computerized fire-control system that finally seemed to have been debugged. The first Delta Dart flew on 26 December 1956. It had very similar lines to its predecessor, with a length of 70 feet 9 inches, a wingspan of 38 feet 4 inches, and a maximum take-off weight of about 41,830 lb. The J75-P-17 engine was almost identical to that of the F-105, and provided the Delta Dart with a maximum speed of 1,525 mph, or in excess of Mach 2.3, at its operational ceiling. The armament comprised an AIR-2A Genie or 2B Super Genie missile with a nuclear warhead and four AIM-4F/G Falcon air-to-air missiles. The armaments system was initially unconvincing, but the USAF nonetheless acquired a total of 277 single-seat F-106As and 63 two-seat F-106Bs that entered service in October1959. Over the years the Delta Dart proved to be an effective, reliable plane and remained in service as an interceptor with the National Guard until the second half of the 1980s. The Dart has subsequently been used as a radio-controlled target drone after numerous planes had been converted to QF-106 specifications.

While the USAF acquired a notable variety of advanced fighters to perform various roles, the US Navy was lagging behind in this process of modernization. A number of carrier-based combat aircraft such as the AD-1 Skyraider were actually still propeller driven, while the fighter ranks had seen a proliferation of different models, all of modest performance. Early in the 1950s, the Panther, Cougar and Fury were joined by aircraft such as the Vought F7U Cutlass (first flight in 1948), the McDonnell F3H Demon (1951) and the Douglas F4D Skyray (1951). These planes shared innovative technical features, but also problems regarding the efficiency and performance of their engines, which were frequently underpowered. None of these fighters had supersonic capabilities.

Things began to change with the advent of the Grumman F11 Tiger, a development of the Panther and Cougar line. The aircraft was, however, completely new and adopted a thin-profile wing, swept tailplanes and a more powerful engine. The first prototype took to the air on 30 July 1954, but in January 1955 a second prototype was introduced. This had an afterburner-equipped J65 engine that made the Tiger the Navy's first supersonic fighter, with a top speed of 752 mph. The F11F entered service in March 1957, but its operational career was fairly brief. It was immediately overshadowed by a brilliant new design: the Vought F-8 Crusader, the US-Navy's first true superfighter.

The F-8 Crusader resulted from a requirement issued by the US Navy in 1952 for a new supersonic carrier-borne plane. Vought won the competition (designated VAX) the following year, and received an order to build two XF8U-1 prototypes in keeping with its design. The aircraft was muscular single-seater characterized by a high

153 bottom Four F-106As belonging to the New Jersey National Guard in formation. This fighter remained in service with the Guard until 1988 when it was replaced by the F-16.

swept wing with a variable angle of incidence that could be adjusted during the take-off and landing phases so that the pilot could avoid giving the plane an excessively nose-up stance. The prototype first flew on 25 March 1955. It was equipped with a Pratt & Whitney J57-P-11 engine producing 14,800 lb of thrust, and immediately proved to be an excellent, fast, and maneuverable airplane. A total of 318 F8U-1s were ordered, and serial production was rapidly initiated. The first F8U-1s were already in service by March 1957. The plane was gradually improved to give it more power, strength and firepower as well as all-weather and reconnaissance capabilities.

The definitive version, the F8U-2NE appeared in 1961. (From 1962 on, following the rationalization of US military designations, the Crusader became the F-8 and the F8U-2NE model became the F-8E.) The F8U-2NE was powered by a J57-P-20A turbojet producing 18,000 lb of thrust on afterburn and could reach 1,120 mph, the equivalent of about Mach 1.7. It had a maximum length of 54 feet 5inches, a wingspan of just over 35 feet and a maximum take-off weight of 33,995 lb. The weapons system retained the original four 20-mm cannons – which, with the advent of the F-4, earned the plane the nickname of "The Last of the Gunfighters." To this armament was added the possibility of mounting up to four AIM-9 Sidewinder air-to-air and bombs or rockets on the two underwing pylons for a maximum total of 5,000 lb. A total of 1,259 Crusaders were built, and the plane fought successfully in Vietnam. In RF-8G form, the Crusader remained in service until the early 1980s. A number of Crusaders were sold to the Philippines air force, while 42 F-8Es (FNs) were delivered to the French navy in January 1965. These were actually the last operational Crusaders, though they were withdrawn from service only in December 1999.

154 top and center This famous sequence from 3 June 1967 shows the shooting down of a North Vietnamese MiG-17 by a USAF F-105 using a 20-mm cannon.

154 bottom The MiG-17 appeared in 1950 as a direct development of the MiG-15 and was the first Soviet fighter armed with missiles. Shown here is a fully restored MiG-17 taking off from an American airfield.

The aircraft discussed above were the principal American designs of the 1950s. However, aeronautical engineers on the other side of the Iron Curtain were similarly prolific. Following the success of the MiG-15 fighter, the MiG factory (led by the designers Mikoyan and Gurevich) soon began work on a new design that would eradicate the defects identified in the earlier plane, particularly with regard to its aerodynamics. The I-330 prototype appeared in January 1950. In terms of overall lines, the I-330 resembled the earlier design, but introduced a longer fuselage, larger tailplanes, a 45° swept wing and, of course, a more powerful engine. The plane went into production in 1951 as the MiG-17, reaching the first operational squadrons the following year. A Klimov VK-1A engine producing 5,950 lb of thrust enabled the I-330 to reach a top speed of 690 mph. It had a length of 36 feet 8 inches, a wingspan of 31 feet 5 inches, and maximum take-off weight of 13,400 lb. The first model (known to NATO as "Fresco-A") was followed by the MiG-17P, which had a VK-1F engine fitted with an afterburner that produced 7,450 lb of thrust, giving the plane to reach a top speed of 693 mph. The MiG-17 was more heavily armed than the MiG-15. It fielded three 23-mm cannons, one 37-mm cannon, and free-fall weaponry weighing up to 1,100 lb. Next followed the MiG-17FP, which was equipped with Izumrud interception radar; then came the MiG-17PFU, which was stripped of its cannons but could carry up to four AA-1 "Alkali" radar-guided air-to-air missiles.

155 top The MiG-19 was the first supersonic Soviet interceptor, thanks to its 55° swept wing and twin Tumansky engines with afterburners. It entered service in 1954.

155 bottom The Mikoyan-Gurevic MiG-21 was the most popular Soviet jet fighter, with almost 12,000 being built. It was used by roughly 50 armed forces around the world. Shown here is one belonging to the Finnish air force, which withdrew it from service in 1998.

The MiG-17 remained in production in various forms up until 1958, with about 9,000 planes being built (about 3,000 of these were built under license in China as the J-5 and in Poland as the LIM-5). Simple and robust, the MiG-17 served with the air forces of all the Soviet bloc countries, as well as the USSR, in some cases up to the present day.

The next development was the MiG-19, an aircraft born out of the I-350 project that was intended to provide the Soviet Union with a supersonic airplane capable of competing with the new American fighters. The prototype, which first appeared in 1953, had a strong 55° sweep to its wings, new tailplanes with low-set horizontal surfaces. The MiG-19 was equipped with two AM-5 engines with no afterburners. This plane first flew on 18 September 1953, and went into service in 1954 with afterburner-equipped AM-5F engines. (NATO designated it "Farmer-A.") However, the plane was not yet fully developed and demonstrated serious instability at high speeds. The MiG-19S version was consequently developed, featuring new, completely mobile tailplanes. This version was followed by the SF, powered by new Turmansky RD-9BF engines producing 3,300 lb of thrust, sufficient to propel the SF to a top speed of 901 mph. The plane was 39 feet 7 inches long, had a wingspan of slightly over 30 feet, and a maximum take-off weight of just over 20,000 lb. It was armed with three 30-mm cannons and up to 1,100 lb of external weaponry.

The MiG-19P all-weather interceptor version was introduced in 1958, equipped with radar and two 23-mm cannons. This was followed by the MiG19PFM, which featured radar and four AA-1 air-to-air missiles, then the PF interceptor with two 30-mm cannons, then the MiG-19R reconnaissance version, and finally the two-seater MiG-19UTI trainer. About 2,500 MiG-19s appear to have been built in the USSR, while double that number were built in China, which acquired a production license for the MiG-19S. China built the MIG-19 and exported it numerous countries as the J-6/F-6 and, in two-seater form, as the FT-6. These last and their derivatives are still in service in a number of Asian countries.

The introduction of the MiG-21A gave Soviet aviation a decisive technological and operational boost. The new MiG was the Warsaw Pact countries' first aircraft capable of Mach 2; it came as a disagreeable surprise to the West, one with similar impact to that of theMiG-15's introduction. Work on a new, lightweight, supersonic fighter had begun as early as 1953. The first configuration (E-2) was the logical development of what had gone before and featured a swept wing. Subsequently, however, the E-5, which introduced a delta wing, was preferred. It first flew on 9 January 1956. Out of this experimental aircraft came the E-6; in effect the prototype MiG-21, which first flew on 20 May 1958. While featuring a delta wing, the jet retained strongly swept traditional tailplanes and had a nose air intake with an antishock cone to feed the afterburner-equipped Turmanski R-11 turbojet. Put into production in 1959 as the MiG-21F ("Fishbed-C" to NATO), the new fighter boasted excellent level flight and climbing performance and was sufficiently agile, but was only lightly armed with two 30-mm cannons. It was 43 feet 10 inches long and had a wingspan of 23 feet 3 inches, and a maximum take-off weight of 16,700 lb. The engine produced 12,765 lb of thrust, giving the plane a top speed of about 1,245 mph at its operational ceiling.

The following year, the MiG-21F's first mass-produced version, the MiG-21F-13 was introduced. This plane was equipped with radar and two K-13 (AA-2 "Atoll") air-to-air missiles, the Soviet copy of the AIM-9B Sidewinder. Over the years, this aircraft was continuously developed and improved. The four principal versions followed. The MiG-21PF featured a more powerful engine, larger fuel tanks, and new radar system): the MiG-21PFM (from 1964 on) featured a new system of flaps, an enlarged fin, new weaponry and avionics. The MiG21R, the reconnaissance version, featured a pod and larger dorsal fairing; and the MiG21S and SM both featured a more powerful engine, new radar, four underwing pylons and an internal cannon. The MiG21S had an export version designated MiG-21MF). Then in 1972 the MiG21BIS ("Fishbed-L") appeared, introducing a larger dorsal avionics pack, new R-13M, R-55 and R-60 air-to-air missiles, and a new Soyus-Gavrilov R-25-300 engine producing around 16,535 lb of thrust, sufficient to provide the plane with a top speed of 1352 mph, or Mach 2.05. The plane's maximum take-off weight rose to 22,928 lb, with 2,650 lb accounted for by the external weapons load.

The MiG-21 design gave rise to various two-seater training versions, starting with the original MiG-21U (based in the MiG-21F-13 and designated "Mongol-A" by NATO), and proceeding to the MiG-21US and lastly to the MiG-21UM. Over 11,950 MiG-21s were produced in the USSR and, under license, in China (as the J-7/F-7), Czechoslovakia, and India. During the 1960s and 1970s, it was in service with all the Soviet bloc countries, in close to fifty air forces.

In the 1990s, two projects appeared for the modernization of the still active MiG-21, through extensive measures improving the airframe, avionics, and armament. Russia (Mapo-MiG) developed the MiG-21-93, which was adopted from India, while Israele (IAI) offered the MiG-21-2000, the version that was chosen by Romania, under the name Lancer. Although obsolete, the MiG-21's performance is still quite valid for an aircraft at 50 years from its maiden flight; and many Fishbeds are still in service today.

Chapter 12

The Cold War

The precarious alliance between the United States and the Soviet Union lasted only until the defeat of their common Nazi enemy. Even during the conflict, despite the incessant flood of military aid dispatched by America to the USSR (planes, trucks, tanks and prime materials in vast quantities), Stalin was never particularly friendly toward his western Allies, either politically or militarily. A number of American bomber crews who had made emergency landings behind the Russian lines in 1944 and ad 1945 were, in fact, jailed as common prisoners and only released many months later; their recognized that the USSR's undeclared aim was that of exploiting the war to take control of as much of Europe as possible.

In the United States, the attitude toward the Soviet Union was still one based on trust, but this changed after the President Roosevelt's death in April 1945, when the Potsdam Peace Conference brought to East, in a perennial state of confrontation and tension. The conflict, though never openly declared, came to be known as the Cold War.

Apart from political and ideological propaganda and international espionage, the principal Cold War weapons were aeronautical; in particular, strategic aircraft,

planes, especially the B-29 bombers, were confiscated by the Russians. During the battle of Warsaw (August and September 1944) Russian strategy was made clear to the western Allies. Following the Warsaw Uprising against the Germans, the Russian troops failed to advance from their positions just a few miles outside the Polish capital. This effectively allowed the rebels to be caught and massacred – because the Polish Government-in-Exile in London had refused to accept the authority of Moscow. The Allies' attempts to supply Warsaw from the sky were unsuccessful, resulting in heavy losses. Churchill, a convinced anti-Communist, light the progressive deterioration of the Allied- Soviet relationship. Other factors also came into play in this gradual dissolution of the wartime alliance. These included the United States' possession of the atomic bomb (1945), followed by the Soviet Union's (1949); creation in Eastern Europe of satellite governments subject to the Soviet Union; formation of the NATO military alliance (1949); the outbreak of the Korean War (1950); and formation of the Warsaw Pact (1955). All these events contributed to the creation of two distinct and contrasting power blocs, the so-called "Free World" in the West and the Communist states in the nuclear bombers, in-flight refueling tankers, and reconnaissance planes.

From the outset, Cold War tactics hinged on bombers with a nuclear capability. The first aircraft of this type was the Boeing B-29, the bomber used in the attacks on Hiroshima and Nagasaki. After the war, beginning in 1947, the B-29s began to be replaced by an advanced version known as the B-50, which featured structural modifications designed to increase both its weapons load and its range. As early as 1950, however, the B-50 had already been rendered obsolete and was relegated to less demanding reconnaissance and

In the meantime, in 1948 the Strategic Air Command (SAC), responsible for nuclear bombers and their in-flight refueling tankers, came under the command of General Curtis LeMay. He was a hardliner who had distinguished himself during the war, above all during the bombardment of Japan. By means of grueling and unrelenting training exercises and thanks to the acquisition of new planes, he managed to revive the forces from the dangerously low level of efficiency to which they had sunk, making of the SAC an efficient, reliable and fearsome unit. The United States' Cold War strategy demanded that armed bombers with a global strike capability be in flight at all times, ready to respond rapidly and decisively to a nuclear attack by the USSR. The introduction of jet-powered bombers and in-flight refueling tankers made this capability possible.

refueling duties. Convair built the first true strategic bomber in response to a USAAF requirement issued in 1941 calling for an intercontinental bomber capable of carrying a weapons load of about 10,000 lb for a distance of 5,000 miles. Owing to the technical difficulties involved and the ongoing Second World War, project development was slow, and the prototype did not fly until August 1947. At 162 feet in length and with a wingspan of 230 feet, the B-36 Peacemaker was a truly enormous aircraft; it was powered by six 3,800-hp Pratt & Whitney radial engines and had a maximum take-off weight of over 165 tons.

The plane's performance was disappointing, however, and 1949 saw the introduction of the B-36D, fitted with four J47 turbojets each producing 5,200 lb of thrust. That same year also saw the introduction of the first strategic reconnaissance version, the RB-36D. This plane's definitive version was the B-36J, which first appeared in 1953. It had a maximum take-off weight of 410,000 lb, a maximum range of 6,800 miles, a weapons system composed of 86,000 lb of bombs and sixteen 20-mm cannons, and a crew of 15 men. A total of 383 Peacemakers were built, and the aircraft remained in service until the end of the 1950s.

156 The Avro Vulcan strategic bomber was developed after 1947 to provide the RAF with a nuclear-capability carrier and to make Britain militarily independent of the US. The Vulcan, which entered service in 1956, had a typical delta wing and was the second in the triad of British nuclear bombers, the other two being the Vickers Valiant and the Handley Page Victor.

157 top The strategic superbomber, the Convair B-36, was developed following futuristic specifications drawn up in 1941 that required an airplane able to fly over 10,000 miles at a speed of at least 244 mph. Shown here is the B-36J, which combined six propeller engines with four turbojets, making it capable of reaching 413 mph.

157 bottom The four-engined Boeing B-50, developed from the B-29, entered service at the end of World War II, though only 79 planes were built. It was more powerful and carried a larger payload than its predecessor.

158 top The Boeing B-47 Stratojet, which entered service in 1950, was the USAF' first really modern bomber. It had swept wings, flew at 609 mph and carried up to 10 tons of bombs. The photograph shows a rapid take-off assisted by jet engines.

158 bottom The B-52H, often considered washed up, is still in service more than 50 years after its first flight. It was used recently in the Kosovo and Afghanistan conflicts.

159 top In-flight refueling of a Boeing B-52 Stratofortress by a Boeing KC-135 Stratotanker over the mountains of the Sierra Nevada in California in 1958. This combination gave the USAF a strategic component that could span the globe.

159 bottom A B-47 flies above the clouds with a three-man crew (the pilot and navigator are in tandem in the nose cockpit). The Stratojet remained in service until 1966. Roughly 1,800 were built.

The B-36's role had already been revised by the appearance of the first jet-powered strategic bomber, the Boeing B-47 Stratojet. This project first saw the light of day in 1944 when Boeing, together with Convair, Martin and North American, set to work on the creation of a jet-powered bomber. After considerable research, initially focused on straight-wing models, the Air Force accepted the Boeing Model 450, which featured an innovative wing with a 35° sweep (derived from acquisition of German know-how), a single-track undercarriage below the fuselage aided by two underwing stabilizers and six engines slung from underwing pylons. The first prototype, the XB-47, took to the air on 17 December 1947, powered by Allison J35 turbojets, which in the second prototype were replaced by more powerful General Electric J47 units. The first production B-47A flew on 25 June 1950, the very day war broke out in Korea. The most significant version of the Stratojet was the B-47E, which appeared in 1951; a total of 1,614 were built, of which 255 were the RB-47E reconnaissance version. The B-47E had a length of 109 feet 5 inches, a wingspan of 115 feet 9 inches, and a maximum take-off weight of almost 200,000 lb. It was armed with two 20-mm cannons 22,000-lb bomb load. Powered by six J47-GE-25 turbojets, each producing 7,200 lb thrust, the plane had a top speed of 606 mph and a range of 4,000 miles that could be extended through in-flight refueling. The three-man crew were provided with ejector seats. The B-47 was a much smaller than the B-36, but far more up-to-date, faster and better suited to the changing operational scenario. This model formed the basis for the RB-47E/H/K reconnaissance variants, and remained in service with the USAF until 1967.

In the meantime, with respect to ultra-heavy bombers, Boeing had been working since 1945 on the replacement for the B-36, an aircraft that had never been wholly satisfactory. The new aircraft was equipped with a 20° swept wing and was powered by turboprop engines. Given the B-47's success, however, and the availability of the new J57 jet engines (powerful, reliable, and less thirsty), Boeing decided to redesign the plane. It did so, making the new bomber a kind of "big brother" to the Stratojet. The new bomber retained a more highly developed version of the 35° swept wing, engines slung from underwing pylons, an in-flight refueling capability, a fuselage-mounted main undercarriage and (on the prototype at least), a cockpit with tandem seating. Thus was born the Model 464-67 or XB-52, a plane that flew for the first time six months after the YB-52's maiden flight on 15 April, 1952. The B-52 Stratofortress was another gigantic machine, 60 feet 9 inches in length with a wingspan of 184 feet 8 inches, and a maximum take-off weight of over 220 tons. It was powered by eight jet engines and could carry a load consisting of about 22 tons of nuclear or conventional weapons. The first production model, the B-52A, was introduced in 1954 and was followed by a series of uprated versions that culminated in the B-52H of 1961. This last model introduced the new Pratt & Whitney TF33-P-3 turbofan engines producing 17,000 lb of thrust, had a maximum take-off weight of about 485,000 lb, a top speed of 595 mph, a range of about 9,950 miles, a weapons load of 60,000 lb, and a crew of five.

Other significant versions had appeared earlier, including the B-52G in 1958, which incorporated a new wing and a lower fin (also fitted to the B-52H) and the B-52D in 1956, which had a greater maximum take-off weight than the preceding B and C variants. Moreover, 1955 saw the introduction of the RB-52B reconnaissance version. In total, the USAF ordered 744 of the various versions of the Stratofortress. Over the years this aircraft, at times criticized and written off as obsolete, has always proved to be up to the job. As of 2012 there are still two Air Force flight formations flying with up-to-date versions of the B-52H, an aircraft that has already survived two programs of cuts calling for its retirement from service, and 60 years on from its maiden flight, it still enjoys excellent health. The USAF plans on keeping this exceptional aircraft in service, through continuing updates and structural interventions, until 2040.

With the B-47 and the B-52, the USAF, or rather the Strategic Air Command, could count on a highly effective nuclear bomber force, but the organization's weak point was in-flight refueling.

Owing to the nature of their missions, the bombers required constant service from airborne tankers: this was an essential element of the overall strategy. In the mid-1950s, however, the USAF was still flying obsolete propeller-driven aircraft on refueling flights, the KB-50 and above all the KC-97 whose slowness obliged the bombers to refuel perilously close to their stalling speed. It was again Boeing that presented a valid response to the SAC's dilemma. In the early 1950s, the Seattle-based company had begun work on the first large passenger jet, the 707 (broadly based on the B-47 bomber), a model that was to become an extremely popular airliner. In order to cover itself against the risk of eventual commercial failure, however, Boeing wanted to recoup its development costs before launching the civilian 707 by selling a military version, the Model 367-80, to the government as a refueling tanker and transport plane. The USAF approved the design that featured a narrower diameter fuselage (lacking most of the windows), an upper deck for passengers and cargo and a lower section that housed the fuel and the refueling gear. Boeing also designed a refueling system employing a rigid probe that could be actuated by an operator located in the tail so as to facilitate the coupling operation. The first KC-135A Stratotanker flew on 31 August 1956, and deliveries of the production version began just three months later. The aircraft was 134 feet 6 inches in length, had a wingspan of 130 feet 11 inches, and a maximum take-off weight of 316,000 pounds. It was powered by four Pratt & Whitney J57-P-59W engines producing 13,750 lb of thrust, sufficient to provide it with a top speed of 584 mph and allow it to carry about 25,000 gallons of fuel.

The Stratotanker represented an enormous advance for the SAC, improving on the KC-97's performance from every point of view, including the fuel transfer rate. Naturally, the KC-135 also proved to be of great use for refueling the Tactical Air Command's (TAC) fighters and fighter-bombers. The USAF purchased 820 Stratotankers, many of which (incorporating updated E and R specifications) still constitute the backbone of its tanker fleet. The most recent version currently in service is the KC-135R, which first appeared in 1982. This variant is in reality an update of the previous models, with a strengthened airframe, new main systems and new CFM International F108-CF-100 turbofan engines producing 22,000 lb of thrust. The 135R can carry more than 6.5 tons of additional fuel, and thanks to the new engines, has improved take-off characteristics, a greater range, and the capability of transferring 150% more fuel to other aircraft, compared to the original KC-135A.

160 A USAF KC-135 refueling an F-15A fighter.

160-161 In-flight refueling by a KC-135 photographed from the cockpit of a USAF twin-seat F-15B Eagle over the Pacific Ocean.

161 bottom A formation of F-106A interceptors being refueled in flight by a KC-135. This procedure has been almost routine for American pilots since the 1950s.

80793
23554
U.S. AIR FORCE
U.S. AIR FORCE

Needless to say, the Soviet Union was also active in developing new bombers, although its progress was slower than that of the United States. After the four-engined propeller-driven Tupolev Tu-4 (a copy of the Boeing B-29), the first jet bombers to appear in the USSR were not aircraft with strategic capabilities, but rather tactical twin-engined planes. These included in 1947 the Tupolev Tu-14 (code-named "Bosun" by NATO) and in 1948 the Ilyushin Il-28 "Beagle," this last being particularly successful, with production of more than 10,000 planes. The Il-28 entered service in the USSR in 1950, and adopted by many Soviet bloc countries, remaining in service until the 1960s.

The USSR's first jet-powered strategic strike bombers were the Myasishchev M-4 and the Tupolev Tu-16. The M-4 was a four-engined plane with swept wings and tailplanes, designed in 1949 in response to Stalin's requirement for a bomber capable of attacking the United States. The first jet bomber appeared in 1954, and was designated the "Bison" by NATO. While incorporating interesting specifications, the M-4 was never a great success: when fully laden it was unable to provide the required performance. The M-4 was 154 feet 9 inches long, had a wingspan of 165 feet 4 inches, and a maximum take-off weight of about 350,000 lb. It was propelled by four Mikulin AM-3D turbojets producing about 21,000 lb of thrust and could reach a top speed of about 620 mph at its operating ceiling. It could carry almost 20,000 lb of bombs, was armed defensively with ten 23-mm cannons, and had an eleven-man crew. The M-4 was in service, it was diverted to less demanding duties such as reconnaissance and in-flight refueling, a role it performed through to the 1980s.

The Tupolev 16 derived from the Tupolev Works' prototype No. 88, which had been initially developed from the four-engined prop-driven Tu-4. The project was, however, radically modified, introducing a swept wing and tailplanes, with engines buried in the roots of the wings and improved aerodynamics. In general terms the plane resembled the M-4. It completed its maiden flight on 27 April 1952, and production of the definitive Tu-16 version began a year later. The aircraft made its public début in 1954, and was codenamed "Badger" by NATO. The Tu-16 remained in production in the Soviet Union for about ten years and some 1,500 were built (together with those built in China as the Xian H-6). The Tu-16 was powered by two Mikulin AM-3M turbojets producing 21,000 lb of thrust (engines virtually identical to those used in the M-4), had a top speed of 609 mph, and a range of over 4,500 miles.

162 top The crew of a Soviet Ilyushin Il-28 bomber embark for a mission. This twin-engined plane was the Soviet Union's first jet-engined bomber.

162 center A Soviet air force Tupolev Tu-16 on a reconnaissance mission is escorted by two F-4 fighters as it flies over the USS Kitty Hawk aircraft carrier during an exercise in the Pacific in 1963.

162 bottom A pair of Tu-16 bombers on a surveillance mission is intercepted by an F-4 and F-8 having landed on an American aircraft-carrier in the north Pacific in January 1963. During the Cold War, these close encounters were common.

163 top and bottom The Tu-22M3 had a swing-wing, could reach 1,250 mph and carry up to 24 tons of bombs and missiles, both conventional and nuclear. It entered service in 1975, and 500 were built. This one belongs to the Ukrainian air force.

Its maximum take-off weight was about 167,000 lb; its armament comprised a bomb load of about 20,000 lb and seven 23-mm cannons. The Tu-16 was 118 feet 9 inches in length and had a wingspan of 107 feet 11 inches. The Tu-16 was developed in a number of variations for anti-shipping strike roles (equipped with AS-2 "Kipper" or AS-6 "Kingfish" missiles), reconnaissance, electronic warfare and in-flight refueling. A number are still in service with the Chinese air force today.

The Tupolev Tu-20 was built at the same time and designed in response to the same requirement as the Myasishchev M-4. It made its first flight on 12 November 1952. The Tu-20, however, reflected a different approach to the problem of meeting the great range requirement; it was fitted not with jets but with more economical turboprop units. The Tu-20 went into service in 1956, and was soon adapted to carry the AS-3 "Kangaroo" nuclear missile. This made it a fearsome adversary for NATO, which named it the "Bear." The Tu-20 was 161 feet long, with a wingspan of 167 feet. Its maximum take-off weight was 340,000 lb (increased in successive versions up to about 408,000 lb); its weapons system comprised 44,000-lb bomb load and six 23-mm cannons. Four Kutznezov NK-12MV engines, each producing 14,795 hp and equipped with contrarotating propellers, powered the Tu-20 to a top speed of 528 mph and provided it with a range of about 7,500 miles. It was crewed by ten men. Production of the Tu-20 continued until 1965, and was then revived in 1983 with the redesigned Tu-95MS version capable of launching cruise missiles. A number of reconnaissance, strategic, electronic warfare and maritime patrol and spotting versions of the Bear were developed, and the Tu-142 variant was built for anti-submarine and communications duties. A total of over 300 Tu-20s were built; many are still in service.

The Soviet Union developed its first supersonic bomber, a Tupolev, in the mid-1950s, and the prototype completed its maiden flight in 1959. The plane, initially designated the Tu-105, eventually became the Tu-22 (known as the "Blinder" to NATO), a sleek, highly aerodynamic design with strongly swept wings and tailplanes, and with two turbojet engines located at the base of the fin. The new bomber's afterburner-equipped engines each provided a maximum of 27,000 lb of thrust and gave it a top speed of 932 mph. It had a range of 1,400 miles, and carried a crew of five men. The Tu-22's armament consisted of a 23-mm cannon for short-range defense, and the plane carried a bomb load of nearly 20,000 lb. Some 250 Tu-22's were built, including reconnaissance, training and missile assault versions. Both the Libyan and Iraqi air forces used the plane.

Some years later, the Tupolev Works developed a new bomber based on the Tu-22 design; NATO code-named it the Tu-26 "Backfire, but it was officially designated the Tu-22M. This aircraft, which flew for the first time on 30 August 1969, had a number of new features, including engines located within the fuselage with lateral air intakes and a new variable-geometry wing. The Tu-22M was a very superior aircraft. Its two Samara NK-25 turbojets produced about 55,000 lb of thrust, and it was capable of a top speed of about 1,245 mph, with a range of about 3,750 miles, carrying an offensive load of up to 53,000 lb of nuclear and conventional bombs and missiles. The plane entered service in 1975; some 500 were built, a number of which are still flying today.

This Cold War chapter would not be complete without mention of two aircraft that have become symbols of the era, the United States' Lockheed U-2 and SR-71 spy-planes.

Lockheed designed the U-2 in the early 1950s in order to provide the American government (or rather the CIA) with an aircraft capable of completing strategic reconnaissance missions deep into Soviet territory. Led by Kelly Johnson, engineers employed in the "Skunk Works" (the California-based company's secret R&D facility) came up with a hybrid machine, mid-way between a glider and a powered aircraft. The fuselage housed the pilot, the electronic systems and the engine, a Pratt & Whitney J57-P-13 producing 11,200 lb of thrust. The U-2 had a relatively immense wing span of almost 80 feet that guaranteed the plane lift at altitudes of over 65,500 feet, where it would be beyond the reach of enemy anti-aircraft fire. The U-2A completed its maiden flight on 1 August 1955 and entered operational service the next year. On 1 May1960, a Soviet SA-2 missile downed a U-2 flown by Gary Powers. This U-2's loss revealed the plane's vulnerability and led to a revision of strategic reconnaissance tactics. Subsequently, numerous improved versions of the aircraft were introduced, with NASA and the USAF operating flights in addition to the CIA. The new models included the U-2C electronic surveillance version and the U-2R, which had a wingspan extended to over 101 feet, new wing pods for surveillance gear, an increased maximum take-off weight, and a new J75 engine producing close to 17,000 lb of thrust. In 1981, the TR-1A was introduced; it as in effect a variation on the U-2R, which was dedicated to radar reconnaissance and used primarily in the European theater. Last, in 1994, the U-2S entered service. It was equipped with the most advanced electronic surveillance gear, a satellite data-link, and a new General Electric F118-GE-101 engine producing 19,000 lb of thrust. Ever since Lt. Gary Powers had been shot down in 1960, the United States had continued to employed the U-2 in numerous crisis-and-war theaters to collect data and information. However, the Powers' episode had convinced the American authorities of the need for an aircraft with far superior performance and features, one that could operate safely on a global scale, even above hostile territory.

Once again the Lockheed Company was commissioned, and it produced from

164-165 *The Lockheed SR-17A Blackbird strategic reconnaissance plane was one of the most secret and interesting aircraft in history. It first flew in 1962 and remained in service until 1997.*

165 top *An SR-71A being rolled out at dawn for a mission. The size of the two Pratt & Whitney J58 engines is clear. Together, with the afterburners, they provided almost 30 tons of thrust.*

165 center *A Lockheed U-2S reconnaissance aircraft on a training flight over the Californian mountains. Thanks to constantly updated models, this airplane, which first flew in 1955, is still in service today.*

165 bottom *Thanks to what were science-fiction techniques for the period, Lockheed was able to produce the SR-71A, an aircraft able to cruise at over 2,000 mph at heights up to 85,300 feet.*

scratch a truly exceptional airplane. The form, dimensions and materials employed (titanium, for example) were exotic and unusual, dictated by the extreme conditions of use and the heat developed during supersonic cruising at over 1,865 mph. In order to achieve such performance, two enormous Pratt & Whitney J58 engines producing 32,500 lb of thrust on afterburn were used. They were designed to operate at maximum power for hours rather than for a few short minutes, as was normally the case with fighters. A special fuel named JP-7 was developed, with an extremely high ignition point that required a tetraethyl borane additive (which explodes on contact with air), in order to start up the engines. The first prototype, initially designated the A-12, took to the air on 30 April 1962, from the ultra-secret Area 51 Base in the Nevada desert. All the A-12's early flight-testing (which was entrusted to the 1129th SAS, a CIA-controlled squadron) was conducted from this base. Subsequently, the SR-71A Blackbird (as the production A-12 was designated) was flown by the USAF's specially constituted 4200th Strategic Reconnaissance Wing based at Beale Air Force Base in California, and by its operational units in Japan and Great Britain. The Blackbird was 107 feet 3 inches in length and had a wingspan of 55 feet 5 inches, and had a maximum take-off weight of 170,000 lb. The Blackbird could be equipped for each mission with interchangeable systems packs with photographic, infrared and electronic surveillance gear. Crewed by a pilot and navigator, the plane had a top speed of over 2,000 mph (Mach 3) at altitudes of over 85,300 feet, while its maximum range of about 3,000 miles could be extended with in-flight refueling. KC-135A tankers were modified to KC-135Q specifications in order transfer the special JP-7 fuel. Some 50 Blackbirds were built (including the prototypes and experimental versions such as the A-12B, the YF-12, and the M-21) served successfully until 1990, with NASA also using it. It was in service from 1995 to 1997, when it was finally withdrawn without being replaced by an aircraft of similar capabilities.

Chapter 13

Helicopters

What to some may today appear to be a "peculiar" way to fly, taking off and landing vertically, even hovering stationary in mid-air, was for many European children from the 14th century onward a popular and highly enjoyable game. The toy that was involved (which probably originated in China) consisted of a kind of spinning top equipped with leather blades that would rise vertically into the air as the result of a cord being tugged. In about 1480, Leonardo da Vinci investigated the underlying principle of vertical lift. In one of his notebooks, he drew an aircraft equipped with a mechanically actuated spiral air screw that appeared to be capable of lifting itself into the air, thanks to an autonomous propulsive thrust.

The principle of vertical flight became increasingly clear to many, but from 1700 until the beginning of the 20th century, pioneers and researchers such Cayley, Phillips, Pènaud, Forlanini and Edison were limited to building only scale models of vertical take-off flying machines. This limitation was due largely to lack of an engine capable of

166 top What may be considered as the first vertical flight project in history was unquestionably Leonardo da Vinci's "vite aere" (air screw).

166-167 center The Frenchman Etienne Oehmichen was one of the pioneers of vertical flight machines. Here he is at the controls of one of his first designs in 1923.

166 bottom Paul Cornu's helicopter before take off at Lisieux, France, 13 November 1907. This was the first such attempt with any machine of this kind. Despite its success, Cornu abandoned the project.

developing sufficient power to allow a full-size, manned prototype to fly.

The first aircraft to lift a man vertically was Breguet-Richet's Gyroplane No. 1, a complex machine equipped with a 32-blade rotor driven by a 40-hp Antoinette engine. On 29 September1907, this machine lifted itself about 2 feet off the ground for one minute. The aircraft was not a true helicopter as the passenger on board had no control whatsoever over it. Nonetheless, Breguet and Richer had taken an important step in the right direction. Paul Cornu then took Louis Breguet's work further, and on 13 November he successfully flew a similar but more sophisticated machine. In 1909, another pioneer of vertical flight arrived on the scene when Igor Sikorsky, aged 20 and fresh from his technical school studies in Russia, began work on his first helicopter. He devoted himself enthusiastically to research and development on his designs No. 1 and then No. 2, but both suffered from problems of instability, vibration, power and control that could not be solved at the time. Sikorsky thus turned to airplane design, a field in which he enjoyed notable success, and many years passed before the supremely talented Russian designer went back to helicopters.

Another three interesting designs appeared in the 1920s. In the United States, George de Bothezat designed a machine featuring a 38-blade rotor driven by a 220-hp engine; in France, Raul Pateras Pescara, a Spaniard, built a helicopter with two twin-blade rotors; and Etienne Oemichen, a Frenchman, built a controllable aircraft with six large rotors. All three machines managed to fly successfully, but technical obstacles meant that they were fragile, dangerous and unsuitable for any serious use at a time when the relatively fast and reliable airplane had already made enormous progress.

167 top The Gyroplane I designed by Louis Breguet and Richet was first vertical flying machine that succeeded in lifting a man. The flight took place in Douai on 29 September 1907 but was not considered a free flight because of the attached safety lines.

167 bottom Raoul Pescara's 16-blade helicopter as it lifts off in Issy-les-Moulineaux, France, on 23 January 1924 during a duration-record attempt.

Thanks to the airplane and the talent of the Spaniard, Juan de la Cierva,the helicopter principle was able to be developed in a new direction. De la Cierva, an enthusiastic aviation pioneer since his youth, recognized that the problem of in-flight safety (in particular the phenomenon of stalling) could be resolved by a horizontal rotor that could provide the

necessary lift at any speed. Far from looking to develop the helicopter, de la Cierva was instead interested in creating a hybrid machine, an airplane that was also equipped with a rotating wing. De la Cierva patented his idea in 1920, calling it an autogyro. In effect, the solution of a rotor free to revolve above the vehicle gave rise to the phenomenon of auto-rotation, which guaranteed a continual source of lift even in the absence of engine power and therefore torque to be counteracted. In order to overcome the problem caused by the difference in lift provided by the freewheeling blades, de la Cierva constructed two counter-rotating rotors installed one above the other. His new aircraft, named the C.1, was ready to fly in October 1920, but like successive models suffered from severe problems of stability in balancing the lift. After repeated studies and experiments. De la Cierva managed to solve the problem by adopting blades that were free to oscillate and adapt themselves to the forces in the various rotation phases. On 9 January 1923, he flew his C.4 model for the first time, and found that it behaved perfectly.

The autogyro was propelled by an engine driving a front-mounted propeller and flew like other aircraft, taking off when the rotor blades provided the necessary lift, but could land vertically, with a minimal runway. The aircraft was truly astounding and was warmly received abroad. In fact, after careful evaluation, the British government acquired the autogyro, while demonstrations were performed in various European countries. De la Cierva's invention was soon a commercial success and by the late 1920s his autogyros were being constructed under license in Germany, Russia, France, Japan and the United States. Technical development continued, and in 1933 De la Cierva introduced the "jump start," which allowed the rotor to be actuated by the engine. This permitted vertical take-off, with the angle of incidence of the blades controlled by the pilot. Initially the implications of this innovation were unclear, but it soon became fundamental to the evolution of the true helicopter. While ingenious, the autogyro lacked the payload-carrying capabilities that

commercial and military applications required, and its performance was by no means exceptional.

The Italian Corradino D'Ascanio was another pioneer of vertical flight. He was an aeronautical engineer who left the fixed-wing field in 1926 and constructed his first helicopter in that same year. In 1927, he designed a new helicopter powered by an 85-hp engine. The machine was designated the DAT-3, partly in honor of his financier, Baron Pietro Trojani. The helicopter, which was equipped with two coaxial rotors and control surfaces at the tail, was presented to the Air Ministry, which in 1928 commissioned a prototype for experimentation. The aircraft's maiden flight was staged in Rome on 8 October 1930, and was a great success. The DAT-3 set world records for altitude, range and duration for that type of aircraft, and the future of the first Italian helicopter briefly appeared to be rosy. However, the Fascist regime's technical experts, military personnel, and politicians were skeptical about the helicopter's potential, and the project was deliberately ignored. The same fate awaited the PD.2, a two-seater equipped with a main rotor and an anti-torque rotor at the tail, built in 1939 by D'Ascanio and the Piaggio Company. Italy's opportunity to take a lead in helicopter design thus evaporated.

168 top Two autogyros flying over New York in 1930. At the start of the 1930s, Harold Pitcairn bought the US construction rights for the Cierva autogyro, but the craft did not catch on.

168 center The autogyro invented by the Spaniard, Juan de la Cierva, was a compromise between a plane and a vertical flight machine and opened the way to the development of the helicopter.

168 bottom The Italian engineer, Corradino d'Ascanio, dedicated himself in 1926 to designing a helicopter; he constructed a craft with two counter-rotating blades on the same axle but the Fascist regime had no faith in the new machine.

169 top The Cierva C.30 autogyro abandoned conventional wings in favor of just the rotating blade to provide lift.

169 bottom The German Focke-Achgelis Fa.61 helicopter was derived from the autogyro even though its two rotors were powered by an engine, as in a helicopter.

Such was not the case in Germany. Henrich Focke, a First World War pilot who had subsequently founded the Focke-Wulf aeronautical company, had been removed from the head of the firm in 1933 because of his anti-Nazi views. He then devoted himself to the construction of the autogyro under license, and soon began to explore the machine's potential for development, primarily in the application of power to the rotor to create a true helicopter. On 26 June 1936, his first aircraft, the Fa-61, took to the air. Within a year it had proved to be so effective that it broke all standing records and attracted the interest of that famous aviatrix, Hanna Reitsch, a darling of the Nazi régime. Reitsch made numerous demonstration flights in Germany and the Nazi government approved the formation of a new company, Focke-Achgelis, which it commissioned to build a new, more powerful helicopter capable of lifting a payload of 1,500 lb. The new machine, the Fa-223, was ready by 1940 and featured a 1,000-hp engine, two tri-blade rotors mounted side-by-side and a closed, four-seat cabin. It could achieve a top speed of 115 mph, climb to a ceiling of almost 23,000 feet and lift a load of just over a ton slung beneath the fuselage. Focke-Achgelis continued to develop the aircraft, but the bombing of the factory by the Allies prevented production, though at the end of the war, the British were able to capture a Fa-223 there.

Anton Flettner was another German engineer working on helicopters; he built his first one in 1930. In 1937, after developing various experimental designs, including one for an autogyro. Flettner built the Fl-265, a true helicopter equipped with two contrarotating rotors. The machine attracted the attention of the German navy, and in 1938 the admiralty ordered a small batch. In 1940 Flettner's company produced the improved Fl-282 Kolibri version that could fly at about 108 mph and carry a total load of about 800 lb. The navy approved the design and placed orders for 1,000 Fl-282s. About 20 pre-production Fl-282s saw operational use in the Mediterranean theater from 1943, but the planned mass-production program was cancelled as a result of the Allies' continuous bombardment of Germany.

On the other side of the world, however, another famous engineer was about to lead the helicopter to maturity. Having abandoned Russia after the end of the revolution, Igor Sikorsky settled in the United States in 1919 where, after overcoming a series of problems, he had founded the Sikorsky Aero Engineering Corporation on Long Island. The company initially concentrated on manufacturing successful fixed-wing air- and seaplanes, but economic success meant that Sikorsky was able to devote himself to his passion for vertical flight. The opportunity to apply his research came in 1938, when he began work on an aircraft that featured an invention he had patented in 1935: a single rotor combined with an anti-torque rotor mounted vertically at the tail. On 14 September 1939, his design, the VS-300,

170 top The VS-300 of 1939 was first helicopter Igor Sikorsky built. This is the second version, with two horizontal tail rotors.

170 center The American Arthur Young was one of the most famous pioneers of the helicopter. Here he is shown with members of his team during a full-load test flight of his Model 30, the first helicopter produced for Bell.

170 bottom Before developing actual size aircraft, Young experimented with scale models like this radio-controlled craft of 1938, with a 20-hp motor.

began to make its first cautious hops lasting just a few seconds. After a seemingly endless series of adjustments and modifications, the VS-300 made its true maiden flight on 13 May 1940. By the end of the year, it was competing with other designs for a US Army contract for a new military helicopter, designated the XR-4. In December 1941, with control problems resolved thanks to the adoption of a cyclical system, the VS-300 was ready to fly in its definitive configuration. Subsequently, the slightly larger XR-4, equipped with a 185-hp engine and incorporating all the advances made with the earlier model, undertook its first flight on 14 January 1942. The American army approved Sikorsky's design and ordered the XR-4 into serial production as the R-4. Two larger models were commissioned during the war. By 1945, the Sikorsky Company had produced over 400 helicopters, placing it at the forefront of the field.

By the end of the Second World War it was clear that the helicopter had a brilliant future and some engineers like Sikorsky himself, predicted that helicopters would come to cost little more than cars and would enjoy widespread popularity. In 1947, in the United States alone, more than 70 companies were working on the development of the new technology; few however managed to establish themselves with successful products. Among those that did was Piasecki, which developed the tandem rotor layout with its PV-3, HRP-1 and H-21 (Flying Banana) models, and Bell, which entrusted its rotating wing sector to an up-and-coming engineer, Arthur Young. Young's first project was the Model 30, a single-engined, single-rotor design with architecture similar to that of the Sikorsky VS-300. It flew for the first time in June 1943. Numerous improvements and modifications were made to this first helicopter, and in the December of 1944 it was redesignated as Model 47 and soon went into serial production. The Bell 47's strong suit was the civil certification obtained in 1946 that accelerated its diffusion in the commercial field. This machine had a characteristic side-by-side two-seat cabin in the form of a transparent Plexiglas bubble, lift being provided by a single twin-blade rotor responsible for the unmistakable noise that led to the "chopper" nickname. The Bell 47 was extremely versatile, and was well suited to numerous civilian and military transportation, search-and-rescue and air-ambulance roles. It was the first modern helicopter. The USAF acquired 28 Bell 47s in 1947 and the US Army acquired 65 in 1948, while further orders flooded in from other forces in America and throughout the world. Over 5,000 Bell 47s had been built before production ceased in 1973, some of them under license in Italy by Agusta and in Japan by Kawasaki. The Bell 47G variant was powered by a 270-hp Lycoming engine and had a top speed of 121 mph and a ceiling of 10,500 feet. With a main rotor diameter of just over 32 feet and a maximum take-off weight of about 2,850 lb, the helicopter could carry two passengers.

171 top One of the first Bell Model 47s as soldiers prepare to load wounded soldiers during the Korean War. This aircraft marked the beginning of the modern helicopter era.

171 bottom Frank Piasecki invented the tandem rotor configuration featured in his PV-3 model. Because of its shape, this craft was soon called the "flying banana." Shown here are three Piasecki HRP-1s landing troops during the Korean War.

172 top The concept of the tandem rotor was developed by Piasecki with new, larger and more reliable models like the H-21. Two USAF H-21s are shown during a matériel transportation exercise.

Two other brilliant American designers were Stanley Hiller and Charles Kaman, who established their reputations in the late 1940s when founding companies carrying their own names. In the meantime, Sikorsky launched the first true cargo helicopter in 1948 in response to a U.S. army requirement, the S-55 prototype being given the military designation YH-19. This aircraft adopted an all-new configuration in which the engine was located in the nose and the cockpit and the transmission system were placed above the cargo hold, which was thus spacious and located close to the ground. The helicopter flew for the first time on 10 November 1949, and soon went into production as the H-19. It was 42 feet 4 inches long and its tri-blade main rotor had a diameter of 52 feet 10 inches. Powered by a Wright R-1300 piston engine producing 800 hp, the H-19 could achieve a top speed of 112 mph. With a maximum take-off weight of about 7,900 lb, the aircraft could carry a crew of two and up to ten soldiers or eight stretcher patients. The air force and the army, and subsequently the US Navy, the Marines, and other operators throughout the world ordered the H-19. Almost 1,300 H-19s were built, some under license by Mitsubishi and Westland.

War broke out in Korea on 25 June 1950, with the western democracies, above all the United States, being drawn into the conflict. This war saw the establishment of the helicopter as extremely versatile tool for the armed forces. Initially, however, the principal duty entrusted to the rotary wing aircraft, one that it performed admirably, was that of the air ambulance: Bell 47s equipped with two external stretchers were supplied to the army's famous MASH (Mobile Army Surgical Hospital) units. The Sikorsky S-51 used by the USAF and the Marines was also adapted for use as air ambulances. With the arrival in Korea of the H-19 and the H-21, however, the use of helicopters became more widespread. First the Marines and then the army began to exploit the helicopter's air mobility capabilities, using it for the rapid transportation of troops, particularly in emergency situations. Operation Bumblebee on 11 October 1951 saw the first large-scale operation in which a battalion of Marines (almost 1,000 men) was dispatched

172 bottom left A USAF Sikorsky H-19 during operations at an advance camp in Korea in March 1951.

172 bottom right The first helicopters led inevitably to the first accidents. Here a Sikorsky S-51 Dragonfly, in service in the Korean War, hits the island of a British aircraft-carrier and falls into the sea.

173 top Helicopters were soon employed on rescue missions. Here a USAF Sikorsky S-55 saves the crew of a US Army tow-boat sinking off Okinawa in 1956.

173 center The French army rapidly acquired helicopters for their armed forces. Here an S-55 (H-19 for the armed forces) evacuates wounded men from the outskirts of Dien Bien Phu, Vietnam, March 1954.

173 bottom An S-51 belonging to the Marines takes off with a wounded soldier during fighting near Seoul on 25 September 1950. Evacuation of the wounded was one of the first duties assigned to helicopters.

aboard a dozen H-19s to relieve another front-line unit located on a mountain top.

The decade that followed the Korean War saw the most intensive period of helicopter development. Designs were refined and differentiated to meet the requirements demonstrated by the military and civilian sectors, and the turbine engine was introduced--a technical and operational revolution that promised to increase both performance and on-board comfort.

The first of the new generation of helicopters was a Kaman K-22 that had been modified in 1951 to accommodate a 177-hp Boeing YT-50 gas turbine engine. However, the first helicopter to successfully exploit the new turbine units was actually French rather than American.

The Sud-Est Company entered the helicopter field following the Second World War, thanks in part to the consultancy of Heinrich Focke, who was attempting to improve on his Fa-223 design with its SE 3000 model. Sud-Est began to establish its reputation with the SE 3120 Alouette, a two-seat helicopter with a single engine and a single rotor. The Alouette completed its maiden flight on 31 July1952. This model was used as the platform for the installation of a Turbomeca Artouste I turbine developing 360 shp (shaft horse power). The turbine engine offered enormous advantages when used in helicopters; it provided significant power while weighing only half as much as an equivalent piston engine. The new helicopter, the SE 3130 Alouette II, flew for the first time on 12 March 1955. Once it had obtained civilian certification, it aroused immediate interest and was a great commercial success. In 1957, following the merger of Sud-Est and Sud-Aviation, the model was redesignated the SE 313B. This helicopter had a top speed of 115 mph, could carry up to three people, and had a maximum take-off weight of 3,527 lb, compared with a dry weight of 1,973 lb. Subsequently, the SA 318C variant was created with the introduction of the more economical Astazou IIA engine. The Alouette II series remained in production until 1975, with about 1,300 being sold around the world. Sud-Aviation (later known as Aérospatiale) expanded its family of French helicopters with the introduction of the SA 316 Alouette III (in 1959), the SA 321 Super Frelon (in 1962), the SA 330 Puma (in 1965), and the SA 315 Lama (in 1969). These aircraft helped make Aérospatiale one of the world leaders in the helicopter field.

In the United States, however, Bell had by no means been resting on its laurels. On the strength of experience gained during the war, the company immediately developed a multipurpose turbine-powered helicopter in response to a US Army requirement issued in 1955. The resulting Bell Model 204 was to remain a milestone in the history of rotary wing aircraft, winning acceptance in June of the same year, even before it had made its maiden flight. That flight took place on 22 October1956, and the helicopter demonstrated its huge potential. It featured the classic Bell layout of a single, twin-blade main rotor with an auxiliary rotor at the tail. The fuselage housed the two-man crew in the nose, with the loading bay behind it, capable of holding six soldiers or two stretchers accessed via two large sliding doors. In 1959, when it entered service with the US Army as the UH-1A Iroquois, the helicopter was powered by an Avco Lycoming T53 turbine developing 710 shp, but this output was soon increased to 1,115 shp with the UH-1B model, and eventually reached 1,419 shp with the Fuji-Bell 204B-2. This helicopter was almost immediately followed by an improved version, the Model 205, offered to the US Army as early as 1960. To all intents and purposes a stretched version of the 204, the Model 205 could accommodate 14 soldiers or 6 stretcher patients, for a maximum payload of 4,000 lb. The prototype flew on 16 August 1961, and the aircraft was in service by August 1963, designated the UH-1D. It was powered by the same engine as the UH-1B and had a maximum take-off weight of 9,500 lb. This version was soon followed by the UH-1H powered by the same 1,419-shp engine as the Model 204-B-2. Like the 204, the 205 was also built under license by Agusta and Fuji, and in all, more than 10,000 of the two versions were built.

Mainly used for military duties, the UH-1 series reached the height of its fame during

the Vietnam War. It became something of a symbol of the American cause and was used extensively for air-borne cavalry and troop transportation, these being the great innovations of the Southeast Asian conflict. Vietnam also saw another helicopter produced by Bell, the Model 209, which privately developed by the American company to provide the army with a dedicated combat helicopter. The 209 used the 204's mechanical elements, incorporating them within a narrower fuselage that was equipped with two tandem seats and wing stubs to which the armament was attached. After the 209 had completed its maiden flight on 7 September 1965, the US Army ordered an initial batch, designating the model the AH-1G Cobra. It entered service in 1967 and was immediately dispatched to Vietnam. The lightweight, single-engined Bell Model 206, also designed for military use as a reconnaissance helicopter, completed its maiden flight on 10 January 1966; it also enjoyed great commercial success in the civilian field. In fact, the 206 is still in production after a long sequence of versions and modifications, and over 9,000 have been built. The Bell multipurpose helicopter series continued with other machines, whose design can be traced back to the 204/205. These were the twin-engined Model 212, the Model 412 with a four-blade main rotor, the Model 214, and the Model 222.

174 top The French Aérospatiale SA.315B Lama was developed for Indian forces that needed a helicopter able to operate at high altitudes and low temperatures, shown here with Pakistani colors.

174 bottom left The Agusta-Bell 205 (this is a Greek army model) was developed from the Model 204 and was one of the most popular military helicopters.

174 bottom right A Sud-Est SE.313 Alouette II belonging to the Belgian army. This machine made its maiden flight in 1952 and the turbine version arrived in 1955.

174-175 bottom The heavy Sud-Aviation SA.321 Super Freon was produced to meet French military requirements and first flew in 1962. Here, commandos disembark on the flight deck of a French aircraft carrier.

175 top A formation of three Bell AH-1 Cobra attack helicopters. This aircraft was derived from the UH-1 transport model for use in the Vietnam War.

175 center An Aerospaziale SA.330 Puma (built by Westland) flying British army colors. It first flew in 1965 and was the result of experience gained from the Super Freon.

175 bottom right A Sud-Aviation SA.316 Alouette III of the Portuguese air force flying at low altitude. This helicopter made its maiden flight in 1959, and more than 1,800 were produced.

Sikorsky, a still active American colossus, concentrated largely on satisfying the requirements of the US Navy. In the late 1950s, following the S-58 model of 1954 (still powered by a piston engine and designated by the military as the HSS-1 Seabat and then--from 1962--as the H-34), the Sikorsky company introduced the S-61, a truly remarkable helicopter still widely used throughout the world. The company developed the S-61 in response to a US Navy brief calling for an anti-submarine helicopter with both search and assault capabilities. Sikorsky designed a large twin-engined machine equipped with a five-blade main rotor, folding tail boom and a large fuselage capable of housing the crew and the mission systems. The prototype first flew on 11 March 1959, and the helicopter, designated as the HSS-2 ("Sea King") proved to be an immediate success. The first production examples went into service

177 top The Sikorsky S-65 (designated the H-53 from 1964) was the American forces' largest helicopter. Here a US Marines' CH-53A is on a transport mission in South Vietnam.

177 bottom The Sikorsky H-60 is widely used by the US Navy and the Marines. The photo shows a VH-60N "White Hawk" operated by HMX-1, the Marines section involved in presidential transport.

in September 1961. Mitsubishi, Westland and Agusta acquired production licenses, and the powerful and robust aircraft was subject to continued development. The year 1960 saw the introduction of the S-61L passenger version for the civilian market, while an amphibious version dedicated to search and rescue duties made its début in 1963. This version, designated the S-61R by the company, had a much larger fuselage and was equipped with a rear-loading door. It was initially produced as the HH-3E Jolly Green Giant and adopted by the USAF; the more advanced HH-3F Pelican model was introduced later. In total, the Sea King, since 1962 known to the American armed forces as the H-3, had a production of over 1,300 planes in almost 40 different versions. It is still in use throughout the world.

In the early 1960s Sikorsky developed yet another large helicopter in response to a US Navy requirement. The Model S-65 was designed for the Marines corps as a heavy transport helicopter, capable of transporting troops and munitions from ship to shore during landing operations. While retaining the basic mechanical elements of the S-64, Sikorsky redesigned the fuselage along the lines of that of the HH-3E. Designated the CH-53 Stallion, the new helicopter began entering service in 1966, and was immediately sent to Vietnam, where there was great need for such an aircraft. Other versions were developed from basic S-65 for minesweeping duties and for the USAF (for search and rescue and special operations), as well as for other armed forces. In all, more than 730 S-65s were built in over 20 versions.

The most recent Sikorsky success story has been the S-70, which flew for the first time on 17 October 1974. In this case, however, the helicopter resulted from a requirement issued by the US Army, which sought a new multipurpose tactical transport helicopter to replace the UH-1 series. The Sikorsky model was declared the winner of the competition, and its production model was designated the UH-60A Black Hawk in December 1976. The first Black Hawks began to enter service in 1979. Subsequently, numerous different versions were developed from the basic model; these included the naval SH-60B, F and R (and the export S-70B and C variants), and the MH-60, EH-60 and HH-60 designed for rescue, electronic warfare, and special operations. The US Army and the USAF, as well as more than 30 countries around the world, used these versions.

176-177 Westland produced the Sikorsky S-61 under license as a naval helicopter, the Sea King, derived from the SH-3A, for the US Navy. It was also built under license by the Italian Agusta company.

176 bottom An RAF Sea King HA3 on a search and rescue mission in extreme weather conditions.

The S-70 series is still in production and has attracted over 2,600 orders to date. The company's most recent projects have included the multipurpose S-76 model and the S-92 heavy transport helicopter, which in 2019 was also chosen, in VH-92A form, for the program "Marine One," the new helicopter destined to carry the President of the United States.

Vertol (later known as Boeing-Vertol) was another American company that established a reputation in the field of large transport helicopters. Within a few years Vertol introduced two notable models still widely used today. The Model 107, a twin-turbine helicopter with two rotors mounted in tandem, designed for commercial transportation, was presented on 22 April 1958. The H-46, as the 107 was designated, was powered by two General Electric CT58 turbine engines, each developing 1,400 shp. It had a maximum take-off weight of over 21,000 lb. The CH-46's success, however, was largely as a result of its acquisition by both the US Navy and the Marines. In all, 624 H-46s were ordered during the 1960s.

At much the same time, Vertol built a new helicopter that retained the same layout as the H-46, from which it clearly derived, but was scaled up in size. The Model 114 completed its maiden flight on 21 September 1961. Its characteristics made it particularly well suited to the requirements of the US Army, and the Army adopted it as the CH-47 Chinook in place of the smaller CH-46. The Chinook was powered by two Avco Lycoming T55 engines, each developing 3,800 shp, and had a maximum take-off weight of about 38,500 lb. Around a dozen different versions of the CH-47 were developed and were used by about 20 different countries. More than 1,200

179 center Two CH-46s turning before landing on the amphibious assault ship USS Peleliu in the Arabian Sea. This was during Operation Enduring Freedom in November 2001.

179 bottom A pair of US Army UH-60A Black Hawk helicopters during Exercise Bright Star in Egypt in 1983. The Black Hawk entered service with the American army in 1979 and has been used by more than 20 countries.

178-179 The Boeing-Vertol CH-46 Sea Knight has long been the Marines' standard transport helicopter. Only recently has it been replaced by the new V-22 Osprey tiltrotor aircraft.

178 bottom A Boeing-Vertol CH-47 Chinook, here displaying RAF roundels, prepares to lift an armored vehicle. The Chinook first flew in 1961 and has been used by around 30 countries throughout the world.

179 top The SH-60 Sea Hawk is the naval version of the Black Hawk and was developed at the same time for the US Navy. Here an SH-60 from HS-4 squadron is shown firing a torpedo.

180 The AH-64D, the penultimate version of the Boeing Apache, is equipped with the Longbow radar, seen above the main rotor. This one belongs to the British army.

181 top A formation of National Guard UH-60As and AH-64As flying over Kuwait in 1998 during Operation Southern Watch.

181 center A Boeing AH-64A Apache assault helicopter launching air-to-ground missiles during a shooting exercise. The Apache is armed with a single 30-mm cannon and a mixture of rockets and anti-tank and air-to-air missiles.

Chinooks were built, with production licenses being acquired by Agusta Leonardo and Kawasaki.

The latest models are the CH-47F and the MH-47G, used for special operations.

Subsequently, Boeing and Sikorsky joined forces to create a new armed reconnaissance and assault helicopter for the American army, the RAH-66 Comanche, an extremely advanced machine that completed its maiden flight on 4 January 1996. The helicopter is currently in development and is due to enter operational service in 2006.

The Hughes Aircraft Company had been among the major American manufacturers until it was absorbed by McDonnell Douglas, which was in turn was absorbed by Boeing. In the early 1960s, Hughes introduced the Model 369 in response to a US Army requirement for a new reconnaissance and scouting helicopter. It flew for the first time on 23 February 1963, and designated the OH-6A Cayuse. In 1965, the US Army began ordering OH-6As. The civil version was known as the Hughes 500. These two helicopters marked the beginning of an enormously successful program that through development of numerous and gradually improved versions resulted in the manufacture of about 5,000 helicopters to date, some being built under license in various countries. One of the most recent versions, the 520N NOTAR, is characterized by the absence of an auxiliary tail rotor. Powered by an Allison 250 turbine producing 450 shp, it is capable of a top speed of 175 mph. With a dry weight of about 1,600 lb and a maximum take-off weight of 3,850 lb, the 520N can carry up to five people, including the pilot. The most famous Hughes helicopter is, however, the AH-64 Apache, which in 1976 the US Army selected as its new assault and combat helicopter. A powerful, robust and large aircraft, the Apache flew for the first time on 30 September 1975. After McDonnell Douglas absorbed Hughes, the AH-64 began to enter service in its "A" specification in 1986. Built by Boeing today, the AH-64E Guardian (a version equipped with the Joint Tactical Information Distribution System, or JTIDS) was characterized by a classical layout with a tandem two-seat cockpit in the nose together with the navigation and weapons sensors, and well-separated engines in a higher, better protected location. It was fitted with a four-bladed rotor driven by two

182 top Two Russian Mil Mi-24s landing near the Chechen capital Grozny in 1999, during fighting to repress the local rebellion.

182 bottom A Mil Mi-24 belonging to the Czech Republic flying in an air show. The Mi-24 has a maximum speed of 210 mph and can carry eight armed soldiers.

General Electric T700-GH-701C engines, each developing 1,890 shp, providing the helicopter with a top speed of 226 mph. Maximum take-off weight is 23,000 lb, and the armament comprises anti-tank missiles, air-to-air missiles, rockets and a 30-mm cannon. The American army uses around 750 examples of the aircraft out of a production total of more than 2,500, with the model also having been acquired by the armed forces of Australia, Saudi Arabia, South Korea, Egypt, the United Arab Emirates, Japan, Greece, Great Britain, India, Indonesia, Israel, Kuwait, Morocco, Poland, Qatar, Singapore, and Taiwan.

Along with the United States and France, the Soviet Union for obvious reasons also invested heavily in the design and construction of its own series of helicopters. Two factories in particular have specialized in this sector, Mil and Kamov. The first was founded by Mikhail Mil in 1947 and in the following year presented its first design for a light, multipurpose helicopter equipped with a 575-hp Ivchenko piston engine, a tri-blade main rotor, and an auxiliary rotor at the tail. This helicopter, subsequently designated the Mi-1, was later developed into the Mi-2 that used two Isotov turbine engines producing 450 shp. This model appeared in 1961; more than 6,500 were built, and production licenses were granted to other manufacturers. Mil then produced a number of other designs for increasingly powerful and more specialized helicopters. These included, among other models, the Mi-6 in 1957 (then the world's largest helicopter, with a 13-ton payload capability) and the Mi-24 in 1969, a combat helicopter of notable strength and power

that strongly cemented its reputation during the Soviet invasion of Afghanistan. Conceived virtually as a flying combat vehicle, the Mi-24 carried a two-man crew and considerable firepower, but also featured a loading bay that could accommodate up to eight soldiers or stretchers for the evacuation of wounded personnel. Built in over 20 different versions, the Mi-24 (like the Mi-25 and Mi-35 versions) was exported to over 30 countries in the former Communist bloc, in all, more than 2,300 of these helicopters were built. Such are the merits of this plane that a great many are still in service, and a number of programs have been developed to update the basic model platform with advanced avionics and armaments.

The Mi-8, which took its maiden flight on 7 July 1961, warrants special mention. A multipurpose aircraft, more than 17,000 units of it were built in over 35 versions, and it is in service in more than 80 countries around the world. Among the most representative versions are the Mi-8T, Mi-17, and the recent Mi-38, practically a new helicopter, though already flying in 2003, entered service, both civilian and military, from 2019.

The most recent Mil combat helicopter is the Mi-28, which flew for the first time on 10 November 1982. The definitive prototype, a tandem two-seat assault helicopter, did not appear until 1988; however the economic and oganizational difficulties resulting from the Soviet Union's collapse, in addition to the aircraft's lack of all-weather capabilities, led in 1993 to cancellation of the program. Thus the version Mi-28N (with nocturnal capability) was developed, taking its maiden flight on 14 November 1996 and entering into service from 2006.

The Kamov design department built its reputation with the creation of a broad range of helicopters built for naval use, especially in anti-submarine roles. Distinguished by a characteristic layout with dual contrarotating coaxial rotors, the Ka-25, the first in Kamov's line of small naval helicopters, made its maiden flight on 26 April 1961. A number of different versions followed, designed for anti-shipping missile, search and rescue, mine sweeping and intercontinental ballistic missile data transmission duties. Then came the Ka-27, 28 and 29 models for transportation and assault purposes, the Ka-31 air-borne radar helicopter and the Ka-32, a multirole helicopter used in both the military and civil sectors. The most recent and best known Kamov design is the Ka-50 Black Shark attack helicopter; it retains the typical Kamov dynamic layout, but combines it with a slim, streamlined fuselage with a single-seat cockpit equipped with an ejector seat.

The aircraft completed its maiden flight on 17 June 1982, but owing to the Soviet crisis, a mere 12 units of the series were built between 1991 and 2009. From this machine, however, the side-by-side two-seater Ka-52 Alligator derived in 1994, and in production since 2008, with more than 170 examples having been ordered, along with 32 of the Ka-52K naval version. The Ka-52, equipped with sophisticated avionics, has a top speed of no less than 217.5 mph, thanks to the thrust provided by Klimov TV-3 turbines. The maximum takeoff weight is 25,000 lb, and the maximum armament payload of missiles, bombs, rockets, and guns of various types, amounts to 6,173 lb. The improved Ka-52M came into service in 2022..

Two other European companies have established solid reputations in the helicopter field over the years, Westland in Great Britain and Agusta in Italy. The former has, among

183 top right A Mil Mi-17 of the Czech Air Force. Over 17,000 units of this family of helicopters have been produced since 1961.

183 bottom The Russian Kamov Ka-50 Werewolf attack helicopter has a dual coaxial rotor and an ejection seat for the pilot. It can fly at 219 mph.

by two Pratt & Whitney PT6C turbines giving 1531 shp each, for a top speed of 193 mph. Maximum takeoff weight is 15,000 lb, and the spacious cabin can carry up to 15 passengers, along with 1 or 2 pilots. In Italy, it was acquired from 2008 by the Financial Police, the Coast Guard, the Air Force, the State Police, the Fire Service, and the Carabinieri, while the Navy also uses a former Financial Police example. Later AW.169 and AW.189 models appeared, in addition to the military model, AW.149 all sharing many features. The first is a machine in the five-ton class that flew for the first time on 10 May 2012 and has currently attracted more than 350 orders. The aircraft is suitable for civil use in air ambulance, transport, supply, and search and rescue roles, as well as for police and military purposes (AW169M), thanks to the possibility of armaments and mission equipment. In Italy, it has been ordered by the Army, the Carabinieri, and the Financial Police.

The AW149 helicopter instead took its maiden flight on 13 November 2009 and was conceived for the military market in the eight-ton class as a troop transport aircraft (up to 18 soldiers), but with multi-role capabilities thanks to on-board sensors and the possibility of being fitted with armaments. It is powered

186 top The naval version of the NH-90 is designated NFH; in this shot, an SH-90A of the Italian Navy, which uses 46 examples, along with 10 multi-purpose MH-90As.

186 center Developed from the AW149, the AW189 is a transport helicopter in the eight-ton class dedicated to the civil sector.

186 bottom The AW149 is a mid-range transport and multi-role helicopter conceived by Leonardo for military use. (Photo: Leonardo)

by two General Electric/Avio CT7-2E1 turbines and can achieve a maximum speed of 192.5 mph. It has been acquired by the armed forces of Thailand, Egypt, and Poland and is the favorite in the New Medium Helicopter program of the British Ministry of Defence, with a contract for 23 examples which was signed in 2026.

The AW189 model is the largest in the family and completed its maiden flight on 21 December 2011. It entered service a little over two years later. This is a helicopter in the eight-ton class derived from the AW149 model and can carry up to 19 people. It is suitable for offshore supply, search and rescue and transport roles. 2017 saw the introduction of the AW189K version equipped with Turboméca Aneto 1K turbines each developing 2,500 shp, permitting the ceiling to be raised to 15,000 feet and the maximum takeoff weight increased to 18,960 lb.

The lightest of the Leonardo helicopters is the AW09, a light single-engined machine in the three-ton class, suitable for various roles in the civil and governmental fields. Originally conceived by the Swiss firm Kopter, the AW09 is powered by a Safran Arriel 2K turbine developing 1,006 shp, can carry up to eight people plus the pilot, and has a cruising speed of 161.5 mph.

Lastly, the most recent machine produced by the Leonardo Helicopter Division is the AW249, the future combat helicopter designed for the Italian Army, which will probably enjoy commercial success abroad too. It is larger than the AW129 it replaces, and provides clear improvements in terms of performance, capabilities, and avionics, placing it at the forefront of the sector. Weighing 18,298 lb fully laden, the AW249 is powered by twin General Electric/Avio CT7-8E6 turbines, each developing 2,583 shp, good for a maximum speed of 192 mph. The AW249 flew for the first time on 12 August 2022 and is expected to enter service in 2027.

Airbus Helicopters is the other major European manufacturer, boasting a wide range of machines from the smallest, the H125, through the H130, H135, H140, H145, H160 and H175, up to the H215 and H225 models, all part of an extended family with multiple common features and available in both civil and military versions. The most recent success story is the H160M Guépard, a military model derived from the H160, which completed its maiden flight on 13 June 2015. This is the first of a new generation of machines and has been chosen by the French government to equip all three of its armed forces. A multi-role helicopter in the six-ton class, it can carry up to 12 fully-equipped soldiers (in addition to the two pilots) and can be fitted with various types of weaponry. The model can reach a cruising speed of 163.4 mph and is powered by twin Turbomeca Arrano turbines, each developing 1,100 shp. The first production H160M is due to enter service in 2026.

China has also become a major helicopter manufacturer, after years of building machines under French and Italian license. Changhe Aircraft Industries was founded in 1969 as a state-owned company and has developed, thanks to joint ventures with AgustaWestland (producing the CA109, the Chinese version of the AW109) and Sikorsky. The Changhe Z-8 is a three-engined helicopter derived from the SA 321 Super Frelon by the French company Aerospatiale, which was subsequently developed into the Z-18 version, which flew for the first time in 2014 and is used by the Chinese army and navy.

The company then developed the CAIC WZ-10, a combat helicopter that flew for the first time on 29 April 2003 and entered service five years later. This is a six-ton-class machine, powered by two Pratt & Whitney Canada PT6C-67C turbines, each developing 1,531 shp, giving it a maximum speed of around 186 mph.

Harbin instead made its debut in 1981, constructing the Aerospatiale SA 365 Dauphin under license, with the model designated as the Z-9 in China. This was then used as the basis for the Z-9B, of which over 100 were produced for the army, the Z-9C, the Z-9W, and the Z-9G versions from the 1990s, for a total of more than 350 machines. Finally, an attack version of this helicopter was developed from the Z-9W version, known as the Harbin Z-19 and featuring a front fuselage redesigned to accommodate a tandem cockpit for the pilot and gunner. The Z-19 has been in service since 2012 and more than 200 units are thought to have been produced. The Z-19E export version was launched in 2017. This machine, with a maximum weight of 9,370 lb, is powered by twin 940 shp WZ-8C turbinesand can reach a maximum speed of 174 mph. Its armament includes rockets, pod cannons and anti-tank, anti-ship and air-to-air missiles. Another machine produced by Harbin is the Z-20, which flew for the first time on 23 December 2013 and entered service six years later. This is a helicopter in the 10-ton class, very similar in shape to the Sikorsky S70, a machine used by the Chinese army since 1984.

187 top The lightest helicopter offered by Leonardo is the AW09, which completed its maiden flight in 2014 as the Kopter SH09.

187 bottom The AW249 Fenice is the most recent model produced by Leonardo and is destined to take up the legacy of the AW129 Mangusta as the army's combat helicopter.

Chapter 14

Trainers

The claim could be made that the most significant aircraft in a pilot's life are trainers: they are the aircraft that enable novices to complete that all-important first flight, to gain experience and obtain a license, whether civil or military. In fact, the trainer is the airplane that allows any budding pilot to realize the dreams of flight. Yet, strange though it may be, at the dawn of aviation, when the need was particularly acute, there were no planes designed specifically for pilot training. Intrepid pioneers frequently began their careers perched behind their instructor, barely able to reach the joystick yet expected to pick up at least rudimentary notions on how to control an aircraft. The earliest aviators of all were of course self-taught, taking off alone with scarce or non-existent theoretical notions – and with results that are all too easily imaginable.

Pioneering aviation was however a field reserved for a few adventurous enthusiasts willing to try anything. The outbreak of the First World War was to transform flying radically, from an elitist sport to an activity of the masses. Military demands and rapid technological development provided enormous impetus to powered flight, and during the conflict the belligerents had to produce tens of thousands of aircraft and, naturally, tens of thousands of pilots capable of flying them. Initially, practical approaches were limited, and training was conducted with aircraft no longer effective as front-line military machines, generally spotters and bombers such as the French Farman, the German DFW B.I and the early Avro 504s. From 1916, however, airplanes expressly designed for the delicate task of pilot training began to appear. These machines were characterized by dual controls and "forgiving" handling.

Among the first great trainers from this period, the British Avro 504J was of particular note. The original model, a two-seater biplane with a front-mounted engine and control surfaces at the tail, completed its maiden flight in July 1913. The British War Ministry then ordered it for use as a spotter and light bomber. The aircraft, which was also used as a fighter for a time, soon proved to be obsolete in terms of front-line combat, and in 1916 Avro decided to revive its fortunes by developing the 504J version as a purpose-built trainer. The aircraft enjoyed notable success, so much so that to overcome problems with standardization of the engine, Avro developed the 504K version. This featured an engine mount that could accept three different types of power unit in use at the time. The 504 had a wingspan of 35 feet 7 inches, a length of 29 feet 2 inches, and a maximum take-off weight of about 1,830 lb. The principal power unit was a Le Rhône rotary engine producing 110 hp, providing the plane with a top speed of just over 90 mph. Also built under license in Australia, Belgium, Canada, Denmark and Japan, the 504 continued to be used well after the end of the war. The 504N version was introduced in 1925; it featured a new undercarriage and a radial engine producing 162 hp. The Avro 504 remained in service with the Royal Air Force as a trainer up until 1932, but was widely used in the civilian sphere until the outbreak of the Second World War. Over 10,000 504s were produced, and the aircraft was employed in over 25 different countries around the world. It remains one of the most significant aircraft in history thanks to the

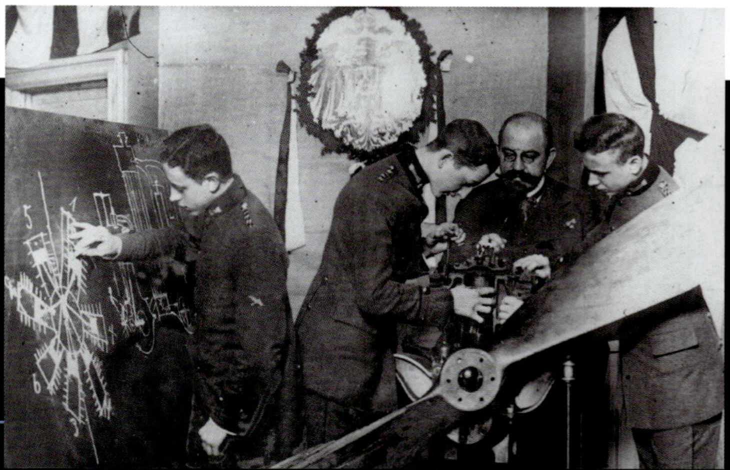

188 top The Curtiss JN-4 Jenny was one of the commonest and most famous airplanes between the two World Wars. More than 6,500 were built and many were used in civil aviation.

188 bottom left Training simulators came into use before the 1940s. Here, in 1938, a Royal Navy cadet is undergoing a flight preparation session at the RAF school at Netheravon.

188 bottom right German air force cadets during a lesson on engine mechanics in February 1914.

189 top In this RAF propaganda photograph from 1940, RAF student pilots receive last instructions on flying in formation before a mission.

contribution it made to the diffusion of flying.

Another "giant" in this sector was the American Curtiss JN, of which more than 6,500 were built. The most famous of the series was the JN-4, which resulted from development of the JN-2 and JN-3 models, and was affectionately nicknamed "Jenny" by its pilots. The JN-4 flew for the first time in July 1916. Britain's RAF immediately purchased 105 planes. Shortly thereafter, in view of the need to produce ever greater numbers of pilots, the American army also began to order the plane. The JN-4D, which incorporated improvements made to JN-4s built in Canada and the USA, was introduced in June 1917. This aircraft had a wingspan of 43 feet 6 inches, a length of 27 feet 2 inches, and a maximum take-off weight of about 1,920 lb. The Jenny was powered by a Curtiss OX-5 engine producing 90 hp, sufficient for a top speed of about 75 mph. Wartime production forced output to ever higher levels, and the plane was also built by a further six American firms.

In 1917, Curtiss launched an advanced training version, the JN-4H, which was powered by a 132-hp engine and could be armed with bombs and machine guns. Subsequently the slightly improved JN-6 was introduced. In the post-war years, a number of Jennies remained in service, being subjected to numerous refits, but a considerable proportion of the military aircraft were sold on the civilian market and made a notable contribution to the diffusion and development of flying. The Jenny's military career continued up until 1927, but the plane continued to be a mainstay of civil aviation for much longer, and even today a number of Jennies can be found in airworthy condition.

189 center A student pilot is prepared for ejection using a Martin-Baker seat. The year is 1949, and with the advent of new equipment, training activities were expanding.

189 bottom Britain's Avro 504T was one of the first real training aircraft in aviation history. It flew in 1913 as a spotter but was used as a trainer after the outbreak of the Second World War.

Italy also produced a notable training aircraft during the First World War, the SVA 9, one of the successful SVA series of planes. Production of these aircraft (designed by engineers Savoia and Verduzio of the Ansaldo company, hence the initials SVA) began in 1917 with the appearance of the SVA 4, a biplane originally developed as a fighter for the Italian army, but soon relegated to reconnaissance duties, given its limited maneuverability. Later in 1917, the SVA 5 bomber version spawned the SVA 9, a two-seater with dual controls for pilot training. This aircraft had a wingspan of 28 feet 10 inches, a length of 26 feet 6 inches, and a maximum take-off weight of 2,315 lb. It was powered by an in-line SPA 6A engine developing 265 hp, and had a top speed of about 124 mph. The SVA 9 began entering service early in

190 top The Caproni Ca.100 – better known as the 'Caproncino' – was a popular and common Italian touring plane and trainer between the wars. It first flew in 1929 and its 130-hp engine could take it to 116 mph.

190 center and 191 top Shown here are two Avro Tutors. The photograph on the right shows a restored Avro Tutor still flying today. The Armstrong Siddeley Lynx radial motor produced 243 hp and a top speed of 123 mph.

190 bottom The Avro 621 Tutor first flew in 1929. Designed to replace the Avro 504 in the RAF's flying schools, it remained in service for ten or so years until the outbreak of World War II.

1918, and proved to be unquestionably more successful that its predecessors. Adopted by a dozen countries, the SVA biplanes remained in production until 1928, with more than 2,000 being built. The most famous feat involving a SVA 9 was undoubtedly Lieutenant Arturo Ferrarin's 1920 expedition from Rome to Tokyo with a group of 6 two-seaters, a journey of 9,500 miles completed in less than three months. The year 1928 also saw the introduction of another famous Italian trainer, the Caproni Ca.100, nicknamed the "Caproncino." About 700 Ca. 100s were constructed between 1930 and 1937, and the plane proved to be a well-designed, versatile machine. It was, in fact, able to accept about fifteen different engines (from the 86-hp FIAT A50 through to the 132-hp Colombo S63), and was also transformed into a seaplane and was fitted with skis to that it could fly from snow-covered runways. Exported to numerous countries for civilian and military uses, the Caproncino also gave rise to the Ca.100bis version, which had a 136-hp Gipsy engine and featured a closed cabin.

The RAF flying schools' retirement of the Avro 504N in 1930 coincided with the Air Ministry's purchase of the new Avro 621 Tudor biplane as its training aircraft. Though about 800 Tudors were built, the plane was soon competing with the de Havilland DH.82 Tiger Moth, another two-seater that was to enjoy far greater success. This aircraft, which completed its maiden flight on 26 October 1931, was developed from the civil Gipsy Moth, but featured new wings, a new engine and a reinforced frame. The RAF immediately ordered the Tiger Moth, and it began to enter service in 1932. Not long thereafter, Avro introduced the DH.82A version, which was equipped with the more powerful Gipsy Major engine, producing 132 hp. The Tiger Moth was to become one of the most widely used and successful trainers in aviation history. About 8,000 were built and production licenses were granted for Australia, Canada, New Zealand, Sweden, Norway and Portugal. For many years, and well past the end of the Second World War, the Tiger Moth was to

191 bottom left The de Havilland DH.82 Tiger Moth was without doubt one of the most successful and commonest trainers. It first flew with the British forces in 1931 but was then licensed for construction and used in many countries around the world.

191 bottom right The Polikarpov U-2 biplane first flew in 1928 but, thanks to its sturdiness, remained in service for many years and many purposes, including night-time bombing raids during the Second World War.

remain the standard military basic trainer for all the Commonwealth countries. Still widely used in the civilian field for recreational flying, the Tiger Moth is now considered to be a classic.

The Soviet Union, which was a leader in aeronautic design in the 1920s and 1930s, also produced a particularly interesting trainer. It is probably still the most widely constructed aircraft of its type, with more than 33,000 being built. The merit for this success goes to the designer Polikarpov, who radically modified the U-2TPK design (for an aircraft that was actually never built) to produce a robust, conventional biplane that was easy to produce and fly. The U-2 flew for the first time on 7 January 1928, and proved to an excellent plane. It was immediately ordered into mass production.

During the Second World War, the U-2 was used in diverse roles such as reconnaissance, tactical support, night bombing, target towing, transport and air-ambulance, and remained in production up until 1952. On Polikarpov's death the aircraft was renamed in his honor as the Po-2.

The early 1930s saw the birth of another great aircraft, this time in Germany There, with the support of the Nazi regime, aviation was rapidly regaining ground lost after the restrictions imposed by the Treaty of Versailles. The newly formed Buckner firm made its debut in 1934 with the Bu-131 Jungmann, a two-seat biplane of compact dimensions, built from wood and aluminum covered in canvas, and featuring a moderate sweep to the wing. The Bu-131, powered by an 82-hp Hirth engine, made its maiden flight on 27 April. In this case too, the aircraft achieved sales success in both the military and civilian fields, being adopted the nascent Luftwaffe, by aero clubs, and purchased by numerous other countries. The Bu-131B model was powered by a Hirth HM 504 engine developing 106 hp, and had a wingspan of 24 feet 3 inches, a length of 21 feet 7 inches, a maximum take-off weight of 1,500 lb, and a top speed of about 113 mph. The Jungmann was also produced in Czechoslovakia and Japan, with 1,254 being built in the latter nation alone. The total production figure is unknown but is certainly in the order of some thousands. A single-seater, the Bu-133 Jungmeister, was also developed from the Bu-131. It was basically a scaled-down version but equipped with a more powerful Hirth engine producing 137 hp. In the C version, this engine was replaced by a Siemens Sh 14 engine, producing 162 hp. This truly agile and lively aircraft was used as an advanced trainer for fighter pilots. Finally, in 1939 Buckner produced the last of its famous planes, the Bu-181 Bestmann, a basic trainer with modern features. A low-wing monoplane with a closed two-seater cabin, the Bu-181 had the same Hirth engine as the Bu-131 but a top speed of about 134 kph. During the Second World War, some thousands of Bestmanns were produced in Germany, Holland and Sweden and subsequently in Czechoslovakia and Egypt.

Another trainer to earn a place in the history of aviation was the American Stearman 75 Kaydet, which flew for the first time in 1935 and was used to train thousands of American pilots during the Second World War. Developed from the X-70 prototype, the Model 75 was initially ordered by the US Army and it began to enter service in 1936 as the PT-13. In 1939, Stearman was bought out by Boeing, a firm capable of meeting the enormous demands that the army and navy made for aircraft with the outbreak of war. In 1940 the PT-17 (known as the N2S to the US Navy) was introduced; this most successful version of the aircraft had adopted the

Continental R-670 engine, producing 220 hp. The PT-17 had a wingspan of just over 32 feet, a length of 24 feet 11 inches, and a maximum take-off weight of 2,710 lb, and was capable of a top speed of about 125 mph. By 1945 Boeing had produced 10,436 PT-17s, and the plane continued to fly in great numbers in the military and civilian spheres for many years after the war.

The General Aviation Company (shortly to become North American) introduced another great military trainer, the T-6 Texan, which was used in the United States and then throughout the world from the 1940s to the 1960s and beyond. The plane, which was redesignated the NA-16, flew for the first time in April 1935. The US Army appreciated features such as the aircraft's all-metal construction, monoplane architecture, two-seat tandem cockpit and 405-hp Wright engine, and ordered an initial batch of 42 that year, designating the plane as the BT-9. This aircraft was then the basis for the BC-1 (the SNJ-1 for the Navy), introduced in 1937 and featuring a retractable undercarriage, a more sophisticated airframe and a new, more powerful engine developing 550 hp. In 1940, the BC-1 was redesignated the AT-6. Output of the airplane increased constantly as war loomed and finally broke out. The basic design was gradually improved and adapted to meet diverse needs with the introduction of specific versions. Baptized by the British as the Harvard, the AT-6 was known to the Americans as the Texan, given that it was built in a new factory in Dallas. The AT-6A was equipped with a Pratt & Whitney R-1340-49 engine producing 607 hp, and was capable of a top speed of 210 mph. It had a wingspan of 42 feet, a length of 28 feet 10 inches, and a maximum take-off weight of about 5,150 lb. Designated the T-6 in the post-war period, the plane was built under license in Australia, Sweden, Spain, Canada and Japan. Taking into account all the versions of the T-6 (over 260), it appears that a total of more than 21,000 were built. After the war, far from being withdrawn from service, the T-6 was adopted by over 60 different air forces and used throughout the world as a standard military trainer, in some cases until the 1980s.

193 top The US Navy called the T-6 Texan the SNJ. Shown are two splendid restored SNJs in their official colors before the US entered the Second World War in December 1941.

193 center left The Bearing-Stearman 75 Kaydet, another famous trainer, was used by the flying schools of the USAAC and US Navy from 1936, and after the war by many other forces. Over 10,000 were built.

193 center right Germany too built excellent trainers during the 1930s, including the Bucker Bu-133 Jungmeister, a small acrobatic, single-seat biplane with a 160-hp engine and top speed of 137 mph.

193 bottom Designed to be a touring plane, the Bucker Bu-181 Bestmann first flew in 1939. It was very advanced for its time and, after the war, was used by various countries, including Egypt, flying the one depicted here.

192 top A North American T-6A Texan flown by the Canadian air force. Built under license, it was given the name Harvard H4M in the Commonwealth countries.

192 bottom A T-6A flown by the Portuguese air force. The Texan was a great success and was sold to many of America's allies after World War II. More than 21,000 T-6s were built and used in over 60 countries.

Perhaps the most significant expression of technological progress in aviation to emerge from the Second World War was the jet engine and the aircraft powered by it: the Messerschmitt Me-262, the Arado Ar.234, the Gloster Meteor and other experimental designs. In the post-war years it became clear that with the rapid spread of jet aircraft, traditional propeller-powered trainers were no longer adequate. A new generation of jet-powered trainers was thus introduced, some of which were actually used for the basic training of student pilots. The first to appear and the most famous was (like others in that period) derived from a fighter, the Lockheed F-80. On the basis of this excellent jet, the California-based manufacturer created a two-seater model that conserved the architecture of the original plane. Designated the TP-80C, the airplane flew for the first time on 22 March 1948, and was immediately approved for serial production. The TP-80C had a wingspan of 38 feet 9 inches, a length of 37 feet 9 inches, and a maximum take-off weight of about 14,450 lb; it was equipped with an Allison J33 turbojet that produced 5,400 lb of thrust, sufficient to give it a top speed of 600 mph. The TP-80C trainer was soon renamed the T-33A Silver Star and enjoyed exceptional sales success: Lockheed and other firms that manufactured it under license produced a total of 6,557 planes. These were purchased by the armed forces of more than 40 different countries around the world. Such was the quality and balance of the design that the last of these planes, in operational, service with the Bolivian Air Force, was only retired in 2017.

Following the United States' lead, jet-powered basic trainers began to be designed in Europe too. Apart from the British de Havilland Vampire T, derived from the well-known fighter, the first true trainer from the Old World was designed in response to a requirement issued by the French Air Force. This was the Fouga CM-170 Magister, which completed its maiden flight on 23 July 1952. The Magister boasted highly streamlined, elegant lines with a straight wing, a tandem cockpit layout with no ejector seats, a characteristic "V" tail and two Turboméca Marboré turbojets buried in the wing roots, each developing about 880 lb of thrust. The plane had a wingspan of 39 feet 7 inches, a length of 32 feet 10 inches, a maximum take-off weight of 7050 lb, and a top speed of 444 mph. It could be armed with two 7.62-mm machine guns and up to 220 lb of external weaponry in the form of bombs or rockets. As it was one of the first jet trainers to be produced in series, the Magister sold very well and was used by no less than 22 different countries. It was also built under license in Germany, Finland, and Israel. A total of 929 Magisters were produced, some of which are still flying today. Fouga also produced the CM-175 version for the French navy as a trainer for deck-landing operations.

Another two significant aircraft appeared two years after the Magister. The first to fly, on 16 June 1954, was the British Hunting P84 Jet Provost, developed from the propeller-driven Provost trainer in the hope of meeting the RAF's requirements for an instructional jet. Maintaining the wing and the tailplanes of the Provost, Hunting designed a new fuselage featuring side-by-side ejector seats and two air intakes in the wing roots that fed a Bristol Siddeley Viper turbojet producing 2,500 lb of thrust, sufficient to provide the plane with a top speed of 440 mph. The Jet Provost had a wingspan of 35 feet 1 inch, a length of 33 feet 9 inches, and a maximum take-off weight of about 9,200 lb. The RAF ordered 201 of the T. Mk.3 version, which was followed by a further nine production versions, many of which were exported. In all, about 530 P84s were built, and the aircraft was adopted by seven air forces. It remained in service with the RAF through until the late 1980s.

On 12 October 1954, the Cessna Model 318 made its maiden flight. It was the company's response to a USAF requirement for a new jet-powered basic trainer. The plane, which won the USAF contract, was a compact twin-engined design, with parallel seating and a low undercarriage. The Model 318 then received the military designation T-37A. It had a wingspan of 35 feet 9 inches, a length of 28 feet 3 inches, and a maximum take-off weight of about 6,395 lb, and was powered by two Continental J69 engines, each producing 920 lb of thrust. In 1974, after 534 T37-As had been constructed and the plane had begun to enter service, the T-37B version with more powerful engines (1,025 hp) was introduced. Then followed the T-37C and the A-37A and B Dragonfly versions, equipped with new engines, a maximum take-off weight of 14,000 lb, armament and wingtip tanks. In the main, these planes were sold to Allied air forces as light strike aircraft. Subjected to an updating program and prolongation of its operational life, the T-37 (with over 1,230 built) was the USAF's basic trainer up until 2009, when it was definitively replaced by the Beech T-6A Texan II.

194 top left The Lockheed T-33A Silver Star, derived from the F-80 fighter, was the first US Air Force jet trainer. It first flew in 1948.

194 bottom left A pair of RAF Hunting P.84 Jet Provosts. This plane was a jet-engined development of the propeller-driven Provost trainer.

194 right Two Cessna T-37As belonging to the Pakistani Air Force. With two side-by-side seats and straight wings, this plane first flew in 1954 and over 1,800 were made.

195 The French Fouga CM.170 Magister was one of the first airplanes designed to train new pilots on jet aircraft; it first flew in 1952. This one is flown by the Irish air force.

The Aermacchi MB.326, another trainer that was to be sold successfully throughout the world, completed its maiden flight in Italy on 10 December 1957. With a traditional layout and tandem seating, the 326 was a simple, reliable and robust aircraft. However, it featured a number of innovative features, and in 1962 the Italian air force (AM) began ordering it for use by its flying schools. The plane had a wingspan of 34 feet 5 inches, a length of 34 feet 9 inches, and a maximum take-off weight of 8300 lb. and was powered by a Rolls-Royce Viper turbojet producing 2500 lb of thrust, providing a top speed of 500 mph. A number of important production contracts granted in South Africa, Australia and Brazil confirmed the MB.326's sales success. A total of 763 planes were manufactured, and they were used in twelve different countries. The MB 326 formed the basis of a single-seat strike version, the MB.326K. In 1972, in order to create a 326 series replacement for the AMI, Aermacchi began work on the MB.339. This was a development of the preceding model, equipped with a more powerful Viper engine producing 4000 lb of thrust (the same unit used in the MB.326K), and a completely redesigned nose with a tandem cockpit featuring stepped seats. This aircraft flew for the first time on 12 August 1976. It was a commercial success; the air forces of nine countries adopted the plane, and it was produced in around 237 examples in a number of different versions.

The Soviet bloc countries had their own basic jet trainer: the Czech Aero L-29 Delfin, a single-engined plane with

a tandem cockpit, a straight wing and a T-tail that flew for the first time on 5 April 1959. This plane had a Motorlet M701 engine producing about 1,960 lb of thrust, sufficient to provide it with a top speed of 407 mph. It had a wingspan of 33 feet 10 inches, a length of 35 feet 5 inches and a maximum take-off weight of about 7,230 lb. Over 3,500 of the L-29s were built, with 2,000 acquired by the Soviet air force for its flying schools. In the 1960s, Aero decided to produce a successor to the Delfin, but initially the L-39 Albatros project encountered difficulties, apparently due to the combination of the new airframe and the Soviet Ivchenko engine. The Albatros completed its maiden flight on 4 November 1968, and entered operational service in 1974. The Albatros had a wingspan of 31 feet 1 inches, a length of 39 feet 7 inches, and a maximum take-off weight of about 10,360 lb. Powered by an Ivchenko Progress engine that produced about 3,790 lb of thrust, the plane had a top speed of 470 mph. The Albatros was extremely successful; over 2,800 were built and exported to over 18 different countries, the majority to the Soviet Union. The L-39 basic training version was followed by ground-attack version and, in 1986, by the improved L-59 with uprated avionics and engine. The L-139, a collaboration with the American aviation industry, which provided the power unit and avionics, made its maiden flight in 1993, and the L-159, a single- or two-seat fighter based on the L-59, was introduced in 1997.

The American firm Northrop produced the first supersonic trainer: its two-seat N-156 made its maiden on 10 April 1959. The USAF then took delivery of the plane in 1961, designating it the T-38A Talon. The plane was a slim tandem two-seater, with twin engines, it featured a short, moderately swept wing. The USAF used it as an advanced trainer to which pilots graduated after basic training with the T-37. The T-38 had a wingspan of 25 feet 2 inches, a length of 46 feet 3 inches, and a maximum take-off weight of about 13,000 lb. Its two General Electric J85 turbojets each produced 3,850 lb of thrust and powered the plane to a top speed of Mach 1.23. A total of 1,187 Talons were built; many went into service with the US Navy, NASA and the air forces of Portugal, Turkey, South Korea, and Taiwan. The original model formed the basis of the AT-38B gunnery trainer. The T-38 was modernized, with the C version allowing it to remain in service until it was replaced by the new T-7A.

196 top The Northrop T-38 Talon was the USAF's first supersonic trainer. It entered service in 1961 and was used for advanced training. Shown here is a Talon belonging to the Turkish air force.

196 center The Anglo-American McDonnell-Douglas (Boeing)/Bae T-45A was developed from the BAe Hawk for the US Navy. Though it still has the general lines of the Hawk it was in fact an entirely new airplane.

196-197 The Alpha-Jet, introduced in 1973, was the result of a French-German collaboration to build a multi-role aircraft that could act both as a light fighter-bomber and as a training airplane.

197 top The Aermacchi MB.339A, which first flew in 1976, was a direct successor of the MB.326. This model is the MB.339C belonging to the New Zealand Air Force and is shown during an acrobatic display.

197 bottom left The Aero L-29 Delfin first flew in 1959 and was for many years the standard training aircraft of the Warsaw Pact countries and allies of the USSR. More than 3,500 were built.

197 bottom right The Aermacchi MB.326 was the greatest Italian success after the Second World War. This plane belong to the South African acrobatics team.

In Europe in the mid-1970s two other successful modern trainers came into being. Together with the MB.339, they represented the new generation. The three were the Franco-German Dassault-Breguet/Dornier Alpha Jet and the British Hawker Siddeley (later British Aerospace) Hawk. The former was introduced in 1969, and was the outcome product of combining the French need for a trainer and the German need for a light fighter-bomber. The resulting Alpha Jet was, however, closer to the French requirement than to the Luftwaffe's: it was a tandem two-seater with a high swept wing and two low-mounted engines that were not equipped with afterburners. The Alpha Jet E trainer had with a wingspan of 29 feet 10 inches, a length of 40 feet 4 inches, and a take-off weight of about 16,500 lb. The plane's twin Larzac 04 turbofan engines produced 2975 lb of thrust, giving it a top speed of 572 mph. The Alpha Jet completed its maiden flight on 26 October 1973, and since then over 500 have been produced and sold in eleven different European, Asian and African countries. In 1982, the Alpha Jet 2 followed, equipped with modern avionics and specific ground-attack weaponry.

The BAe Hawk, which flew for the first time on 21 August 1974, also enjoyed considerable sales success. The plane was an advanced trainer with tandem seating and a low wing, and was designed to provide the RAF with a replacement for its Folland Gnats and Hawker Hunters. The fast, maneuverable Hawk entered service in 1976, and soon proved to be a fine strike aircraft. Its export success was largely due to this versatility. The Hawk T Mk.1 had a wingspan of 30 feet 10 inches, a length of 38 feet 9 inches, and a maximum take-off weight of about 17,000 lb. Powered by an Adour turbofan engine producing 5,200 lb of thrust, the plane was capable of a top speed of 646 mph.

After 1980, with the introduction of various export versions (the Mk. 50, 60 and 100 series), the Hawk was sold to about fifteen European, Asian, African and North American countries. The Mk. 200, a single-seat version expressly designed for combat roles, was introduced in 1986, and the Hawk Mk. 60 was chosen as the basis for McDonnell-Douglas/BAe joint venture advanced trainer for the US Navy. Designated the T-45A Goshawk, this plane flew for the first time on 16 April 1988, and entered service in 1992. The US Navy has placed orders for a total of 174 of the T-45A's, which in 2003 ended up replacing the TA-4 Skyhawk and, in 2008, the T-2 Buckeye. The T-45A is a substantially different aircraft from the Hawk; it features significant avionics, structural and aerodynamic modifications enabling it to respond to the US Navy's particular demands, including those of carrier-borne operations. Finally, the last series of Hawks were those produced as Lead-In Fighter Trainers (LIFT), advanced trainers for the latest generation fighters, boasting digital avionics, new wing and fuselage structures, and Adour 951 engines with 6500 lb of thrust. These versions entered into service beginning in 1997, with the Mk.127 for Australia, followed later by other versions for the UK, South Africa, and India. Total production for all the various types of the Hawk had exceeded 1,000 examples when the model was discontinued in 2022.

198-199 The Italian Frecce Tricolori, one of the three most appreciated acrobatic teams, has flown ten Aermacchi MB.339s since 1982. The Frecce, known as the 313th Group, are also trained as a fighter-bomber unit.

199 top The Patrouille de France, another of the world's top acrobatics teams, uses nine Dassault-Breguet/Dornier Alpha Jet E. The French use this plane only as a training aircraft, but the Germans also use it as a light fighter-bomber.

199 bottom The RAF Red Arrows acrobatics team uses nine BAe Hawk trainers. The Hawk first flew in 1974 and can be used as a light attack aircraft carrying bombs and rockets. It has been exported to a dozen or so countries as a trainer and light, multi-role fighter. A single-seat version, known as the Hawk Mk.200, was also developed.

The Italian SIAI Marchetti SF-260, designed as a recreational aircraft, flew for the first time on 15 July 1954. The aircraft was derived from two designs of the engineer Stelio Frati, the F.8 Falco and the F.250. SIAI Marchetti acquired and developed the F.250, and adapted it for serial production. After its initial success in the civilian market, civil and military flying schools took the plane into consideration, thanks to its notable capabilities as a trainer. As a result, toward the end of the 1960s, the SF.260M was developed for military applications; this version bwas followed by the SF.260 Warrior, which could be lightly armed. The airplane featured very clean, elegant lines with a laminar profile wing, side-by-side cockpit seating and a retractable undercarriage. It was powered by a Lycoming AIO-540 engine producing 260 hp and could reach a top speed of 215 mph. It had a wingspan of 27 feet 5 inches, a length of 23 feet 4 inches, and a maximum take-off weight of about 2,645 lb. A TP turboprop version SF.260 was developed, and overall the SF.260 has been a great commercial success: more than 900 have been produced, and the military version has served in 31 different countries.

In the 1970s a third type of training aircraft fitted with turboprop engines began to become established. The turboprop combined the high performance of jet-powered alternatives with lower running costs. Along with the Marchetti SIAI T-34C and the SF.260TP, the most successful trainers of this type were designed in Switzerland and Brazil. In the 1960s, the Swiss company Pialatus, which already produced the P-3 piston-engined aircraft, decided to revitalize its range with the introduction of a turboprop-engined plane. The new plane, designated the P-3B, flew for the first time on 12 April 1966, but the definitive PC-7 Turbo Trainer version did not appear until 1978. It had a wingspan and length of 33 feet 1 inches and a maximum take-off weight of about 7,050 lb. The PC-7 was powered by the classic 700-shp Pratt & Whitney Canada PT6A engine, which provided a top speed of 345 mph. The PC-7 entered service with the Swiss air force in 1982; 15 or more countries have adopted it, and more than 600 PC-7s have been built. It also formed the basis for the PC-9, which flew for first time on 7 May 1984. This version featured a more powerful PT6A engine (950 shp), better performance, and ejector seats for the crew. More than 160 PC-9s were built; the model in turn formed the basis for the Beech Mk.II, which in 1995 won the JPATS competition for a new trainer for the American armed forces. Designated the T-6A Texan, the plane, practically a new one, is built by the American Beechcraft. It has a 1250 shp engine (and other modifications, such as new avionics and pressurized cockpit) and to date more than 1,000 units have been produced for the USAF and the US Navy, as well as for the air forces of Argentina, Canada, Colombia, Germany, Japan, Great Britain, Greece, Iraq, Israel, Mexico, Morocco, New Zealand, Thailand, Tunisia, and Vietnam. The model AT-6B has also been developed, for light attack and counterinsurgency.

The latest evolution of the family of Pilatus trainers is the PC-21, which took its maiden flight on 1 July 2002. Although similar to its forerunner the PC-9, the PC-21 is a completely

new aircraft, characterized by digital avionics and smaller, swept wings to better approximate the flight characteristics of jets. It is powered by a Pratt & Whitney Canada PT6A-68B turbine giving 1600 shp to an advanced-profile five-blade propeller. Maximum takeoff weight is 9400 lb; top speed is 425 mph; wingspan is 30 feet; and overall length is 37 feet. The aircraft is currently in production (more than 300 examples have been built to date) with orders for the air forces of Switzerland, Saudi Arabia, Australia, Canada, United Arab Emirates, France, Jordan, Qatar, Singapore, and Spain.

202-203 bottom The SIAI Marchetti (today Aermacchi) SF.260 was one of Italy's most successful postwar light trainers. Introduced in 1964, it is still in production, and more than 900 have been sold to 30 or more countries.

203 top The Swiss Pilatus PC-7, introduced in 1978, is another successful turboprop trainer. More than 600 have been sold to 16 countries, including Malaysia, which owns the three shown here.

203 center Captured in flight during an exercise, a Brazilian air force A-29A Super Tucano and a KT-IP of the Peruvian air force, two of the most modern turboprop trainer/attack aircraft in service.

203 bottom The latest Pilatus turboprop trainer is the PC-21, which flew for the first time in 2002. It has already been ordered by several air forces.

In the wake of the success of the formula and the demand registered, other turboprop trainers, very similar to the Pilatus in appearance and characteristics, have been produced by the aviation industries of various countries. KAI, South Korean aviation manufacturer has introduced the KAI KT-1, which took its maiden flight in 1991, entering into service in 2000) and in Turkey, where it provided the basis for the TAI Hurkus, a development of the KAI T-1, which completed its maiden flight on 12 December 2013.

In the mid-1970s, the Brazilian manufacturer Embraer focused on development of new turboprop trainer, producing a prototype that flew for the first time on 16 August 1980. The plane, named the EMB-312 Tucano, has a wingspan of about 36 feet, a length of 32 feet 3 inches, and a maximum take-off weight of about 7,000 lb. The Tucano was powered by the Pratt & Whitney PT6A engine, capable of producing 760 shp but restricted to 590 hp, giving the plane a top speed of 270 mph. The plane entered service with the Brazilian air force in 1982 and has been produced in 504 examples for 17 countries. It has been successfully exported, in particular to two countries notoriously reluctant to import aircraft, Great Britain and France, both of which assigned the

204-205 The Alenia Aermacchi M-346 is the most recent advanced trainer to appear on the market and has already attracted considerable approval and sales success in nine countries, including Italy, Singapore, and Israel.

204 bottom The Beechcraft T-6A Texan II, a modern turboprop trainer produced for needs of the USAF and the US Navy, was derived from the Pilatus PC-9.

205 bottom The new basic jet trainer from Leonardo is the M-345 HET, here seen in the colors of the Italian air force with which it has been in service since 2025.

Tucano to their flying schools. The Tucano is still in production as the improved EMB-312H assault and training version (denominated A-29 by the Brazilian Air Force), with around 300 units delivered to 20 countries throughout the world.

The panorama of advanced trainers (Lead-In Fighter Trainer or LIFT) has lately been enriched by two new aircraft, designed to better prepare young pilots in managing the latest generation of fighters. In 1999 the South Korean industry KAI, in collaboration with Lockheed Martin, created the project of a new advanced trainer that also has combat capabilities. Designated the T-50 Golden Eagle, the aircraft took its maiden flight on 20 August 2002, and is intended to replace the older T-37s and T-38s in service with the South Korean Air Force. The Golden Eagle, which entered into service in 2005, is clearly inspired by the aerodynamic lines of the Lockheed Martin F-16, especially the wing. The aircraft was also developed in the T-50B version, assigned to the acrobatic Black Eagles team; the TA-50 version, with assault capability; and the FA-50 version, a light fighter with all-weather capability, destined to replace the F-5E/Fs in Korea. The trainer is powered by a General Electric F404 with afterburner that delivers a thrust of 17,700 lb, making the aircraft supersonic, with a top speed of Mach 1.4 (about 932 mph). Maximum take-off weight is 29,700 lb; wingspan is 30 feet; and overall length is 43 feet. In recent years, the T-50 has been one of the very few candidates on the market for the role of LIFT, competing above all with the Alenia Aermacchi (today Leonardo) M-346. Built in more than 300 examples, the plane has to date enjoyed sales success in its various versions with the air forces of South Korea, the Philippines, Indonesia, Iraq, Malaysia, Poland, and Thailand.

The other advanced trainer dominating the scene of today and tomorrow is the M-346, an aircraft originating from the Yak/Aermacchi 130 project of the 1990s. The two companies parted ways in 2000, and Aermacchi developed the M-346 Master, improved and almost completely redesigned, beginning with the engines, installations, structure, and aerodynamic features. The first prototype flew on 15 July 2004, while the first unit of the series soared on 31 March 2011. The M-346 is powered by two Honeywell F124 turbofans, each providing 6250 lb thrust. Its maximum take-off weight is 20,900 lb; wingspan is 32 feet; and overall length is 38 feet; top speeds can reach about 684 mph. The Master boasts advanced aerodynamic lines, redundant performance, and it may also be used as a combat aircraft thanks to the development of the FA version. Currently on order with the Italian Air Force (120 examples of the T-346A) and with the air forces of Austria, Greece, Israel, Nigeria, Poland, Qatar, Singapore, and Turkmenistan, the M-346 is competing in various tenders, including that of the US Navy, for a new advanced trainer to replace the T-45. In 2024, the T-346 was chosen to equip the Italian National Aerobatic Team, the "Frecce Tricolori."

Italy's Leonardo has developed another jet trainer, intended for the initial training of military pilots, to replace the MB.339A used by the Italian Air Force flight school. Designated M-345 HET (High Efficiency Trainer), this is an evolution of the M-311 project, itself derived from the SIAI Marchetti S.211, which has been completely revised and redesigned to meet modern requirements and offer operating costs equal to those of modern turboprops, while maintaining the performance of jet aircraft. The M-345 features a new 1,540 kg Williams FJ44-4M-34 turbofan engine (soon to be upgraded), new engine air intakes, a redesigned and optimized airframe, new onboard systems, new Martin Baker Mk.16 ejection seats and, above all, a complete state-of-the-art avionics suite, derived from the M-346's already highly advanced systems. The first prototype of the new aircraft made its maiden flight on 21 December 2018. The first two production aircraft were delivered to the Italian Air Force in December 2020, but operational issues that were encountered forced the company to make a series of modifications which then required a new series of approval tests by the Armed Forces. The model for the Italian Air Force, designated T-345A, with 18 units ordered so far, was then introduced into service in December 2024, with the first pilot-training course beginning in July 2025.

The Russian firm Yakovlev, a former partner of Aermacchi, for its part, has built the Yak-130, which took its maiden flight on 30 April 2004, with series production of the aircraft beginning in 2012. The Russian Air Force uses more than 100 examples and the aircraft is currently in production with exports to Algeria, Bangladesh, Bieorussia, Myanmar, Iran, Laos, and Vietnam.

For its part, China produces the Hongdu L-15 Lie Ying, an aircraft developed in collaboration with the Russian firm Yakolev and which in fact bears a striking resemblance to the Yak-130. This aircraft flew for the first time on 13 March 2006 and was developed in training (L-15A) and attack (L-15B) versions. Entering service in 2018, the L-15 is powered by two Russian-built Ivchenko-Progress AI-222K-25F turbofans, each delivering 9,260 lb of thrust, and has a maximum takeoff weight of 20,944 lb. Employed by the Chinese Air Force and Navy, the L-15 has also been ordered by the United Arab Emirates, Zambia, and Zimbabwe.

Another emerging country in the military aviation sector is Turkey, which has developed the TAI Hurjet, an ambitious advanced trainer that flew for the first time on 25 April 2023. Powered by a General Electric F404-GE-102 turbofan engine with afterburner, delivering approximately 17,650 lb of thrust, the Hurjet has supersonic capabilities but has yet to enter operational service. 16 examples have already been ordered by the Turkish Air Force and it has been selected by the Spanish Air Force, which is expected to acquire 30.

Finally, the United States has also committed to developing an advanced training jet to replace the USAF's aging fleet of T-38s. The project selected by the Pentagon in 2018 was the Boeing T-7 Red Hawk, developed in collaboration with Sweden's SAAB. This aircraft is also powered by a General Electric F404 engine with afterburner and can reach a speed of Mach 1. The first prototype flew on 20 December 2016. 351 aircraft have been ordered, with an option for a further 125, while initial operational capability was scheduled for 2020. Little publicity was given

to the flight tests of this aircraft, which appears to have encountered difficulties, but the first production T-7A aircraft rolled out of the St. Louis, Missouri, factory on 28 April 2022, with USAF testing beginning on 8 November 2023. There were initially five test aircraft assigned to Edwards Air Force Base, with four more due to be added in 2026. Also in 2026, the first contract for seven aircraft, which had been scheduled for 2025, will be signed. Initial operational capability will be further delayed and has been moved back to November 2027.

206 The Yak-130 by the Russian firm Yakolev is an aircraft born out of a joint venture with Aermacchi but then developed independently, entering service in 2012. (photo: Niccoli Archive)

207 top The Hongdu L-15 is an advanced trainer in service with the air force of the People's Republic of China and derived from the Yak-130 design.

207 center The Turkish aviation industry has also produced an advanced jet trainer with combat capabilities. The TAI Hurjet flew for the first time in 2023.

207 bottom To replace its aging T-38, the US Air Force has chosen the Boeing T-7A Red Hawk, which first flew in 2016 but has yet to enter service.

Chapter 15

General Aviation

The term "general aviation" is relatively recent, being coined as the result of the ever-increasing differences and specialized categories that were becoming established in the world of powered flight. Prior to the Second World War, in many cases there was no clear distinction between aircraft for civil use and those designed for military purposes. At that time sectors such as training, passenger services and transport and even bombing saw the use of aircraft that had never been designed with an exclusive role in mind; civilian aircraft were frequently converted with only minor modifications for military purposes and vice versa. In the post-war era, with the advent of jet engines and increasingly sophisticated technology in aerodynamics and avionics, aircraft could be designed to respond more closely to end-users' specific demands, and the differences between them became increasingly well defined. In the 1920s in America, the Curtiss JN was used by both military flying schools for training and by civilian pilots for recreational purposes; nowadays the differences between the military T-37 or T-45 jets and the Cessna 172s flown by numerous American aero clubs are self-evident.

General aviation thus comprises all those aircraft for private use that do not fall into the military, commercial and airliner categories. In practice they are light training, tourism, or business planes, the highest expression of which are the business or executive jets which the largest corporations and most rushed VIPs use to carry their staff around the world.

Among the companies representing the pillars of this sector are the American firms Piper and Cessna, responsible for introducing millions of people around the globe to the world of aviation.

In this context, the Piper J-3 Cub, which first flew in September 1930 as the Taylor Cub, may be seen as the progenitor of modern general aviation planes. This was a lightweight aircraft covered in canvas with a high, braced wing, a closed, two-seat, tandem cabin, a tricycle undercarriage with a tail-wheel and a 41-hp Continental engine. In 1937, the newly founded Piper company took over development and production. It increased the Cub's engine power output to 66 hp, a modification that increased the plane's popularity. In 1949, Piper decided to introduce an improved, more robust version, the PA-18 Super Cub, the relative weight and performance figures of which were increased. The Super Cub was initially equipped with a 91-hp engine, but a number of variations of the Cub theme were produced; some models equipped with skis and floats and powered by engines of up to 152 hp. The PA-18-150, equipped with a 150-hp Lycoming unit, had a maximum take-off weight of 1,750 kg and a top speed of 130 mph, but

208 The Piper Pa-28 Cherokee is one of the commonest tourist flight planes. Since January 1960, more than 30 versions have been built and the plane is still widely used today.

209 top A Piper Pa-18 Super Cub at low altitude over Maryland. The Pa-18 was the natural evolution of the J-3 and was developed from 1949 on.

209 center left A Piper J-3 Cub, the progenitor of a long line of modern touring aircraft. Designed in 1929, the Cub originally had just a 41-hp engine and weighed less than 1,100 lb on take-off.

compact dimensions: a wingspan of 32 feet 10 inches and a length of 19 feet 7 inches.

Military versions of both the Cub and the Super Cub were constructed for personnel transport and observation purposes (the L-4, L-18 and L-21 models) and were used by the armed forces of numerous countries. Albeit with a number of interruptions, the Super Cub remained in production through to the mid-1990s and together with the Cub, production exceeded 26,000 planes. A heavier tourist airplane, powered by a 150-hp Lycoming engine, completed its maiden flight on 14 January 1960: this was the PA-28 Cherokee, an attractive low-wing monoplane with a fixed undercarriage and a four-seat cabin.

The PA-28 went into production in 1961 and is still available today, having enjoyed an enormous commercial success that led to the construction of over 19,000 planes to differing specifications. Among the more than 30 thirty models that have been produced are the Archer, Arrow, Dakota and Warrior series, variously equipped with carburettor, fuel-injected and turbocharged engines, retractable undercarriage systems, and power outputs of up to 235 hp.

In October 1977, Piper introduced the PA-38 Tomahawk, a new single-engined, two-seat light trainer characterized by a T-tail and a fuel-efficient 114-hp engine. Although carefully designed with advice from some 10,000 flight-instructors, the Tomahawk never enjoyed commercial success; up until 1982, fewer than 2,500 had been built.

Another of the Piper company's major strengths is twin-engined aircraft, the first being the PA-23 Apache, a low-wing monoplane with a retractable undercarriage, a four-seat cabin, and engines faired into the wings. The Apache made its first flight on 2 March 1952 and remained in production until 1965, when the Aztec replaced it. The new model featured a larger cabin, a greater carrying capacity, and two 250-hp engines. Next came the Turbo Aztec, which was powered by two 253-hp turbocharged Lycoming units, had a top speed of 253 mph, and a maximum take-off weight of 5,600 lb. It had a wingspan of 37 feet 5 inches, a length of 31 feet 1 inches, and could carry five passengers and a pilot. Over 2,800 PA-23's were built.

209 bottom left The Pa-38 Tomahawk was produced by Piper in 1977 to equip flying schools with a new light, single-engined plane, but it did not meet with the hoped-for success.

209 bottom right The four-seater Pa-23 Apache was the first twin-engined plane designed and built by Piper. It made its maiden flight in 1952 and remained in production until 1965.

The PA-31 Inca completed its maiden flight on 30 September 1964; three years later this model became the Navajo. This was a larger, heavier airplane that, with its eight-seat capacity and 300-hp engines, was closer to the business aircraft and air taxi sector. Piper developed various versions of the PA-31, offering a pressurized cabin and turbocharged engines, and in 1972 introduced the Navajo Chieftain, which had an elongated fuselage and even more powerful engines. The company also introduced another new plane in 1972, the PA-42 Cheyenne, powered by turboprop engines. The plane was developed in various more advanced versions. The Cheyenne III had a wingspan of 47 feet 6 inches, a length of 43 feet 4 inches, and a maximum take-off weight of 1,200 lb, and was powered by two 730 shp engines guaranteeing a top speed of 342 mph. The Cheyenne IV featured eleven seats, a power output of 1,014 shp per engine, and a top speed of about 398 mph.

Yet another successful Piper model was introduced in 1972, the PA-34 Seneca developed from the single-engined Cherokee Six, This airplane featured typical Piper styling with a low wing, a six- or seven-seat cabin and two 230-hp Lycoming engines. Then came the Seneca II, equipped with turbocharged engines and contrarotating propellers, and then the Seneca III. The latest in this line is the Seneca IV powered by twin 220-hp Teledyne Continental engines that provide a top speed of 224 mph. The model has a wingspan of 36 feet, a length of 26 feet 2 inches and a maximum take-off weight of 4,740 lb. The latest version is the model V, certified in 1996; it is equipped with 220 shp Continental TSIO engines, providing a top speed of 227 mph.
The wingspan is 39 feet; the length is 23 feet; and the maximum take-off weight is 4,750 lb.

210 top Pa-31-350 Navajo Chieftain introduced in 1972 was one of the planes in the Pa-31 family. It had a lengthened fuselage and 335-hp engines.

210 center The Piper Pa-34 Seneca appeared in 1972 as a development of the single-engined Cherokee Six. Shown here is the 1975 Seneca II with turbocharged engines.

210 bottom The Piper Pa-31 Inca, renamed the Navajo in 1967, was the basic model from which 20 versions were developed. It had turbocharged engines, a pressurized cabin and a lengthened fuselage and wingspan.

More than 5,000 units of the Seneca were built, before it went out of production in 2019. Piper continues to produce the twin-engine Seminole models and the single-engine models, the PA-28 Archer, PA-28 Arrow, and PA-46 M-Class.

The other American colossus in this sector, Cessna (founded in 1927), launched its most successful product in 1948 with the introduction of the Model 170, a four-seater with a high, braced wing and a fixed tail-wheel tricycle undercarriage derived from the earlier Model 120. True success came, however, some years later, first with the introduction of the 170 B (powered by a 147-hp engine and equipped with Fowler slotted flaps) and then, on a larger scale, with the Model 172 of 1955 that introduced a nose-wheel tricycle undercarriage. A total of 1,170 planes in this series were sold in 1956 alone. Since then the range has been continuously improved and developed, with numerous different 172, 175 and 182 versions being introduced with turbochargers and a retractable undercarriage (RG). The Cessna 172 had a wingspan of 35 feet 9 inches, a length of 26 feet 3 inches, a maximum take-off weight of 2,500 lb, and a Teledyne Continental engine producing 213 hp, sufficient to provide the plane with a top speed of about 152 mph. A total of 730 of the 172s were built to T-41 military trainer specifications. Around 70 years later, the 170 and its successors are still in production (albeit after a number of interruptions), and over 58,000 have been built.

In parallel with the Model 170 series, Cessna also produced the Model 180 series, in practice a heavier and more powerful version of the 170B that used the same wing. The 180 Skywagon first flew in 1953. Seven years later came the six-seat Model 185, which was powered by a 300-hp engine producing a top speed of about 178 mph. The plane had a maximum take-off weight of 3,350 lb.
A military version was also developed from the 185, equipped with a more powerful engine and a nose-wheel tricycle undercarriage. This was the Cessna 206 Super Skywagon, introduced in 1964 and followed in 1969 by the Model 207, which had a stretched fuselage. This general-purpose plane was particularly well suited to airdrop, air taxi and air-ambulance duties. Over 8,000 Model 207s have been built, with some equipped with turbocharged engines.

Another exceptionally successful aircraft was the Model 150, in which generations of pilots learnt to fly. Equipped with the typical Cessna braced high wing, a fixed nose-wheel undercarriage, the two-seat 150 was introduced in September 1957 as a low-cost light aircraft for flying schools and aero clubs. The Model 150 had a wingspan of 32 feet 10 inches, a length of 22 feet 11 inches, and a maximum take-off weight of 1,600 lb. It was powered by a 100-hp Continental engine and could attain a top speed of about 125 mph. The Model 150 enjoyed great commercial success throughout the world, and was also built under license in France by Reims. Model 152 was introduced in 1977; version, the only significant difference being the adoption of the 112-hp Avco Lycoming engine. A total of more than 27,000 of the 150/152 series were built, and the planes are still very common today.

Cessna currently produces the series of single-engined Skyhawk, Skylane, Turbo Skylane, and Turbo Stationair models, all with the classic high-wing configuration typical of the company. For light passenger and goods transport purposes, the twin-engined Skycourier and the

211 top The Cessna 172, perhaps the most famous touring plane of all time. It was first produced in 1955, and more than 60,000 of the three-wheeler front undercarriage planes were built.

211 bottom The Cessna 206 Super Skywagon hydroplane version. It first flew in 1964, and is highly suitable for general work tasks.

212-213 A twin-engined Cessna E310P flying over the sea. Cessna has produced an enormous series of twin-engined touring and business planes, which has culminated in the 425 turboprop model.

single-engined Caravan and Gran Caravan are in production.

Cessna also built a noteworthy series of successful twin-engined planes. The first, the Model 310, made its maiden flight in January 1953. The 953's low-wing design incorporated supplementary wing-tip fuel tanks, a retractable tricycle undercarriage, a six-seat cabin and two 228-hp Continental engines (replaced by 264-hp units when production began).

In February1961, Cessna introduced an aircraft that was to a certain degree revolutionary: the Model 336 Skymaster, characterized by the "push-pull" layout with one engine in the nose and the other in the tail. The plane featured a four-seat cabin, the traditional Cessna braced high wing, a twin-boom tailplane, and twin-rudder assembly. Not only was the plane original but it also handled very well; nonetheless, owing in part to its outdated fixed undercarriage, it never achieved great success. The 337 Super Skymaster was introduced in 1965, featuring a needed retractable undercarriage, an optional belly luggage compartment, and a cabin that could now accommodate six in comfort. The plane had a wingspan of 36 feet, a length of 29 feet 6 inches, and a maximum take-off weight of about 4,630 lb. Two Continental engines, each producing 213 hp, powered the 337 to a top speed of about 206 mph. The 337 was built under license by Reims. Two military versions were developed (the 0-2A and FTB337) for observation and light ground attack duties. In all, by 1980, a total of 2,678 Super Spymaster 337s had been manufactured.

In July 1962, Cessna introduced the first of a long line of traditional twin-engined planes featuring a low wing, a retractable undercarriage and six-, eight- or ten-seat cabins. These Cessna 411 series aircraft were very similar to the earlier 310, but slightly larger and more powerful. The 411 was followed by the more economical 401 and 402. 1965 saw the introduction of the 421 Golden Eagle, developed from 411, and with a pressurized, air-conditioned cabin that could accommodate up to ten passengers. The 421 was powered by two Continental turbocharged engines developing 370 hp, and providing a top speed of almost 300 mph.

In 1968 the Cessna 414 Chancellor was introduced, incorporating the 401's wing type and the 441's fuselage. In the 1970s the company introduced the 441 Conquest and the 404 Titan; they were outwardly similar and both were equipped with positive dihedral tailplanes. The 401 was powered by turboprop engines and could accommodate up to eleven passengers, while the 404 was powered by turbocharged piston engines. These two models were followed by the 425 Corsair, another turboprop plane that combined the 421's airframe with Pratt & Whitney Canada PT6A engines.

In the late 1960s, Cessna took a significant step in the direction of twin-engined business jet, developing the Model 500, which flew for the first time on 15 September 1969. This aircraft, named the Citation, had a classical layout with a straight wing and the engines paired at the tail, and immediately won a position as a valid competitor in the executive jet sector. Over the years, the Citation family has been expanded with the introduction of ever more sophisticated and powerful models. In 1982, the Citation III became the first to be equipped with a swept wing. The next models were the Citation V (replacing the II), then VI and the VII, and finally in 1993, the Citation X. A VIP jet for intercontinental flights that can accommodate up to eight passengers, the Citation X has two Rolls-Royce AE3007C2 turbofan engines delivering 3070 kg of thrust for a top cruising speed of 604 mph, a range of 3,530 miLES, and maximum take-off weight of 36,500 lb. Its wingspan is 69.2 feet and overall length is 74 feet. Today the Citation family is still in production, with the improved Longitude, Latitude, Ascend, CJ3, CJ4 and M2 models, constructed in thousands of examples.

212 bottom This Cessna 321 Golden Eagle has British markings. It can seat up to 10 people and fly at 300 mph.

213 top A German twin-jet Cessna 525 Citation about to land. This plane first flew in 1969 and was one of the first relatively cheap business planes within the reach of a fairly large market.

213 bottom The Cessna 336/337 was undoubtedly a revolutionary plane, characterized by the in-tandem or "push-pull" twin engines. This is a F337F model with retractable undercarriage.

In 1975 SOCATA introduced a new series of more modern airplanes, of which the TB.10 Tobago led the way, completing its maiden flight on 23 February 1977. This low-wing monoplane had a fixed undercarriage, a spacious four-seat cabin and a Lycoming O-320 engine producing 160 hp. This first model was followed by a variant equipped with a180-hp engine. At this point Aérospatiale decided to call the first type the TB.9 Tampico and to assign the TB-10 designation to the more powerful version that could carry up to five people. A third version appeared in 1980, characterized by an even more powerful engine (250 hp) and a retractable undercarriage; it was designated the TB.20 Trinidad. This last model could attain a top speed of 192 mph; it had a wingspan of 29 feet 6 inches, and a length of 22 feet 11 inches, and had a maximum take-off weight of about 3,085 lb. Still later, the XLTB.200 Tobago, powered by a 200-hp engine, and the TB.21 Trinidad equipped with a turbocharged engine, were introduced.

The Jodel/Robin family of light aircraft was another successful series. The first model to make its mark was the Jodel DR.100, which went into production in 1958. This airplane, designed by Jean Delemontez and Pierre Robin, was a three-seat with a fixed two-wheeled undercarriage, a 96-hp Continental engine and a low wing with upturned outer sections, a feature that was to become a characteristic of the series. In 1961 the aircraft was renamed as the DR.1050 Ambassadeur, and various versions were developed with engines of up to 106 hp. Two other firms were engaged in building the planes; Avions Pierre Robin and the Société Aéronautique Normande (SAN). This last built the D.140 Mousquetaire model with a 180-hp engine and up to five seats in the cabin. SAN went out of business in 1969, with Robin taking over its activities. In1972, Robin introduced its most famous aircraft: the DR.400, characterized by its forward-sliding canopy. The DR.400's first model was the DR.400/125 Petit Prince, powered by a 125-hp engine and designed to carry two adults and two children. The Petit Prince was followed in close succession by other models with power outputs of up to 180 hp, four-seat cabins and glider-towing capabilities. The Robins are also popular among with the flying schools and aero clubs, and more than 1,400 have been built. Seven versions of the DR.400 were manufactured with engines producing between 110 and 200 hp, while the DR.401 model still in production.

France also saw the establishment of a major national aircraft manufacturer, namely the Mudry firm, founded in 1958. Its CAP 10, introduced in 1968, was the first of a successful series of aerobatic aircraft. The CAP 10 was a low-wing monoplane with side-by-side seating, a 180-hp engine, and a fixed undercarriage. It had a wingspan of 26 feet 2 inches, a length of 22 feet 11 inches, and a maximum take-off weight of 1,830 lb. It was capable of a top speed of about 168 mph. In 1969 the CAP 20, a competition single-seater, was introduced. The series continued with the CAP 21 (featuring a new wing) and then the CAP 230, 231 and 232.

In the 1960s, Dassault pioneered the way in the business plane sector, introducing increasingly successful models. Today the company remains a leading manufacturer in this sector. The Mystère 20, a particularly elegant twin-engined jet with swept wings and tailplanes, completed its maiden flight on 4 May 1963. Subsequently named the Falcon 20, the plane enjoyed considerable success, and a number of different versions were developed. The Falcon 20 was followed by the smaller Falcon 10 (later replaced by the 200 and 100 models, respectively), while in 1976,the Falcon 50m, a long-range tri-jet with a new wing and tailplanes and a stretched fuselage accommodating up to twelve passengers. In 1984, Dassault introduced the Falcon 900, a tri-jet with an intercontinental range and able to accommodate up to nineteen passengers. In the mid-1990s, the firm rejuvenated its product range rejuvenated with the introduction of the Falcon 2000 (twin engines) and the 50EX and 900EX models. The most advanced and capable model today is the Falcon 7X, released in 2005. This triple-jet eight-seater VIP is powered by Pratt & Whitney Canada PW307A turbofans delivering 6,400 lb of thrust, for a top cruising speed of Mach 0.80, a range of 6,850 miles, and a maximum take-off weight of 69,000 lb.

Today, Dassault's production is based on the Falcon 6X, Falcon 10X, and Falcon 2000LXS, along with the three-engined Falcon 8X, an aircraft that can carry 12 to 16 passengers for distances of up to 7,422 miles, with a ceiling of 51,000 feet and a maximum speed of Mach 0.9. It is powered by three Pratt ‹ Whitney Canada PW307 D turbofans providing around 6,600 lb of thrust each.

Falcons have also been produced for military purchasers, and more than 1,200 have been built.

216 top The latest executive jet built by Dassault is the Falcon 7X, released in 2005. The triple-jet eight-seater has intercontinental capabilities.

216 bottom The fixed wheel DR.400, the most famous plane built by the French manufacturer Avions Pierre Robin. It is shown here with Italian registration.

216-217 Two two-seater CAP.10Bs during an air acrobatics display. This 1968 plane is still one of the best acrobatics trainers.

217 center Two Robin DR.500/200I Presidents shown over the French coast. These aircraft derived from predecessor designs, adopting the typical Jodel/Robin rising wing.

217 bottom The TB.9 Tampico, the TB.10 Tobago and the TB.20 Trinidad, a family of tourism planes, were produced by SOCATA during the 1970s. All were developed around the same cell.

In Italy, SIAI Marchetti was the most famous firm in aircraft manufacturing. Along with the SF.260, Marchetti also produced the renowned S.205/208 series. The first to fly was the S.205 on 4 May 1965. This monoplane, which had a roomy four-seat cabin, could be ordered with various engines (180, 200 or 220 hp), and with a fixed or retractable undercarriage. In 1967 the five-seat S.208, which was equipped with wing-tip tanks and used a more powerful O-540 engine producing 260 hp (the same unit fitted to the SF.260), made its maiden flight. Over 520 planes of the two models were built, and many are still flying in Italy and abroad.

Rinaldo Piaggio, today Piaggio Aeropace, has also built twin-engine aircraft for general aviation and business flights. The first was the P.166, equipped with piston engines and featuring upside-down gull wings, with its first flight in 1957. It was later developed in the DL3 turboprop version, sold primarily to the military sector. Additionally, the P.180 Avanti took its maiden flight on 23 September 1986. The innovative turboprop for the executive aircraft market presented a sophisticated aerodynamic formula with canard wing configuration fixed forward and "T" tail planes, coupled with two turboprop engines equipped with five-bladed propellers in thrust position. Capable of up to nine passengers, the Avanti boasts performance rivaling those of executive jets, at a fraction of the cost. The Avanti II version was released in 2004, with improved engines and digital avionics and has recently been updated as the P.1800 Evo+ model.

The Czech firm Zlin has also been an important light aircraft manufacturer. The first plane in its most successful series was the Zlin 42, which flew for the first time in October 1967. This small two-seater, powered by a 137-hp Avia 137 engine, featured a side-by-side cockpit, a fixed undercarriage and a characteristic wing with a slight negative sweep, ad had been designed to meet the demands of Eastern European flying schools and aero clubs. The Zlin 42 formed the basis of the Zlin 43, a four-seat version, and then the Zlin 142, a two-seater with a 213-hp engine. This latter had a wingspan of 29 feet 6 inches, a length of 22 feet 11 inches, and a maximum take-off weight of about 2,400 lb. It could attain a top speed of about 143 mph. Zlins were produced in the Czech Republic by Morovan through to 2009 when the company revived the historic Zlin name. The top of the range being the Zlin 143L with a 235-hp engine, while the most advanced trainer is the Zlin 242L, powered by a 200-hp engine.

Morovan has also produced a very popular series of aerobatic planes that began with the Zlin 526 (in both single- and two-seat versions) and continued with the Zlin 50L, introduced in 1975. In 1978, the 50L won the world aerobatics championship. The Zlin 50LS powered by a 300-hp engine and with a maximum take-off weight of about 1,345 lb is no longer in production. Other aerobatic planes built in the Eastern Bloc countries include the Soviet Yak-50 and 55 and the Sukhoi Su-26/29 series, all winners of various world championships.

In the business jet field, a number of the major players are still located in the United States. Learjet (founded by William Lear) was the first to produce a modern executive jet capable of offering both speed and comfort. Model 23, the progenitor of the Learjet line) first flew on 7 October 1963, and was an immediate success. It was a small twin-engined jet with attractive, fighter-like slim lines, and it that handled like a fighter. The low, straight wing was derived from a design for the Swiss P16 fighter-bomber (never produced in series), while the engines were carried at the rear of the fuselage, along with a T-tail. The Model 23 formed the basis for the stretched models 24 and 25; the 35 and 36 had an even longer fuselage and greater wingspan.

In 1990, Learjet was acquired by the Bombardier group. The latest models include the Learjet 31A/ER, 45/XR, 55, and 60 (with production terminated in 2007). The Learjet 85 made its maiden flight on 9 April 2014, but although it attracted no less than 60 orders, the program was canceled by Bombardier in 2015 after just two examples had been built. The last of the Learjets was the model 70/75, which flew for the first time in 2012 and was built in 169 examples through to 2022. Also from the Bombardier group are the Challenger executive jets, created in 1978 as the Canadair CL600, later produced in versions 300 (a twin-jet taking its maiden flight in 2001); 605 (built in 2006 as the digital-avionics version of the 604); and 850, a 14-seater jet derived from the CRJ200 passenger transport. The Global Express is a large twin-jet seating 8 to 19 passengers; used in both civilian and military sectors since 1996, over 400 units of the aircraft have been built. Today Bombardier offers six models, the Global 5500, 6500, 7500, and 8000, along with the Challenger 650 and 3500. The twin-engined Global 8000 is the top of the range and the world's fastest passenger jet with its maximum speed of Mach 0.94 (621 mph).

218 top left The Pilatus PC-6 is a robust and austere utility aircraft used all around the world for carrying and launching parachutists.

218 bottom The three-engined Falcon 8X is the most prestigious of the executive jets offered by French firm Dassault Aviation.

It can carry up to 19 passengers for distances of up to 9,200 miles.

In 1965, Grumman began work on a high-end twin-engined executive jet, giving rise to the Gulfstream II, which was derived from the original turboprop model. The Gulfstream II was a large aircraft with transatlantic capabilities and could accommodate up to 19 passengers; the low, swept-wing and T-tail architecture was similar to that of the Learjets. A total of 258 Gulfstream IIs were built. In 1978 Grumman sold the rights to the plane to Gulfstream American, which a year later introduced the Gulfstream III. This plane featured a stretched fuselage and a new supercritical wing profile. In 1985 the Gulfstream IV, which had more economical engines, was launched. Further improvements to the airframe and avionics helped make the Gulfstream IV into a more versatile plane, suited to military and governmental roles as an air-ambulance and a search and rescue plane, as well as for electronic warfare, surveillance and transportation duties. The Gulfstream V, introduced in 1995, is an even larger aircraft with an intercontinental range (over 7,500 miles) and a cabin that can accommodate up to nineteen passengers. It has a wingspan of 91 feet 10 inches, a length of just over 95 feet, and a maximum take-off weight in the order of 45 tons. The two Rolls-Royce BR710 turbofans each produce about 14,750 lb of thrust, and provide the plane with a top speed of approximately 580 mph and a range of over 7,450 miles..

Today, Gulfstream manufactures the G280, the smallest model in the range, the G400, G500, G600, G700 and the G800, which is the business jet with the longest range, as it can fly 9,436 miles at a cruising speed of Mach 0.82.

The original G650 was developed into the G700 model, which boasts a cabin that is almost 10 feet longer, and the G800, which uses the same wing and engines as the G700 but with the cabin of the G650. The G800, which flew for the first time on 28 June 2022, has a maximum speed of Mach 0.93 and can carry up to 19 passengers. It can accommodate up to 10 in nighttime configuration, with sleeping berths. The model is powered by two Rolls-Royce Pearl 700 turbofan engines, each delivering approximately 17,600 lb of thrust, with a maximum takeoff weight of over 104,720 lb. It is priced at around £60 million. To date, more than 660 aircraft in the G650/700/800 family have been built. Several sophisticated variants for military use have also been developed from the Gulfstream G.V and 650 models, designed for surveillance, command and control, and electronic warfare missions.

219 top The Piaggio Aerospace P.180 is an innovative turboprop design with surprising performance that has enjoyed considerable sales success on the civil and military/governmental markets.

219 bottom The Gulfstream G800 is the most recent design from the American company and first flew in 2022. It can carry up to nine passengers on intercontinental routes.

Chapter 16

Military Transports

In the field of military aviation, the importance of transportation aircraft, both tactical and strategic, has grown increasingly over the years. Today they constitute a branch of great, and sometimes crucial, importance in the economy of air operations.

The aircraft that can be considered the archetype of military air transport was the Douglas C-47 Skytrain, a military version of the DC-3 passenger transportation aircraft, described in Chapter 5. During World War II, the C-47 was the "mule" that allowed the Allies to plan and execute their daring assault campaigns, first "Operation Overlord," the Normandy landings in 1944, but also numerous other operations of aerial or amphibious landings, as well as airlift bridges in all the theaters. The C-47, which had its maiden flight on December 23, 1941, was considered one of the 10 most important instruments of war allowing the Allied Forces to win the war against the Axis Powers. In just the period of the Second World War, 10,692 units of the C-47 were constructed. These were primarily used by American, British, and Commonwealth forces (taking the name Dakota for the latter two services), whereas around 2,000 units were built under license in the Soviet Union, under the name Lisunov Li-2. After the war, the aircraft was also developed in the version C-53 for the USAF and the version C-117D for the U.S. Navy. It remained in service for many years, also adapted to other types of missions, such as photographic and electronic reconnaissance (RC-47 and EC-47), search and rescue (SC-47), and VIP transport (VC-47). Perhaps the most famous version was the AC-47 "Spooky," created in 1964 for the purpose of having available a "flying gunship," capable of hitting ground targets with a high rate of fire, owing to its circular flight orbit. The AC-47, which was deployed during the Vietnam War, was armed with three 7.62-mm GAU-2 miniguns with six rotating barrels arranged at the side windows, capable of firing 2,000 rounds per minute. The Skytrain transport could normally carry 28 soldiers or a payload of 5,953 lb at a cruising speed of 160 mph. The maximum flight altitude was 26,247 feet with a range of 1,600 miles. The C-47 had a wingspan of 96.5 feet, a length of 64 feet, and a height of 17 feet. Its maximum takeoff weight was 26 lb, and it was powered by two 14-cylinder Pratt & Whitney R-1830 Twin Wasp radial engines providing 1200 hp each, equipped with three-bladed propellers. The C-47 was adopted by the air forces of about 100 countries throughout the world, and there are a few units, refitted with turboprop engines, still flying to this day.

The successor to the C-47 was an aircraft that became famous as the Flying Boxcar. This

220 A U.S. Navy C-117D in service at the Kenitra Air Base in Morocco in 1975. This aircraft was a derivative of the C-47 Skytrain used for transport during World War II.

221 top In 1995 the Taiwanese Air Force became the last in the world to discontinue use of the Fairchild C-119, known famously as the Flying Boxcar.

was the Fairchild C-119, an improved version of the C-82, developed during the war. The C-119 had a characteristic aerodynamic formula with a double tail boom, leaving free the rear of the fuselage for loading and unloading of materials. Between 1949 and 1955, 1,112 units of the Flying Boxcar were built and developed in various versions; they were exported to many countries and flown until 1995 by the Taiwanese Air Force. For strategic transport, however, the American Air Force adopted the Douglas C-124 Globemaster, which was the last great American cargo aircraft with piston engines. Its maiden flight was on November 27, 1949, and later a total of 450 units of the craft were constructed. Powered by four Pratt & Whitney R-4360 engines, each delivering 3,853 hp, the C-24 could transport up to 200 soldiers or 34 tons (31 metric tons) of cargo; its maximum range was 4,027 miles and its maximum speed was 271 mph.

An aircraft with a revolutionary design, and one that dictated the guidelines for all military transports that followed it, was the Fairchild C-123 Provider. It completed its maiden flight on October 14, 1949, and entered into service with the U.S. Air Force in 1955. Its innovative lines were the fruit of careful studies about military transport needs. The production model C-123B, which derived from the design of the Chase XCG-20, a World War II assault glider, adopted the low carriage with a raised tail, with the wing attachment at the high point of the fuselage, so as to allow installation of a wide tail ramp that could be lowered, on ground or in flight, for easier loading and unloading of the materials and men being transported or air launched. The Provider was powered by two Pratt & Whitney R-2800 Double Wasp engines rated at 2,332 hp each (these were later raised to 2500), and could carry up to 60 fully-equipped soldiers, or a truck towing a 155 mm cannon, or combined loads. The maximum takeoff weight was 60,000 lb. In order to improve performance, the aircraft prototype was, already in 1955, equipped with two small J44 jet engines, each providing a thrust of 1000 lb. In view of the positive results, the production model, called the C-123K, received two J85 engines providing 2,850 lb of thrust, which raised the payloads to 24,250 lb, and the range (at full load) to 1,035 miles. Over 300 units of the C-123 were produced, and these were used by a dozen countries.

221 bottom The Fairchild C-123 Provider was the first military aircraft to adopt the now typical configuration for modern transport airplanes: high tail, fuselage low to the ground, ramp for rear loading.

During these same years, the experience gained and the technological advances led to the design of a transport that made history, and which even today is a major protagonist in the history of flight, the Lockheed C-130 Hercules. The origins of the C-130 date to shortly after those of the C-123, and derived from a USAF General Operating Requirement (GOR) issued on February 2, 1951, calling for the industry to propose a new tactical transport aircraft capable of accommodating 92 persons or 72 fully equipped soldiers; additionally it was to be propelled by the new turboprop engines, which represented the future in transport aviation. The general configuration Lockheed selected for the aircraft made the new machine appear like an enlarged C-123, but the key solution was the adopting of the Allison T56 engines that were especially developed for the new project, which was initially called L-206 and later Model 82. In the end, it was Lockheed that prevailed over some dozen other designs, and the development contract was signed on July 2, 1951. The YC-130 prototype took its maiden flight on August 23, 1954, from the airport at Burbank, California, Lockheed's design studio headquarters. The aircraft went into series production as the C-130A, and it was characterized by three-blade propellers, even though the general configuration of the aircraft was already the definitive one. The wingspan was 133 feet, and the four engines provided 3,750 shp each. Subsequently, a new nose was adopted, containing the navigation radar, and continuing improvements were made to the airframe, the systems, and the engines. The C-130E, which entered into service in 1962, was the first version to be mass-produced. It was further enhanced with the Model H, which remained in production from 1964 to 1996. Powered by T56-A-15LFE engines rated at 4,508 shp, it had a maximum takeoff weight of 175,000 lb, a maximum speed of 404 mph, a ceiling of nearly 33,000 feet, and a maximum range (with 20 tons/18 metric tons of load) of 2,237 miles. The Hercules was an excellent platform, and flexible, also excellent for the fact that dozens of specialized versions were produced, such as the gunships (AC-130), those for electronic warfare (EC-130), those allowing in-flight refueling or for search and rescue (HC-130 and KC-130), for operations in the Arctic and the Antarctic (LC-130), for special operations (MC-130), for reconnaissance (RC-130), and for meteorological reconnaissance (WC-130). In the 1990s, Lockheed Martin decided to revive the project, grafting onto the airframe everything the latest technologies could offer. Thus appeared the C-130J Hercules II, which completed its maiden flight on April 5, 1996. The aircraft had new engines, the Rolls-Royce AE-2100-D3, rated at 4,637 shp with six-bladed propellers in composite materials, digital avionics, with Heads-Up Display (HUD) for the two pilots, and new systems and equipment. Overall, the C-130J assured a 40% increase in range, a 21% increase in speed, and a 41% shorter takeoff distance. The launch customer for the new airplane was the Royal Air Force, which began receiving the first of its 25 units in 1999. And just like its predecessor, the Hercules II was also produced in numerous specialized versions. With the new model, the number of C-130s built exceeded the threshold of 2,600 units. They are operated by about 70 countries throughout the world and production continues.

Because of the needs of the Cold War, and the contingents deployed in Europe and in Asia, in the 1960s the United States needed a transport capable of large loads, but also capable of high speed and long range, in order to redeploy troops and materials in short times, even when at great distance from home. Thus was identified the need for a strategic transport powered by jet engines, and the competition organized by the USAF was won in 1961 once again by a Lockheed project, the L-300, which gave rise to the C-141 Starlifter. An aircraft of imposing dimensions (160 feet wingspan, 145 feet length, 39 feet height), which maintained the design features of military transport aircraft, but introduced the "swept" wing, the T-empennage for the tail, and, most importantly, four Pratt & Whitney TF33 jet engines with 21,000 lb thrust each, allowing a cruising speed of 564 mph, a maximum takeoff weight of 316,584 lb, and a range of 6,040 miles (carrying a load of 15.4 tons/14 metric tons). The maximum loading capacity came to 70,848 lb, and the hold could accommodate up to 154 equipped soldiers or 123 paratroopers. Developed in a short span of time, the C-141A took its maiden flight on December 17, 1963, entering into service already in 1964 and starting almost immediately to operate on the routes to Vietnam, where the U.S. commitment was becoming increasingly more massive. In 1976, approval was given for introduction of the model C-141B, which featured a fuselage that was 23 feet longer. The new model, obtained through conversion of all 271 of the C-141As then in service, entered the lines in 1979, offering a maximum takeoff weight increased to 345,211 lb. In practice, it was as though the USAF had acquired 90 new Model A units. After a long and honored career, the last C-141 was retired from service in May 2006. Although impressive, the Starlifter's abilities were still not enough for the Air Force. In fact they wanted an aircraft capable of transporting even larger loads, and particularly larger in size, like the M60 tank, or the Armored Vehicle-Launched Bridge (AVLB), weighing fully 74 tons (67 metric tons).

After some reconsiderations, in 1964, the initial requirements came out for the CX-Heavy Logistics System (CX-HLS), requiring the design of a four-jet aircraft with a payload of over 89 tons and a speed of Mach 0.75 (500 mph). The cargo bay needed to measure at least 100 feet in length, 17 feet in breadth, and 13 feet in height, and it needed access ramps both front and back. Five companies in the sector responded in the design competition for the aircraft (and three for the engines). In September 1965, the L-500 design presented by Lockheed was declared the winner, prevailing over the Boeing design, primarily for cost reasons.

222 top The Lockheed Martin C-130J Hercules II took its maiden flight in 1996. A derivative of the C-130H, its exterior is distinguished by six-blade propellers of composite material.

222 bottom A C-130H Hercules as U.S. Air Force paratroopers jump from the side doors. Several specialized versions of this aircraft were derived for dedicated applications.

223 top A Lockheed C-141A Starlifter in flight over San Francisco Bay. Two hundred seventy-one units of the aircraft were built.

223 bottom left Shown in flight over Alaska during the 1980s, a C-141B and a C-130H. For over 30 years these two aircraft served as the backbone of USAF transport.

223 bottom right A C-141 in flight at sunset, with streaks of condensation developing from the engines. This aircraft was retired from service in 2006.

At any rate, Boeing derived the B747 Jumbo Jet from its design, which revealed itself to be a huge commercial success. The aerodynamic formula of the Lockheed C-5A Galaxy appears very similar to that for the C-141, except that its fuselage has a larger section and its nose opens upward. At the same time, the General Electric TF39 engine, a revolutionary unit introducing the high-dilution bypass ratio (providing increased power for reduced fuel consumption) was declared winner in terms of propulsion. The prototype took its maiden flight on June 30, 1968, and the aircraft entered into service with the USAF in June 1970. Initially, there were problems in achieving the design specifications, particularly regarding the strength of the wing, and in containing costs; nonetheless 81 units of the C-5A were produced. The aircraft had a wingspan of 223 feet, a length of 248 feet, and a height of 65 feet. Maximum takeoff weight was up to 768,994 lb, with a payload of 220,965 lb. The four TF39-GE-1 turbofan jet engines had a power of 41,099 lb each, assuring a maximum cruise velocity of 553 mph and a maximum range (with a load of 56.2 tons/51 metric tons) of 6,525 miles. In terms of load, the Galaxy could carry up to 75 soldiers on the upper deck and as many as 270 below. In the cargo version, it could carry loads such as two M60 tanks, or three CH-47 Chinook helicopters, or six AH-64 Apache combat helicopters. A load consisting entirely of pallets could arrive to 36 units of the standardized 463Ls.

In 1983, a new, more robust wing was introduced, and from 1985, 50 more units

of the model C-5B were produced, which had more powerful engines, and maximum payload increased to 260,145 lb. Beginning in 1998, an Avionics Modernization Program (AMP) was launched, and then the Reliability Enhancement and Re-Engining Program (RERP), which gave rise to the C-5M Super Galaxy model. A total of 52 units were updated to the new standard, which included the General Electric F138-GE-100 engines powered by 50,045 lb of thrust. The M version went into production in 2009.

The last in order of time of the large military transport aircraft manufactured by the United States, however, was not produced by Lockheed, but rather by McDonnell Douglas. The C-17 Globemaster III (today produced by Boeing) was produced in 1979 as the YC-15, with the purpose of responding to the USAF's C-X program requirements for a new strategic transport capable of substituting the C-141 for a while, and part of the C-130s. The project was declared the winner, and the contract was signed in August 1981. The supply contract was finalized in December 1985, and called for acquisition of 210 units of the aircraft, which would later be reduced to 120 in 1990, owing to economic problems and changing strategic requirements. The C-17A took its maiden flight on September 15, 1991, and entered into operational service in 1993, with the 437th Airlift Wing in Charleston. The Globemaster III was the world's first military transport equipped with completely electric flight commands (Fly-by-Wire), a cockpit with multifunction digital displays, and Heads-Up Display; and a minimum crew consists of only three people. The aircraft's payload is twice that of the C-141; it can operate from short and semi-prepared runways, as well as in proximity of combat zones; and it has a payload of over 159,835 lb, which can include, alternatively, up to 154 equipped soldiers, 102 paratroopers, three Apache helicopters, three Bradley armored vehicles, or 18 standard model 463L pallets.

The external configuration is that of the now classic American military transport, characterized by a highly efficient swept wing, with winglets at the ends, for improved fuel economy. The aircraft is powered by four Pratt & Whitney F117-PW-100 engines with 41,667 lb of thrust each, providing the aircraft with a maximum speed of 597 mph, a practical ceiling of 45,112 feet, and a maximum range of 2762 miles, with a payload of over 78 tons. To date, the USAF has ordered 223 units of the C-17A, but by contrast with its jet-powered predecessors, it has also been sold in small quantities to other countries: Great Britain, Australia, Canada, NATO, Qatar, the United Arab Emirates, and India, together, have ordered 27 units.

224 top A Lockheed C-5M during takeoff. This version of the Galaxy, outfitted with new engines, entered into USAF service in 2009.

224 bottom left The Boeing C-17 Globemaster III transport runs on jet engines, but it can also operate on semi-prepared runway surfaces, as shown here.

224 bottom right The C-5 Galaxy's cargo compartment is truly impressive, accommodating even bulky transports, as displayed with the unloading of this huge front-opening American truck.

225 top Two C-17s assigned to the 62nd Airlift Wing at Joint Base Lewis-McChord. The USAF now has 223 units of the aircraft.

225 bottom A C-17 shown during in-flight refueling operations. On just a single tank of fuel, this aircraft can carry a cargo of 78 tons (71 metric tons) for a range of 2,762 miles.

Obviously the Soviet Union did not neglect development of military air transport, even if for several years after the end of the Second World War, no significant new designs were produced for any aircraft. The first major project realized was that for the Antonov An-12, a four-engine high-wing aircraft with raised tail, for allowing a rear loading ramp. The prototype made its maiden flight on December 16, 1957, and the general configuration of the aircraft appeared to have been strongly influenced by the American C-130 Hercules. The An-12 (the NATO called it "Cub") had a wingspan of 125 feet, a length of 109 feet, and a height of 34 feet. Equipped with four Progress AI-20L turboprop engines providing a thrust of 4,000 hp each, it had a maximum takeoff weight of 134,482 lb and a payload of 44,092 lb. Maximum cruising speed was 416 mph, and the range at maximum load was about 2,237 miles. More than 1,320 units of the An-12 were constructed, also in the civil version, and these were used to equip the air forces of the USSR, all their satellite countries, and numerous other nations. Over the years, the aircraft was produced in dozens of different versions and subversions, and its engines and performance were improved. A copy was produced in China under the denomination Shaanxi Y-8. Recently the An-12s were almost completely retired from service, both in the civil and military sectors. Owing to the experience acquired with the An-12, the Antonov technical office was able to realize a veritable giant of the skies in the early 1960s, at that time, the heaviest aircraft ever constructed. Work started in 1962, and was intended to develop an aircraft capable of transporting large loads over long distances as well as of operating from semi-prepared runways. The prototype's maiden flight was recorded on February 27, 1965, and a few months later, the aircraft was presented to the world at the Paris Air Show, arousing great interest. Denominated the An-22 Antei (and code-named "Cock" by NATO), this aircraft had a classic configuration, except that it adopted a tailboom fitted with tailplane having twin fins. Equipped with four Samara/Kutnetsov NK-12MA turboprop engines providing 14,800 shp each, powering four pairs of four-blade counter-rotating propellers, the An-22 had a wingspan of 211 feet, a length of 188 feet, a maximum takeoff weight of 551,156 lb, a payload of 176,370 lb, a maximum speed of 460 mph, and a range of 3,107 miles at maximum load. Up until 1975, 68 units were produced, and these were used by the Soviet Air Force and by the company Aeroflot, followed by Russia, Bulgaria, and Ukraine.

Considerably smaller was the Antonov An-26 tactical transport (called "Curl" by NATO), which took its maiden flight on May 21, 1969. The aircraft was derived from the model An-24, a twin-engine civil transport of which more than 1,000 units were built, but it introduced new and more powerful engines (the Ivchenko Progress AI-24VT turboprop, providing 2,820 hp of thrust), and a modified fuselage so as to allow for a back-loading ramp. More than 1,500 units of the An-26 were produced between 1969 and 1986; and it has been used by the air forces of more than 50 countries, as well as numerous civilian operators. With a wingspan of 96 feet and a length of 78 feet, the An-26, produced also in China as the Y-7, has a maximum takeoff

weight of about 52,911 lb, a maximum speed of 336 mph, and a range of 1,678 miles.

Also derived from this project were the Antonov An-30 and An-32. The first successful jet transport realized by the Soviet Union, however, was the fruit of the Ilyuschin technical office, which in 1967 responded to a Soviet Air Force requirement for a new air transport destined to substitute An-12s, which could transport at least 44 tons (40 metric tons) of load to distances of 3,107 miles in less than six hours, and could operate from semi-prepared surfaces, even in extreme weather conditions. The prototype took its maiden flight on March 25, 1971, and its general lines were very reminiscent of the Lockheed C-141. The aircraft entered into service with the Soviet Air Force in 1974 as the Il-76 (its NATO code was "Candid"), and two years later with the national airline Aeroflot. The aircraft proved to be exactly appropriate, to such an extent, in fact, that it was developed in numerous versions. Versions produced over the years included the Il-78, with in-flight refueling; the Il-82, the flying Command Post; the A-50 airborne warning and control system; the Il-76PP for electronic countermeasure; the Il-76MD PS for search and rescue; as well as some three dozen subversions for both military and civilian use. In 1995, the elongated prototype version Il-76MF began flying, though only a few units of it were produced. The most advanced version in service is the Il-76TD-90, equipped with Aviadvigatel PS-90-76 turbofan engines providing 38,405 lb of thrust, allowing a maximum takeoff weight of 429,901 lb, a payload of 55 tons (50 metric tons), a maximum speed of 559 mph, and a maximum range of 2,672 miles at full load. The military version is designated Il-76MD-90. The Il-76, with over 950 units produced and in use worldwide, has a wingspan of 166 feet, a length of 153 feet, and a height of 49 feet.

In 1982 the Soviet industry succeeded, once again, in amazing the world by realizing what was at that time the largest airplane ever produced, the Antonov An-124. This machine was born out of the necessity, recognized in the 1970s, of providing the Soviet Armed Forces with a large transport aircraft, on the scale of the American C-5, because the Soviet abilities of strategic heavy transport were limited to some 50 outdated An-22 turboprops. Construction of the first airframe began in 1979, and the maiden flight of the first prototype was recorded on December 26, 1982. The An-124 Ruslan (with NATO code name "Condor") highly resembled the Galaxy, and this also had a nose that opened upward, to allow loading and unloading operations at the same time. But unlike the American aircraft, this one was equipped with a traditional-configuration tail, with horizontal tailplanes in the low position. The aircraft had a wingspan of 240 feet, a length of 226 feet, and a height of 68 feet. Its maximum takeoff weight can reach 892,872 lb and its payload is 330,693 lb. And this also explains why the Ruslan is employed by various civilian companies, for transportation of the most disparate vehicles, including locomotives, yachts, submarines, and even fuselages of other airplanes. The aircraft is powered by four Ivchenko Progress D-18T Turbofan engines, each producing a thrust of 51,645 lb, allowing a maximum speed of 537 mph, and a range of 4,411 miles carrying 99 tons (90 metric tons) of load. About 55 units of the An-124 were produced, but its production has not been terminated. From the Ruslan derived the An-225 Mriya, which even today is the world's largest airplane. It was born to satisfy the need to carry the Soviet space shuttle Buran, and components of its launch system as far as the Baikonur space base. The An-225 recorded its maiden flight on December 21, 1988, and only one unit of it was produced. Equipped with six D-18T Turbofan engines, and with a double-fin tailplane, the Mriya had a payload of 275 tons (250 metric tons), a range of 2,796 miles fully loaded, a wingspan of 290 feet, a length of 276 feet, and a height of 60 feet.

226 top The Antonov An-12, shown here bearing the Ukrainian Air Force colors, was the first major Soviet transport aircraft designed after the end of World War II.

226 bottom An Antonov An-26 of the Slovak Republic Air Force shown while taxiing. This twin-engine transport took its maiden flight in 1969 and remained in production until 1986.

227 top The record for world's largest aircraft goes to the Antonov An-224, produced in the 1980s to carry the Buran space shuttle. Only one unit was ever produced.

227 bottom The Antonov An-124 is the world's largest aircraft built as a series, numbering about 55 units. Here we see an aircraft of the Ukrainian International Cargo company while unloading an armored vehicle.

The production of important military transport aircraft was not, however, the exclusive right of the two superpowers. In Europe after the war, it was primarily Great Britain that worked on this front, realizing the Handley Page Hastings in 1947, the Blackburn Beverly in 1953, and the Short Belfast in 1964; whereas in France in 1950, the Nord 2501 Noratlas appeared, an aircraft quite similar to the C-119, and more than 400 units of that were produced for the air forces of France and Germany. For the specific purpose of replacing the Noratlas, the governments and industries of France and Germany (Aérospatiale, MBB, and VFW) gave birth in 1959 to the Transporter Allianz consortium, which created the design for the Transall C-160. The first of the three prototypes took its maiden flight on February 25, 1963, and mass production began in 1967, for a total of 169 units, for France (C-160F), Germany (C-160D), Turkey (C-160T), and South Africa (C-160Z). The Transall has a maximum takeoff weight of 112,435 lb, a maximum speed of 319 mph, and the maximum range of 1,151 miles, with a maximum payload of 35,270 lb. In 1977, production was revived, to construct 29 units of the C-160NG for France, characterized by new avionics and a probe for in-flight refueling. In the 1990s, the fleets were subjected to measures intended to prolong their operational life, destined to compensate for the delays in the European A400M program for the strategic air transport system of the future.

The Italian industry, as well, tried its hand at creating a tactical transport to replace the American C-119s in the Italian Air Force. In 1963, an agreement was made between the aviation department of Fiat and the Italian Armed Forces, and the first prototype recorded its maiden flight on July 18, 1970. The aircraft, designated G.222, appeared substantially as a small version of the C-130, equipped with two General Electric T64-GE-P4D turboprop engines providing 3,447 shp each. The aircraft had a maximum takeoff weight of 61,729 lb, with a maximum payload of 19,842 lb (e.g., up to 53 equipped soldiers or 42 paratroopers). With a wingspan of 94 feet and a length of 74 feet, the G.222 had a maximum speed of 336 mph, and a range of 782 miles fully loaded. It proved to be a valid aircraft, particularly for operations on small, semi-prepared fields, or for transport of small loads. In all, Italy acquired 52 units of the aircraft, deploying it also for electronic warfare, air signal surveillance and control, and fire control. Fiat's successor Aeritalia exported another 58 units for the air forces of Dubai, Libya, Nigeria, Thailand, Somalia, Venezuela, Argentina, and the United States, where it was deployed for special operations as the C-27A Spartan. In 1996, initially in collaboration with Lockheed Martin, Aeritalia's successor Alenia Aeronautica prepared for development of an evolved version of the G.222, characterized by its adoption of sophisticated avionics (including Heads-Up Display), new Rolls-Royce AE-2100-D3 engines (the same as in the C-130J) with six-bladed propellers, an integrated self-defense system, and a system for in-flight refueling. The resulting aircraft was designated the C-27J Spartan, taking its maiden flight in September 1999. Over 110 units of this model were ordered by the air forces of Italy, the United States, Australia, Greece, and fifteen other countries; currently in development is a version intended for special operations, designated the MC-27J Pretorian. The C-27J boasts a maximum takeoff weight of 70,107

lb, a payload of 25,353 lb (equivalent to 60 troops or 46 paratroopers), a maximum speed of 374 mph, and operational flight altitudes superior to those of the G.222. Although more expensive than the competition, the C-27J is the only modern tactical transport specifically designed for military tasks. The main rival of this craft is built by EADS, the European colossus that inherited the project from the Spanish CASA, and is called the CN-235, an aircraft born in 1983 from an agreement with the Indonesian Nurtanio for development of a twin-engine transport aircraft for both civilian and military uses. In 1995, the company introduced on the market the elongated model C-295, and there were also versions produced for maritime patrol and airborne radar warning. The C-295 is powered by two Pratt & Whitney Canada PW127G turboprops with six-bladed propellers rated at 2645 hp each. It has a maximum speed of 357 mph, a range of 828 miles fully loaded, a maximum takeoff weight of 51,147 lb, and a payload of 20,393 lb (e.g., 71 fully equipped troops). The two models are still in production, with a tally of over 600 orders from civilian and military operators in more than 40 countries.

228-229 One of the Luftwaffe's Transall C-160Ds in flight. This aircraft has been in service since the 1960s.

229 top In service since 1974 and with nearly 1,000 units built, the Ilyushin Il-76 has become the most successful and widely used Russian-built jet-propelled military transport.

229 center One of the Italian Air Force's Alenia Aermacchi C-27Js flying over the Tuscan coast. A derivative of the G.222 transport, the C-27J took its maiden flight in 1999. The air forces of 10 countries have ordered units of the aircraft.

229 bottom The CASA CN.235, today produced by EADS, is a twin-engine medium transport craft, born in 1983 to meet the needs of both civilian and military operators. A stretch version of the plane, called the C-295, later derived from this.

Chapter 17

European Fighters

During the 1950s and 1960s, the USA and the Soviet Union's dramatic Cold War confrontation overshadowed the world. The hostility between the Western and Eastern global superpowers was also evident in military aviation as their aeronautical industries designed and mass-produced some of the world's most effective aircraft, many of which were sold to the two blocs' various allies. Nonetheless, even during those difficult years Europe proved capable of maintaining a degree of independence, and produced a number of aircraft of significant technological specifications and performance.

The French aeronautical industry emerged from the Second World War in a very poor state, but found its inspiration in Marcel Dassault. Born in 1892 (with the name of Marcel Bloch), Dassault was soon caught up in the world of aviation, and as early as 1917 he was building his first industrial aircraft, the SEA4. During the 1930s, his factory produced diverse military and civil aircraft, but with the imposition of the German occupation Bloch was arrested, interred and finally deported to Buchenwald, after having refused to collaborate with the invaders. In 1945, following the liberation, the family changed its name to Dassault and began once again to produce increasingly successful aircraft. The first post-war fighter was the MD 450 Ouragan, which completed its maiden flight on 28 February 1949. This was the first French jet aircraft, and its specifications had much in common with other contemporary jets and engines, including the Rolls-Royce Nene engine, which produced 5,000 lb of thrust.

The MD 450's design featured a straight low wing with tip tanks, air intakes at the nose and cruciform tailplanes. It was armed with four 20-mm cannons and about 1,920 lb of bombs and rockets and flew at a top speed of about 585 mph. In total, 493 of the MD 450s were built; they were used by the air forces of France, Israel and India. The plane had only a brief front-line career as Dassault soon introduced its natural successor, the MD 452 Mystère, which in effect combined the MD 450 Ouragan's fuselage with a new 30° swept wing. The Mystère fighter first flew on 23 February 1951, but the development program led to the creation of diverse variants (with different engines) prior to the introduction of the Mystère IIC production model. In 1952, after 1,250 Mystères had been built, the Dassault introduced the definitive IVA version. This version was powered either by a Hispano-Suiza 250A engine producing 6,283 lb of thrust or by a Hispano-Suiza 350 Verson unit delivering 7,716 lb of thrust. The aircraft had a wingspan of 36 feet and a length of 39 feet 4 inches, and was armed with two 30-mm cannons and 2,200 lb of bombs or rockets.

230 top A Dassault Mystère IV of the French air force. This fighter was introduced in 1952 and perormed well for its era, including its weaponry of two 30-mm cannons and 2,200 lb of bombs and rockets.

230-231 The Dassault Mirage III was an exceptional fighter for its time and produced a family of 1,400 aircraft sold to more than 20 countries worldwide. This is the two-seat Mirage IIIB.

231 top Three Dassault MD.450 Ouragans of the French air force photographed in formation in 1956. The Ouragan was the first French jet fighter.

Maximum take-off weight was about 20,950 lb, and top speed was 696 mph.

By now, the Mystère's similarities to the Ouragan were minimal: the Mystère's fuselage had an oval section, the wing was thinner and stronger and had a 41° sweep, and a different engine had been fitted. A total of 411 Mystères were built and were sold to Israel and India as well as to the French air force. The IVA was followed by a new version designated the Mystère IVB, which first flew in December 1953. The IVB was a significant step forward: it was the first French supersonic jet, thanks to installation of an afterburner-equipped Atar 101G engine.

Serial production concentrated, however, on the Super Mystère B2 that was introduced in March 1955. This was a further development of the basic design, featuring an improved fuselage with an oval air intake in the nose and a thin wing profile with a saw-tooth leading edge and a 45° sweep. The plane had a wingspan of 32 feet 10 inches and a length of 45 feet 11 inches; it carried two 30-mm cannons, a container for 55 65-mm rockets and 2,200 lb of bombs or rockets slung beneath the wings. The maximum take-off weight rose to just over 22,000 lb. This version too was powered by the Atar 101G engine, which produced 9,920 lb of thrust and provided a top speed of Mach 1.12. The Super Mystère was Europe's first operational fighter with a supersonic capability; a total of 180 were built, of which 32 were sold to Israel.

The Super Mystère IVB is fighter had barely completed its maiden flight before Dassault was working on a completely new and interesting concept: a high performance fighter with a delta wing. The MD.550 Mirage, which first flew in June 1955, featured a 60° delta wing and two Armstrong Siddeley Viper engines producing 2,200 lb thrust, along with an auxiliary rocket that allowed a top peed of Mach 1.3 to be attained.

Following the Mirage II project, which used an engine that never actually went into production, the Mirage III flew for the first time on 18 November 1956. This version was powered by an Atar 101G engine equipped with an afterburner and also carried a detachable rocket propulsion unit. After a number of improvements and modifications had been introduced, Dassault finally launched the production version, the Mirage IIIC, that was later joined by the two-seat Mirage IIIB trainer. The Mirage III was 42 feet 7 inches long, had a wingspan of 26 feet 3 inches, and a maximum take-off weight of about 26,000 lb. Its afterburner-equipped SNECMA Atar 9B engine delivered up to 13,225 lb of thrust, sufficient to power the Mirage III to a top speed of Mach 2.1. Armament comprised two 30-mm cannons and an external load of almost 3,000 lb, including air-to-air missiles and supplementary fuel tanks. Thanks to its dramatic aerodynamic shape, the Mirage III caused a sensation; it also provided the French air force with an excellent Mach 2 interceptor that attracted significant orders from Africa and Israel. Well aware of the fundamental quality of the design, Dassault exploited it to the full and soon introduced the IIIE model, which was modified for a ground attack role. Equipped with a slightly more powerful Atar 9C engine, the IIIE completed its maiden flight in April 1961. It featured a new fire control system, an increased fuel load and five weapon-mounting points. Given its versatility, this version was particularly well suited for export.

The Mirage III airframe was also used as the basis for the IIIR reconnaissance version, the IIIS (exported to Switzerland) and the two-seater IIID. Apart from the Mirage 5 version built for Israel but never delivered for political reasons, the last of the line was the Mirage 50 introduced in 1979, but this plane failed to attract the same sales success. A total of more than 1,420 of the Mirage III and 5 were built and exported to around 20 different countries throughout the world; they fought successfully in diverse conflicts in the Middle East, Asia and South America. A number of these planes are still in service, following modernization.

In the early 1960s Dassault was already working on the successor to the Mirage III. In 1964 the firm obtained French government approval, and the new Mirage F.1 was in the air by 23 December 1966, after the parallel F.2 project (a large aircraft weighing almost 20 tons) had been shelved. The F.1 broke with the Mirage delta wing tradition and featured a fuselage very similar to that of the series III together with a swept wing and an Atar 9K engine producing almost 16,000 lb of thrust. The plane was equipped with the new Cyrano IV radar and new Matra air-to-air missiles. The first prototype was destroyed in a crash but had nonetheless demonstrated its potential, and the government authorized construction of three pre-series aircraft. Deliveries of the F.1 C interceptor, the production version began in 1973. The plane had a wingspan of 26 feet 4 inches, a length of 49 feet 3 inches, and a maximum take-off weight of about 35,700 lb. It could attain a top speed of Mach 2.2 and its armament comprised two 30-mm cannons and a load capacity of up to 8,800 lb, carried externally on seven mounting points. The basic design also gave rise to the F.1A ground attack, the F.1B two-seater, and the F.1E multi-role export versions. Subsequently, Dassault produced the F.1C-200 with an in-flight refueling capability, the F.1CR reconnaissance plane, and the F.ICT multi-role version for the French air force. This last plane was delivered by converting 55 of the F.1C-200s. The Mirage F.1 series was also commercially successful; a total of 731 planes were purchased by the air forces of eleven different countries.

232 A Mirage IIIS interceptor of the Swiss air force. The plane given a complete overhaul featuring canard fins on the air intakes to improve maneuverability, and its new aerodynamic profile is clearly shown.

233 A Mirage F.1CJ of the Jordanian air force climbing at high altitude. With this airplane, Dassault returned to a more conventional aerodynamic design that was also more effective in terms of the technology of the period.

In the meantime, across the Channel, Great Britain, another pre-war aeronautic colossus, continued to be a player. In the field of military aircraft, domestic firms produced numerous fighters and strike aircraft for the Royal Air Force and the Royal Navy. The pioneering jets developed during the war were the Gloster Meteor and the de Havilland Vampire, but it was Hawker that gave rise to the first true post-war success, the P.1067, developed out of the Sea Hawk naval fighter. This aircraft was developed in response to a RAF statement issued in 1948 that required a swept-wing aircraft to replace the Meteors; performance was to match that of the American F-86. Named the Hunter, the new aircraft encountered certain teething troubles. However, because of global crisis that the Korean War represented, serial production had been authorized even before the plane's maiden flight on 20 July 1951.

The Hunter featured a long, slim fuselage, a swept wing with an ample chord, and engine air intakes at the root. The F Mk.6 fighter version was powered by a Rolls-Royce Avon Mk. 207 engine producing just over 10,000 lb of thrust, sufficient to ensure a top speed of 700 mph. The Hunter was 45 feet 11 inches long and had a wingspan of 32 feet 10 inches; maximum take-off weight was about 23,800 lb. Because initially fighter and then bomber versions were being developed, the Hunter was armed with four 30-mm cannons and an external bomb and rocket load of 3,000 lb. The Hunter entered service in 1954, but it was no match for the jets flown by the Americans: they already had the supersonic F-100. Nonetheless the Hunter enjoyed a long and successful career with the RAF through until the 1970s. Also built in reconnaissance two-seater training and export versions, the reliable, well-armed Hunter was a great international success. In all, 1972 Hunters were built, and they were sold in 18 countries.

The Hunter's replacement as interceptor was developed by English Electric (later BAC), with the company using a design on which they had actually been working since 1947. The prototype P.1A first flew on 4 August 1954, and featured innovative lines with a stubby swept wing and two Sapphire engines mounted one above the other. In 1957, the P1.B appeared, featuring notable improvement; the RAF ordered 20 pre-series planes. Deliveries of the first production model, the F Mk.1, began in 1959. The Lightning (as the fighter had been named) was a truly exceptional aircraft for its time, capable of flying at twice the speed of sound, with a fantastic rate of climb, good handling and an armament comprising radar-guided Firestreak air-to-air missiles. The principal version of the Lightning was the F Mk.6, which appeared in 1965. It had a length of 52 feet 6 inches, and a wingspan of 32 feet 10 inches, and was armed with two air-to-air missiles and an optional pair of 30-mm cannons in a belly pod; it had a maximum take-off weight at take-off of 50,000 lb. The F Mk.1 was powered to a top speed of Mach 2.27 by two afterburner-equipped Rolls-Royce Avon 30 engines producing 16,300 lb of thrust. Ten RAF air defense squadrons few the Lightning, and in Mk. 53 form it was also exported to Saudi Arabia and Kuwait. Together with the two-seater models of the T series, 334 of these fighters were produced. This figure by no means reflects the quality of the aircraft; however, the Lightning was competing on the global markets with other European planes such as the Mirage III and the F-104, against which it had little chance.

In the meantime, the British aerospace industry had ventured into a wholly new field of great technical complexity: that of vertical flight. A joint venture between Hawker Siddeley and Bristol had been set up as early as 1957 in order to design a fighter with vertical take-off and landing capabilities. The engine-builder came up with the new BS.53 Pegasus turbofan, which was well suited to the particular demands of the project. The first prototype, the P.1127 completed its maiden flight on 21 October 1960. It was a single-seat aircraft characterized by a high,

234 top An RAF English Electric Lightning. The plane, which first flew in 1954, was unusual in shape but very effective. Shown here is a Lightning fitted with auxiliary fuel tanks above the wings and Firestreak missiles in the fuselage.

234 bottom An RAF Hawker Hunter F Mk.6 fighter. This aircraft was a great commercial success for British industry, with more than 2,000 being built and sold to 18 countries.

235 top In the foreground a Sea Harrier F/A Mk.2 armed with missiles. This model was introduced in 1988 and features Blue Vixen radar on the nose.

235 bottom An RAF Harrier Mk.1 firing air-to-ground missiles. This British fighter-bomber entered service in 1969; its vertical take-off and landing capability revolutionized tactical air-support procedures.

swept wing, below which were located four directional nozzles that allowed the plane to fly vertically and to hover. The six prototypes were followed by nine Kestrel pre-series planes destined for evaluation by a combined British-American squadron. The first British Harrier flew on 31 August 1966, and the first production examples, designated the Harrier GR Mk.1, entered service with the RAF as fighter- bombers in April 1969. The Harrier offered new opportunities in the field of tactical support: it could be operated from advanced positions with no airfield infrastructure whatsoever. The Mk.3 version (equipped with laser telemetry systems in the nose) was equipped with a more powerful engine, the Pegasus Mk. 103 engine, producing 21,500 lb of thrust and providing a top speed of about 735 mph. The plane had a wingspan of 22 feet 11 inches and a length of 45 feet 11 inches. Maximum take-off weight was 26,000 lb, of which 8,000 were accounted for by external loads in the form of bombs, rockets, fuel tanks and pods for two 30-mm cannons.

Despite its great potential, the Harrier never enjoyed significant sales success and only 305 were actually built and operated by Great Britain, the US Marines (designated the AV-8A) and Spain (AV-8S), this last purchaser subsequently selling its planes to Thailand in 1997. In 1975, British Aerospace (born out of the merger of various companies including

Hawker) began to develop a naval version of the Mk.3 in response to a Royal Navy requirement for a multi-role carrier-borne fighter with a V/STOL (Vertical/Short Take-Off and landing) capability. The Sea Harrier, as it was named, made its first flight on 20t August 1978. It featured a multi-role radar system, Sidewinder air-to-air missiles and anti-ship missiles. The plane entered service in 1979 as the FRS Mk.1. Two-seater and export versions of the plane (the Mk.51 for India) were also developed, and it distinguished itself in 1982, during the Falklands War, when it shot down a number of Argentinean fighters without suffering any losses. A modernization program initiated in 1988 brought the aircraft up to F/A Mk.2 specification with a new Blue Vixen radar system and the potential for employing AMRAAM and ALARM missiles. In all, 106 Sea Harriers were built, and the last British units were withdrawn in 2006. Beginning in the 1970s, British Aerospace and McDonnell Douglas worked jointly on developing the AV-8B Harrier II, a derivative of the Harrier.

Despite traditional national rivalry, in the mid-1960s France and Great Britain managed to collaborate on the development of a new military jet, an advanced training and strike aircraft. In May 1965, the Breguet 121 was selected as the most promising design, and to further develop it, BAC and Breguet set up a joint company, the Société Européenne de Production de l'Avion d'École de Combat et d'Appui Tactique, better known by its initials SEPECAT. The resulting aircraft had twin engines, a high wing, a slim fuselage and a high undercarriage suitable for operation from improvised airstrips. It completed its maiden flight on 23 March 1969. Great Britain insisted that the aircraft should have increased operational potential and eventually the Jaguar (as the plane was named) developed into a true supersonic fighter-bomber. The Jaguar GR Mk.1, the basic version produced for the RAF, had a wingspan of 26 feet 3 inches and a length of 49 feet 3 inches. It was powered by two Rolls-Royce/Turboméca Adour 104 turbofans delivering about 8,000 lb of thrust on afterburn and was capable of a top speed of Mach 1.6 or about 1,056 mph. The plane had a maximum take-off weight of close to 34,600 lb, of which about 10,000 lb were accounted for by external loads. Two 30-mm cannons made up the plane's armament. The RAF also had a two-seater version, designated the T Mk.2.

The French versions, the Jaguar A and the two-seater E, though characterized by simpler avionics than the British models, were fitted with laser telemetry, a sophisticated inertial navigation system, a mission computer, and HUD equipment. An international version of the Jaguar was also built, and sold to four countries including India, where 91 Shamshers were build under license. The British continued to update the design to new standards with the GR Mk.1A, Mk.1B, Mk.3, and most recently

the Mk.3A, a truly modern and capable aircraft which entered service in January 2000. In total, SEPECAT has to date built 588 Jaguars. In 2005, the French retired their last units from service, and in 2007, the British, as well.

Sweden was the third European country to distinguish itself in building excellent fighter aircraft in the post-war period, primarily through Saab, its principal aeronautical firm. Following the Saab 21, 29 Tunnan and 32 Lansen jets, the company produced the 35 Draken, a highly innovative fighter that was actually the first European fighter capable of exceeding twice the speed of sound. The origins of the design dated back to the early post-war years, when the Swedish air force identified the need for a high- performance interceptor to defend the country against possible raids by enemy jet bombers. The research gave rise to an unconventional design featuring a double delta wing and no tailplanes. Testing was conducted using a 7:10 scale aircraft, the 210 Lilldraken, which first flew in 1952. The first Saab 35 prototype subsequently completed its maiden flight on 25 October 1955, with its futuristic lines and cutting-edge performance causing an immediate sensation. The 35 Draken entered service with the Swedish air force in 1960. In all, including the J35A, B, C, D and F fighters, the E reconnaissance version, and the two-seat Sk35C version, a total of 525 Drakens were built.

In the 1980s, Saab introduced the J35J Draken. This version was basically an improved model F. The J35J had a wingspan of 29 feet 6 inches, a length of 49 feet 3 inches, and an afterburner-equipped Volvo RM6C engine delivering 17,200 lb of thrust and providing a top speed of Mach 2 or about 1,321 mph. The plane's maximum take-off weight was about 27,550 lb, and the armament comprised a 30-mm cannon and up to 6 air-to-air missiles. Although the Draken had exceptional characteristics and performance, for political reasons it was exported only to allied or neutral countries such as Finland and Denmark and, in 1988, to Austria, the last country to retire the plane from service, in 2005. Had the Draken been built in France or Great Britain it would have been a formidable rival for the Mirage and the F-104 in the global markets of the 1960s.

236 top left The Swedish SAAB Viggen, which first flew on 8 February 1967, was the Swedish air force's basic combat aircraft until the advent of the Gripen. Shown here is the improved model AJS37.

236 top right A SAAB 35F Draken fighter flown by the Finnish air force is shown landing. This aircraft was the first European plane to reach Mach 2, and more than 600 were built.

236 center A FIAT G.91R/1 of the Italian air force's 5th Air Brigade in 1959 at Treviso air field. The G.91 was the first true attempt by NATO at standardization but it did not meet with its expected success.

236 bottom A SAAB 350E of the Austrian air force performing acrobatics. The typical form of the Draken's delta wing without tailplanes is clearly shown.

236-237 A SEPECAT Jaguar E of the French air force's 7th Squadron. France introduced the plane to service in 1972, two years before Great Britain did so, and bought 200 of them.

Nonetheless, including export versions, 606 Drakens were built, a highly respectable figure.

During the 1960s Saab was working on a successor to the Draken. This had to be an aircraft exhibiting excellent performance but adapted to the Swedish defense strategy that, in the case of war, featured (and continues to feature) the dispersal of forces in the forests. A robust aircraft was needed, with short take-off and landing capabilities and operable with minimal technical-logistical support. This requirement led to the development of the Saab 37 Viggen, a large single-engined, single-seater featuring a double delta wing with canard control surfaces, with an engine equipped with a thrust-reverser to reduce the runway length required for landing. The Viggen made its maiden flight on 8 February 1967, and the production aircraft entered service in June 1971. The first version to be developed was the AJ37 ground attack model with secondary air defense capabilities. The SF37 reconnaissance version, the SH37 maritime reconnaissance version, and the Sk37 two-seater version followed. In 1979, Saab introduced the JA37 interceptor. It had a wingspan of 32 feet 10 inches, a length of 52 feet 6 inches, and was equipped with a Volvo RM8B turbofan engine producing about 28,000 lb of thrust on afterburn and providing a top speed of 1,368 mph. The plane's armament was a 30-mm cannon, up to six air-to-air missiles, and air-to-ground weapons (bombs, rockets and missiles Maximum take-off weight was about 22 tons.

The potential market for the Viggen was restricted to an even greater degree than with the Draken, and Sweden remains the only country to operate the aircraft. Since 1993 the Swedish air force has been flying the AJS37 version with uprated avionics that make it an extremely versatile aircraft capable of flying interception, attack or reconnaissance missions as required. Saab has constructed a total of 329 Viggens, the last of which was retired from service in 2005. This overview of the most significant European fighters of the 1950s and 1960s could hardly be complete without mention of the Fiat G.91. This plane resulted from one of NATO's first attempts to adopt standardized weaponry among its member states. NATO requirement NBMR-1 (1953) called for a light fighter-bomber capable of operating from advanced bases. The Fiat team, led by led by engineer Gabrielli, developed the G.91, an aircraft with a low swept wing and an air intake beneath the nose, a layout that recalled the general architecture of the F-86K (which Fiat built under license). The first of three prototypes flew on 9 August 1956. Despite the loss of that first prototype, within a few months the design proved to be fully responsive to NATO requirements, and the G.91 was approved for production in January 1958. The G.91 attracted numerous favorable opinions, but because of the usual political-economic maneuvering of national governments, it was never adopted by the majority of NATO air forces, but only by Italy and Germany.

Italy adopted the base model G.91R/1. This model had a wingspan of 26 feet 3 inches, a length of 32 feet 10 inches, and was fitted with a Bristol Siddeley Orpheus engine producing about 4,050 lb of thrust, providing a top speed of 640 mph. It had a maximum take-off weight of 12,125 lb, and was armed with four 12.7-mm machine-guns and up to 2,000 lb of bombs and rockets carried on two underwing pylons. This model formed the basis for the two-seater G.91T and the G.91R/3, which was built for Germany and equipped with improved avionics, two 30-mm cannons and four underwing pylons. The Portuguese air force also flew the G.91, receiving 66 of them from Germany. In total, despite the ostracism of various countries (above all France, whose rival to the Fiat model had been rejected), over 760 of the G.91R and T models were built, and the plane remained in service until the 1990s.

Chapter 18

Vietnam

Although Vietnam had been engaged in guerrilla warfare and fighting since the 1930s, first against the French and then against the Japanese, and then against the French again after World War II, the denomination Vietnam War properly refers to the war fought from about 1960 to 1975. This conflict involved the Vietcong and the North Vietnamese army on one side and the South Vietnamese army, supported by the United States, on the other. From an aeronautical point of view, the Vietnam War was a hugely significant conflict, given that it saw the application of new aircraft, new weaponry, and new tactics that have made their mark on the history of aviation.

The Vietnam War began as a fight by South Vietnam's government against the attempted insurrection urged by North Vietnam's government (with large-scale material support from the Soviet Union). The crisis heightened with the direct involvement of the American armed forces; it culminated in 1968 when the US troops in the area numbered some 550,000. The United States Government made grave political errors: on the one hand global public opinion (and also that of a section of American society) came down in favor of the North Vietnamese, on the other, military operations were ultimately unsuccessful despite the enormous resources expended.

Initially, the United States restricted its intervention to the dispatch of advisors and armaments to the government of South

238 top left A CH-54 helicopter transports a Marine Corps 155-mm howitzer to a hilltop during operations around Khe Sanh in January 1969.

238 bottom American fighters fly over a burning ship during the Vietnam war. The shadow on the water is from a US Navy F-8 Crusader.

238-239 The Vietnam War is remembered as "the helicopter war." Shown here is a formation of Bell UH-1s supporting South Vietnamese troops during an attack on the Vietcong, March 1965.

239 bottom A US Navy Douglas A-4E Skyhawk fighter-bomber drops a Mk.84 high-explosive bomb on a Vietcong troop emplacement in January 1966.

Vietnam. The goal was to form the first nucleus of the South Vietnamese air force, equipped with A-1 Skyraiders, T-28 Trojans, B-26 Invaders, C-47 Skytrains, and C-123 Providers. In Agust 1964 following the Gulf of Tonkin incident (an attack by North Vietnamese patrol boats on the destroyer USS Maddox), the US Navy began to play a direct role in the conflict, conducting bombing raids using its own carrier-borne aircraft. Early in the war, the Navy's aircraft carriers were generally equipped with F-8 and F-4 fighters and A-4 and A-6 fighter-bombers.

The US Navy's most modern fighter at the time was the McDonnell F-4 Phantom II, which

had entered service in December 1960. This plane was designed and built following a naval requirement for a large interceptor (designated the AH-1), derived from the F3H model and to have a three-hour in-the-air time frame and the capability to launch the new Sparrow missiles. The navy ordered two prototypes of the F4H-1 Phantom, the first of which took to the air on 27 May 1958. The twin-engined design featured a trapezoidal wing and tailplanes with negative dihedral, was equipped with radar, and carried a second crewmember to operate the weapons system. The Phantom immediately proved to be an exceptional multi-role fighter, and in December of that year the US Navy ordered an initial batch of 375. In 1962, the fighter was redesignated the F-4, and the first mass-production version was the F-4B. Its two General electric J79-GE-8 engines delivering 17,000 lb thrust on afterburn powered it to a top speed of Mach 2.2. Moreover, the plane also featured an APQ-72 multifunction long-range radar system (which allowed it to fly interceptor missions without the need for ground radar), a redesigned cockpit, and the ability to operate as a fighter-bomber carrying bombs and rockets. While the F-4A and B versions began to fly with the US Navy and the Marines, the USAF also showed an interest in the design, given that it was well suited to its own requirements. McDonnell thus built a specific version of the fighter for land-based operations, the F-4C (initially designated the F-110A), fitted with a different undercarriage and tires, self-starting, J79-GE-15

240-241 A USAF RF-101Cs at low altitude over the My Dug road bridge in North Vietnam, which has just been bombed. Its shadow is reflected on the river.

240 bottom A row of A-4Es waits to take off from the USS Oriskany off Vietnam in 1968. An A-3D Skywarrior has just been launched.

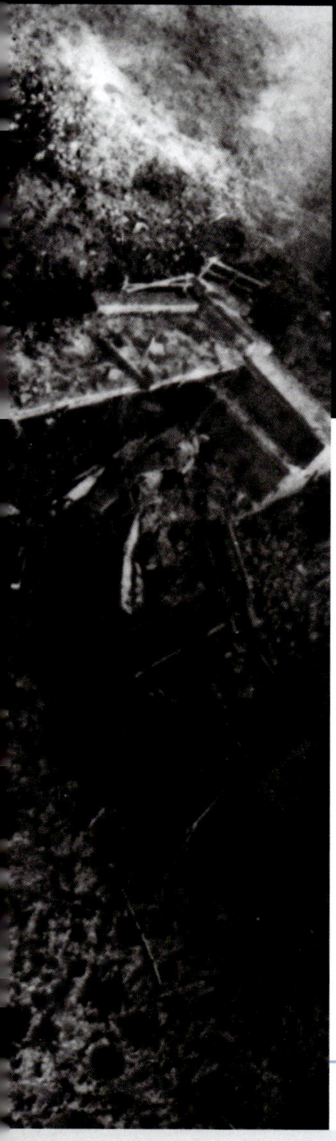

241 top Two Marine A-4Ms firing a burst of rockets at a land target. The Skyhawk was produced up till 1979; in total 2,960 were made.

241 center A pair of USAF F-4E Phantom II's together with a KC-135 tanker just before inflight refueling, on the way back from a mission over North Vietnam.

241 bottom left A US Navy McDonnell Douglas F-4B Phantom II a moment before taking off from aircraft-carrier's deck. The nose-wheel strut was lengthened to improve the aircraft's angle during take-off.

engines, APQ-100 radar, dual flight controls and a rigid-probe in-flight refueling capability. The USAF acquired 583 of the various F-4s between 1963 and 1966. These versions were soon followed by reconnaissance variants (RF-4B and RF-4C), the F-4D for the USAF and the F-4J for the Navy powered by J79-GE-10 engines producing around 18,000 lb of thrust and fitted with AWG-10 fire control systems. The final version for the USAF was the F-4E, introduced in 1966. Its design drew on experience gained during the war and introduced more powerful engines, new wing flaps and a redesigned nose equipped with APQ-120 radar and, finally, the M61 20-mm Vulcan cannon. The RF-4E reconnaissance version then followed. The F-4E had a wingspan of 36 feet, a length of 62 feet 4 inches and a maximum take-off weight of about 61,650 lb. The two J79-GE-17 engines provided about 17,900 lb of thrust on afterburn and propelled the aircraft to a top speed of about 1,485 mph. The weapons system comprised an external load of up to 16,000 lb. A powerful, robust and reliable plane, the Phantom also shone in the arduous Indochinese theater once the air force and the navy had developed appropriate tactics. The F-4 proved to be a solid base for the creation of specialist versions and innovative applications. Thee included the F-4G anti-radar strike version, the F-4K for Britain's Royal Navy, and the F-4N and S, modified and rebuilt versions of the F-4B and J. The Phantom was also sold to numerous allies, NATO states and otherwise, such as Israel, Iran, Japan, Egypt and South Korea. A total of 5,195 of the various F-4 versions were built; many have been modernized and are still in operational use.

The smallest of the carrier-borne fighter-bombers at that time was the Douglas A-4 Skyhawk, a plane privately developed by the Californian firm in order to offer the US Navy a jet replacement for the manufacturer's own A-1 Skyraider. The A-4 was a small, single-engined plane with a delta wing and a cruciform tailplane. It had surprising technical and performance characteristics. As it was built in one piece, the A-4 was notably robust and light; the small wing, for example, did not need to be folded for the aircraft to be parked aboard a carrier. The Skyhawk flew for the first time as the XA4D-1 on 22 June 1954, and with only a few modifications required, it entered service two years later, equipped with the J65-4B engine producing around 7,700 of lb thrust.

In 1962, the Skyhawk was redesignated the A-4, and numerous and continuously improved versions were introduced. The A-4E was the first of a new series that introduced the more efficient J52 engine, producing 8,500 lb of thrust (while consuming less fuel), and was equipped with five underwing pylons rather than three, raising the potential weapons load to 9,150 lb. The F model saw the first appearance of the dorsal "hump" containing new avionics, while the TA-4F was the first two-seater advanced training version.

Fast, maneuverable, robust, and capable of carrying a considerable weapons load, the Skyhawk performed extremely well in Vietnam, operating both from aircraft carriers and the Marines' land bases. The plane remained in production until 1979; 2,960 were constructed and sold to ten countries around the world. Among the last American forces to fly the Skyhawk were the Marines' Aggressor units, which used them in combat training to simulate the agile and deadly MiG-17s and MiG-19s.

242 top A USAF Vought A-7D during the Vietnam War. After the Navy showed interest in this fighter-bomber, the Air Force did so too. It began to replace the F-100 and F-105 after 1970.

242 bottom The Vought A-7E was fitted with a navigation and attack system (the NWDS) that made it most advanced fighter-bomber of its era. This aircraft was in service on the USS Coral Sea.

The US Navy also flew the Grumman A-6 Intruder, which could be considered the diametric opposite of the A-4. The A-6 Intruder was a large, carrier-borne all-weather bomber then considered to reflect the height of avionics sophistication. The basic model, the Grumman G-128, which was approved in 1957, had been designed in response to a requirement for a new carrier-borne strike aircraft. The plane was to reflect the lessons of the Korean War. The G-128 was a twin-jet aircraft featuring a large nose housing the radar system and a two-man cockpit with parallel seating. The prototype made its first flight on 19 April 1960, and the first production planes, designated the A-6A, began entering service in 1963. The Navy and the Marines used them in Vietnam from 1965 on. The Intruder's strengths included a significant payload (fully loaded, the plane weighed in the region of 27.5 tons) and sophisticated strike and navigation systems (APQ-146/148 radar and ASN-92 inertial navigation) that allowed it to find and hit targets even in conditions of zero visibility. Subsequent versions included the A-6B anti-radar aircraft, the A-6C night attack variant, the KA-6D in-flight refueling tanker, and the EA-6A electronic warfare plane.

The definitive version of the Intruder was the A-6E (most versions were built around converted A-6A airframes), which boasted even more advanced avionics. The A-6E had a wingspan of 52 feet 6 inches, a length of 62 feet 4 inches, and a maximum take-off weight of 60,450 lb. It was powered by two Pratt & Whitney J52-P-8B engines without afterburners, producing around 9,300 lb of thrust, and was capable of a top speed of around 650 mph. An updated version of the A-6E was equipped with the TRAM (Target Recognition Attack Multi-sensor) system composed of FLIR (Farward Looking Infra-Red) sensors and lasers for navigation and strike operations. A total of 710 A-6s were built, all for the exclusive use of the US Navy and Marines. The aircraft was withdrawn from service in the mid-1990s.

The Vietnam War saw the use of another two carrier-borne bombers. The first of these was the Douglas Skywarrior, which made its first flight on 28 October 1952. When it

entered service in 1956 as the A3D, the plane was the largest ever built for carrier-borne operations. Equipped with a high wing to which the two Pratt & Whitney J57 jet engines were attached, the aircraft was designed as a nuclear bomber, but was subsequently converted to perform diverse specialist roles such as reconnaissance, electronic warfare, in-flight refueling and training. In 1962 the plane was redesignated the A-3; the B model had a wingspan of 72 feet 2 inches, a length of 75 feet 6 inches and a maximum take-off weight of about 82,000 lb, of which the internally stored weapons load amounted to about 12,000 lb. The engines delivered about 15,615 lb of thrust, giving the plane a top speed of 610 mph. The USAF also used the plane, which it designated the B-66, and beginning in the mid-1950s purchased 290 for the reconnaissance and electronic warfare roles in which it was employed in Vietnam in the 1960s.

The second Navy bomber, the A3J-1 Vigilante (redesigated the A-5 in 1962), was also used for reconnaissance in Vietnam in its RA-5 version. This aircraft had been designed in the mid-1950s as the supersonic successor to the Skywarrior, and featured highly futuristic lines with fully mobile tailplanes and variable-geometry air intakes. The Vigilante had a wingspan of 52 feet 6 inches, a length of 75 feet 6 inches, and a maximum take-off weight of 66,000 lb. It was powered by two General Electric J79-GE-10 engines and could attain a top speed of Mach 2.2.

The pressures of war led to the development of a new fighter-bomber for the US Navy, the response to a requirement for a successor to the A-4. In order to accelerate development time, the requirement specified that the proposed designs should be derived from planes already in service. Ling-Temco-Vought (LTV) developed a project loosely based on the Vought F-8 Crusader, but the resulting plane was in effect all new. It was a subsonic aircraft with a fixed wing and an enlarged fuselage that could carry a notable weapons load on six underwing pylons. The navy declared this plane to be the winner of the competition. Designated the A-7 Corsair II, the new fighter-bomber flew for the first time on 27 September 1965 (ahead of the contract-specified date) and in 1966 entered service as the A-7A; by the following year it was already seeing combat duty in Vietnam. The A7-A was equipped with a TF30-P-6 engine delivering 5,150 lb of thrust; the B and C versions received slightly more powerful units. The USAF also took an interest in the plane. In 1968 the A-7D version developed for the Air Force made its maiden flight. The A-7D featured a TF41 engine producing 14,500 lb of thrust and an M61 Vulcan 20-mm cannon with six rotating barrels (in place of four cannons of the same caliber). It also had more sophisticated navigation and strike systems and a rigid-probe, in-flight refueling capability. The A-7D entered service in 1970 and was also used in Vietnam for some months from October1972 until the withdrawal of the American forces.

In 1967 LTV used the A-7D as the basis for the A-7E model for the US Navy. The A-7E retained the A7-D's avionics but was powered by a TF41 engine uprated to produce 15,000 lb of thrust. A total of 506 A-7Es were built, and the plane made its combat debut in 1970, flying from the carrier USS America. The A-7E was 45 feet 11 inches long, had a wingspan of 36 feet, a maximum take-off weight off of almost 21 tons and a top speed of almost 700 mph. It could carry a 15,000-lb weapons load. The Corsair II was an excellent aircraft, capable of carrying a notable variety and quantity of arms that it could release with precision at considerable distances. However, the workload for a one-man crew was sometimes excessive, and led to a number of accidents. The production total for the A-7 (including the models exported to Portugal and Greece and the two-seat TA-7C and TA-7K trainers) amounted to 1,526 planes.

The US Air Force was also widely involved in the Vietnam War with all types of aircraft. It used a number of new concept models during the conflict years, both for strategic reasons and also to verify their operational capabilities and gain useful experience.

243 first picture *A Vought A-7A Corsair of the VA-147 series on the USS Ranger, before landing. This unit was the first to be used in Vietnam with the Corsair, in 1967.*

243 second picture *The North American A-5 Vigilante was designed in the 1950s as a high-performance bomber for the U.S. Navy. It remained in service until the 1980s. The version shown here has folding wings and rudder.*

243 third picture *The US Navy Grumman A-6A Intruder bomber was brought into the Vietnam War in 1965 and proved itself to be formidable in attacking enemy air defenses.*

243 fourth picture *An A-6A ready for take-off from an American aircraft carrier. The plane's thirty 500-lb bombs are distributed on the wing and body pylons. The A-6 could carry a load of up to 8 tons.*

Along with the F-4 Phantoms and the A-7 Corsairs, the USAF employed other military aircraft with very different characteristics: these included the Northrop F-5 and the General Dynamics F-111 series. The first was a lightweight plane that General Dynamics developed privately, hoping it might interest the American government as a tactical fighter to be sold to allied countries. The company developed two planes: the N-156F and the two-seater N-156T (which eventually became the T-38). The N-156F completed its maiden flight on 30 July 1959. It was a compact, slim supersonic jet powered by two J85-GE-5 engines, each producing around 3,850 lb of thrust. In 1962, the US Defense Department chose the plane as the US contribution to the MAP (Military Assistance Program) and designated it the F-5A Freedom Fighter (the two-seater being the F-5B). The aircraft, which first flew on 31 July 1963, was powered by J85-GE-13 engines each producing around 4,050 lb of thrust with afterburn. It had seven mounting points for up to 6,000 lb of external weaponry, two 20-mm cannons, a reinforced airframe, a maximum take-off weight of about 19,850 lb, and a top speed of Mach 1.4. Even though it was primarily destined for export, the F-5A also went into service with the USAF, which sent a contingent of twelve to Vietnam within the ambit of the Skoshi Tiger evaluation program. These planes were fitted with in-flight refueling probes, new armor, and camouflage paint better suited to the Vietnam conflict theater.

Between 1965 and 1966, the Skoshi Tigers were evaluated in tactical support missions, and acquitted themselves well. In 1967, the US began supplying them to the South Vietnamese air force. Over 1,100 of the F-5A/Bs were built; they were flown by the air forces of 20 countries. In 1969, on the basis of F-5A/B experience gained in the field, Northrop produced an improved version, the F-5E, which flew for the first time in production form on 11 August 1972. This model had a greater weapons load-carrying capacity and improved handling characteristics. Powered by J85-GE-21 engines producing 5,000 lb of thrust, it had a top speed of Mach 1.64. About 1,500 F-4Es were built; of these over 120 (single-seaters and F-5F two-seaters) were acquired by the USAF and the US Navy for use as trainers and aggressor planes in combat training.

244 left The Northrop F-5 light tactical fighter was designed for use by the air forces of the USA's allies. Shown here is a South Vietnamese air force F-5A dropping Mk. 117 bombs in 1966.

244 bottom left A USAF C-130A taking off from Kien Duc Field in December 1973. The Hercules, capable of operating from semi-prepared runways, represented a great step forward by comparison with its forerunner the C-123.

244-245 A USAF F-4C Phantom during launch testing in the United States of the AGM-12 Bullpup air-to-ground missile. This version, with the first type of gray livery, entered into service in 1963.

The General Dynamics F-111 was the result of a very ambitious but never fully realized program. In 1960, the US Defense Department initiated the TFX program, aimed at equipping the USAF and the US Navy with a shared multi-role combat aircraft to replace the F-105, the F-8 and ultimately the F-4. In 1962, it announced that General Dynamics (together with Grumman) had been selected to develop its innovative project for a large, powerful airplane with variable-geometry wings and an ejectable capsule for the two-man crew. A contract was drawn up for the building of 18 F-111As for the USAF and 5 of the carrier-borne F-111Bs for the Navy. The first A-series prototype flew on 21 December 1964, but existing technology proved inadequate for the development of an aircraft truly suited to the widely varying demands. In 1968, the US Congress halted development of the F-111B. In that same year, however, the F-111A entered service and was subsequently used in combat at the tail end of the Vietnam War. After early teething troubles, the F-111A proved itself to be an excellent bomber, above all at night and in critical bad weather conditions. In various versions, the F-111D, E and F, the aircraft demonstrated its capabilities as an air defense and all-weather, low-altitude strike aircraft.

The F model had a wingspan of between 62 feet 4 inches (minimum sweep) and 29 feet 6 inches (maximum sweep) and a length of 72 feet 2 inches. It was powered by two Pratt & Whitney TF30-P-100 engines

producing 25,000 lb of thrust with afterburn, and could attain a top speed of Mach 2.5 at high altitude. It could carry a total weapons load of up to 321,500 lb.

The F model also formed the basis for the development of the EF-111A electronic warfare, the FB-111A nuclear strike, and the F-111C versions. This last version was built for export to Australia, the last country to operate the aircraft in its modernized F-111G form. Australia retired its last F-111s from service in 2010. In all, General Dynamics built 562 of the aircraft.

Another Vietnam War protagonist was the Lockheed C-130 Hercules tactical transport plane, of which various versions were developed, with some used in combat roles. In a region such as Indochina, where the tropical vegetation, the lack of roads and guerrilla warfare made ordinary communications difficult and at times even impossible, transport planes performed a vital role, allowing combat units to be supplied and supported at great distances or when under siege by the enemy. When American involvement in Vietnam intensified,

245 center A USAF B-52 during carpet bombing of a North Vietnamese target, using 750 lb M117 bombs.

245 bottom left An F-111E landing. The plane had a variable-geometry wing and two large Pratt & Whitney TF30 engines able to provide 23,000 lb of thrust and a top speed of 1,625 mph.

245 bottom right The US air forces used the F-5 in limited numbers, mainly as an aggressor airplane, i.e., to simulate enemy fighters during combat. This F-5E of the US Marine Corps' VMFT-401 is flying over Arizona.

the Hercules had already been in service for a number of years in A, B and E forms.

The challenges of the Vietnam War highlighted the need to develop and expand certain specialist activities such as attacks on enemy anti-aircraft installations (Wild Weasel and Iron Hand missions). During the conflict, these had been were conducted with increasing determination, initially with the F-100F, EF-105F and A-6B and then with specialist planes such as the F-105G.

Another important type of mission was Combat Search and Rescue (SAR), undertaken to pluck downed pilots from enemy territory. Here HH-3E helicopters were used along with support and protection aircraft of various types, but primarily with the USAF's ancient propeller-powered A-1 Skyraiders.

Naturally, the Vietnam War also represented a turning point for army helicopter operations, with the chopper becoming one of the symbols of the conflict. The helicopter was irreplaceable for transportation, combat and medical-evacuation uses, and thousands of helicopters of all types were employed. In addition to the introduction of deadly defoliant chemicals intended to deprive the Vietcong of the cover of jungle vegetation, but which did terrible damage to the Vietnamese and the environment, the Vietnam War also saw the introduction of other new weapons. These included anti-radar missiles (AGM-78 Standards and AGM-45 Shrikes) and the first laser-guided bombs, the Paveway Is, which permitted a previously unthinkable degree of accuracy when used correctly.

Along with the US forces' numerous air-ground operations conducted, the Vietnam War also saw aerial duels between the large, advanced American fighters and the old, simple but agile Soviet planes flown by the North Vietnamese air force. From this point of view, the conflict provided a shock for the Americans who were completely unprepared for this type of engagement. In-depth analyses and the institution of new training techniques were required (resulting in the Top Gun and Red Flag courses for the Navy and the USAF respectively) before the F-4 pilots, late in the war, systematically began to get the better of the enemy's MiG-17s and MiG-19s.

The United States' military involvement in Vietnam effectively ended with the ceasefire of 27 January 1973. Between 1961 and 1972, America lost 3,792 aircraft and 4,922 helicopters in Vietnam, not to forget more than 45,000 men. Over seven million tons of bombs and other armaments were dropped and in practice every available weapon was used except those with nuclear warheads. And yet with the conquest of Saigon on 25 April 1975, it was clear to even the most stubborn observers that North Vietnam had won the war. The immovability, fanaticism and strength of those fighting on their own land, for themselves and for their own ideals eventually wore down the world's greatest military power. The Americans had never known or been able to adopt the more appropriate strategies and tactics that would have given them a chance of winning the war. America had lost on both the military and political planes a war that it never felt to be its own, the first war in history to have been largely influenced and decided by the strongest weapons a democracy holds: the media and public opinion.

246 left A squadron of multi-role UH-1 helicopters take South Vietnamese troops the 199th Light Brigade to the landing zone in 1967.

246 bottom right A transport helicopter transformed into an air-ambulance takes American wounded to a field hospital.

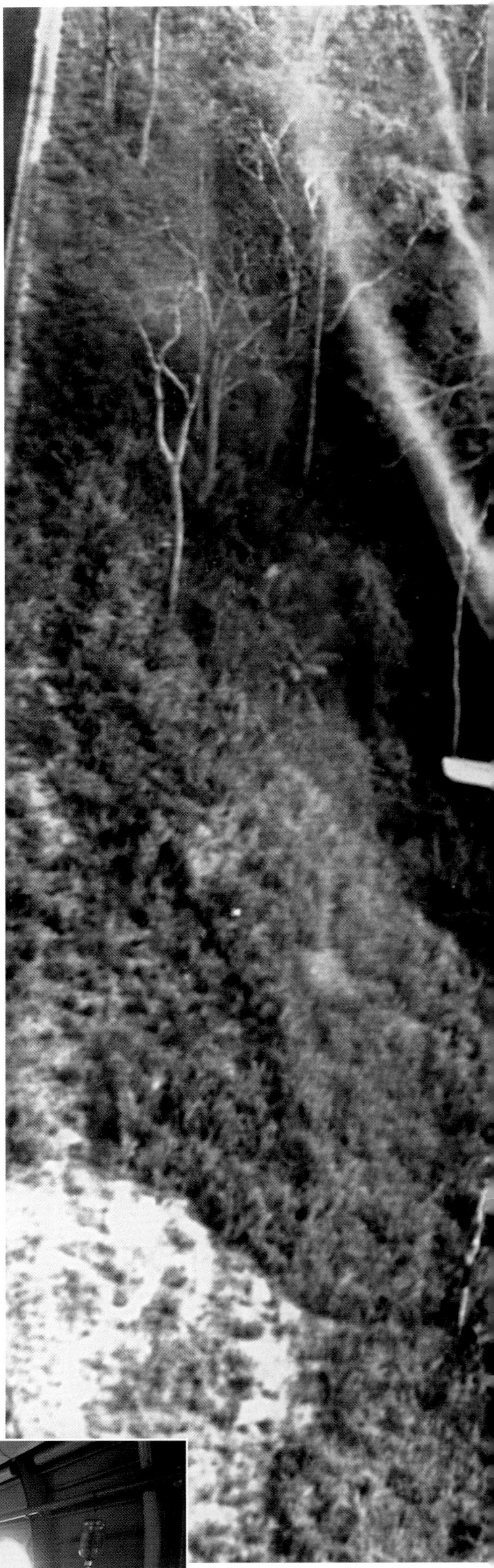

246-247 The Fairchild C-123 Provider, introduced in 1949, was the first tactical transport aircraft to adopt the "high tail, high wing, low undercarriage, rear loading ramp" arrangement still used today. This 1966 photograph shows a C-123 dropping defoliating liquid onto the jungle to prevent Vietcong ambushes.

247 bottom An F-8J Crusader armed with Mk.84 bombs flying over a zone under attack. The Marines used their own aircraft during troop support tasks.

the engineering drawings for the Mirage 5 and its SNECMA Atar 9C engine. The resulting aircraft was known as the Nesher; it was to all intents and purposes a copy of the French plane, and the first in a long line of aircraft derived from the delta-winged Dassault fighter.

The first American-built warplane Israel ordered was the Douglas A-4 Skyhawk, the robust fighter-bombers operated by the US Navy. The Israeli government had placed the order early as 1966, and 48 A-4Hs and two two-seat TA-4Hs were delivered from 1968 onward. At the same time, the Chel Ha'Avir was looking for a new multi-role, high performance fighter. The plane most closely matching the Israeli requirements was the McDonnell-Douglas F-4E Phantom II. This plane offered both fearsome air-to-air armament (with eight radar-guided Sparrow and infra-red Sidewinder missiles and the Vulcan cannon) and a large air-to-surface external weapons load amounting to almost 6 tons of bombs, rockets, and missiles, as well as fuel tanks and electronic warfare pods. The first 50 F-4Es (including six of the RF-4E reconnaissance version) reached Israel in 1969, equipping two squadrons that participated in the so-called War of Attrition, a low-intensity conflict waged against Israel between March 1969 and August 1970, with the aim of wearing down the Israeli air force.

Still smarting from defeat in the Six-Day War, the Arabs, first and foremost the Egyptians, were determined to be the first to strike next time. They spared no expense in acquiring Soviet armaments, including the new MiG-21 (along with numerous "volunteer" pilots from certain Warsaw Pact countries and the USSR), the Su-7 and Tu-16, Mi-8 helicopters. They also acquired a complex and powerful anti-aircraft missile system composed of SAM (surface-to-air) SA-3, SA-6, SA-7 and SA-9 missiles, as well as a vast quantity of armored vehicles. On the eve of the October 1973 conflict, the Arab forces could count on about 730 combat aircraft against Israel's 375.

The Arab attack was disguised by the simple expedient of staging mass exercises. It came out of the blue at 14.00 hours on 6 October 1973, in the middle of the Yom Kippur celebrations, when the Israeli defenses were at their lowest level of alert. The only armed force anywhere near ready for the attack was actually the Chel Ha-Avir, whose commanders had suspected a probable enemy attack. The Arab assault used a pincer strategy, with the Syrians striking to the north on the Golan Heights and the Egyptians staging a mass invasion of the Sinai peninsula, crossing the Suez Canal in operations devised by its engineering corps.

The Israelis concentrated on the attack closest to home. Within two hours of the beginning of the attack, they dispatched as many F-4, Mirage and Nesher fighters as possible, primarily to attack the enemy's armored columns. Israeli losses were extremely high in terms of men, vehicles and aircraft, in the last case because of anti-aircraft fire. After three days of bitter fighting, the Syrians were obliged to retreat; they were now too weak to exploit possible gaps in the enemy defenses. On the western front too, the first few days of the fighting led to grave losses, above all for the Chel Ha'Avir, which in the first four days lost 81 aircraft. After its easy victories in the Six-Day War, the Israeli air force had in fact been taken by surprise by the new SAM missile systems used by the Arabs. In the meantime, the Arab coalition had been reinforced by the intervention of aerial forces from Iraq, Libya and Algeria.

November 9 saw the initiation of Soviet and US airlifts supplying the Arabs and the Israelis respectively. The Soviets sent 100 new fighters to Egypt and to Syria, which alone received over 15,000 tons of supplies. The US sent Israel additional A-4N and F-4E fighters and 22,000 tons of equipment and munitions. These included CH-53 helicopters, Hawk anti-aircraft missiles, and new and more efficient arms such as Walleye and Hobo intelligent bombs and Rockeye fragmentation devices. Important in the aid were systems designed to combat Arab anti-aircraft defenses, namely AGM-45 missiles and electronic warfare pods to confuse the new Soviet SAM systems.

On 14 October, the Egyptians committed

250 top Soviet ground-to-air SA-2 missiles provided effective anti-aircraft defense systems for Arab allies during the Yom Kippur War and offered an unwelcome surprise to the Israelis.

250 bottom The Sikorsky CH-53 is the heaviest helicopter Israel has in service. It flies missions of transport, air ambulance, search and rescue, and special operations.

250-251 Many Israeli missions during the Yom Kippur War were run by A-4 Skyhawks, 53 of which were lost. Shown here are some of the A4Ns and TA-4Js used until 2015 by the Hazerim advanced flight school.

what was to be a costly error: their land forces advanced without cover from their anti-aircraft systems, thus becoming easy prey for Israeli fighter-bombers. The Israeli army then attacked in the Suez Canal area and succeeded in disabling a large part of the Egyptian anti-aircraft defenses. Having crossed the canal, Israeli tanks entered Egyptian territory. The situation was beginning to look bleak for the Arabs. On 20 October, Saudi Arabia played a political card by declaring the suspension of oil supplies to the west. The time had come for a ceasefire that received the backing of the superpowers. It came into force on the evening of 22 October, but the Israelis ignored it and continued with their military operations to encircle the Egyptian 3rd Army in the Sinai. It was only the intervention of the Americans that convinced Israel to accept the ceasefire on 24 October.

251 center An Israeli McDonnell Douglas F-4E from the Bat Squadron landing. The Phantom II was one of the most important aircraft for the Chel Ha'Avir, which bought about 230 from 1969 on. In 2004, Israel retired its last F-4s from service.

251 bottom A MiG-23 with Libyan insignia. During the 1982 war with Lebanon, the Israelis had to combat the Syrian air force's new MiG-23s.

From an aeronautical point of view, in the 19 days of war, the Arabs lost 440 aircraft against more than 120 aircraft lost by the Israelis. Despite their advantage of surprise, the Egyptian and Syrian forces were unable to overcome their enemy and, after having lost the initiative in battle, were obliged to retreat.

The Yom Kippur War highlighted new aspects of aerial warfare that had in part already been seen in Vietnam. First, it became clear that the new Soviet anti-aircraft system's capabilities were far more efficient and dangerous than western analysts had predicted. Second, the importance of measures to counter anti-aircraft defenses became clear, especially the use of electronic warfare systems, dedicated weapons and new tactics; and third, in air-to-surface armaments the new precision bombs proved their worth. Further, the war saw the combat début of RPV pilotless aircraft used for reconnaissance, observation and deception purposes that were to become increasingly important in future conflicts. On a political level, the Yom Kippur War again failed to resolve any of the problems associated with the co-habitation of Arabs and Jews, although militarily it did underline the Israeli forces' superiority in terms of training, combat, and operational capabilities.

After the Yom Kippur War, the two combatants again continued to rearm; the Arabs drawing on the Soviet arsenal, in particular MiG-23 and MiG-25 fighters, while the Israelis continued to rely on America. At home, the Israeli aeronautical industry proceeded with the development of a national fighter based on the original Nesher. With the acquisition of the Phantom, Israel had also had access to the General Electric J79 engine, a far more modern and more powerful unit than the Atar 9C. The Israeli air force thus decided to create a new fighter with improved performance by combining the Nesher airframe and the new engine. Design work on the new fighter probably began in 1969, while the maiden flight of what was to be named the Kfir took place in the summer of 1973, preceded by a development program that also utilized a two-seater Mirage IIIB equipped with the J79 engine. The first Kfir C1s were officially introduced on 14 April 1975, and delivery of the first 40 began that year. Naturally, the Kfir also featured Israeli-produced specialist avionics including telemetric radar, a mission computer and electronic counter-measure equipment. The following year saw the introduction of the C2 model (along with the two-seater TC2), characterized by the appearance of a saw-tooth configuration on the wing's leading edge and two canard wings alongside the air-intakes; these considerably improved the aircraft's maneuverability. The Kfir C2 had a wingspan of 26 feet 3 inches, a length of just over 49 feet, and a maximum take-off weight of more than 32,000 lb. The J79-GE-17 engine delivered 17,900 lb of thrust on afterburn and enabled the plane to achieve a top speed of Mach 2.3 (1,516 mph) without external loads. It was armed with two 30-mm cannons and up to 8,800 lb of external weaponry. The Kfir C7 and TC7 were developed in 1983. The planes featured new avionics and an extra two underwing pylons that allowed the maximum external load to be brought up to 13,225 lb. Israel has built about 240 Kfirs in various versions (some models were created by converting older planes). The total includes those reserved for export: Ecuador, Colombia and Sri Lanka have purchased about 40 Kfirs.

The IAI also offered on the market the Kfir 2000 (also called the C10), equipped with advanced avionics and an in-flight refueling capability, which was adopted by Ecuador (as the KFir CE) and by Colombia (Kfir COA). Used primarily as a fighter-bomber, the Kfir has been joined by two new aircraft, in 1976 by the McDonnell Douglas F-15A/B (and later the C/D), and in 1980 by the General Dynamics F-16 A/B (and later the C/D).

Thanks to these acquisitions, in the 1980s the Chel Ha'Avir became the strongest air force in the Middle East, a predominant position confirmed by three separate military operations. The first was the bombing of the Iraqi nuclear reactor at Osirak, a pre-emptive strike in June 1981 aimed at preventing Iraq from building atomic bombs. The operation was difficult and complex as it involved flying for almost 700 miles beyond the Israeli borders over the territories of Jordan and possibly Saudi Arabia, striking the

252 top An Israeli F-15C Baz during take-off from Tel Nof airbase. The aircraft, in service since 1976, has undergone various updating programs.

252 bottom A Kfir C2 flying over the Sinai. In building this multi-role fighter, which first flew in 1973, Israeli industry drew on experience gained from production of the Nesher fighter, derived from the Mirage.

target and returning to base. The mission was made possible thanks to the recent acquisition of the F-16A. Four years later Israel provided further confirmation of its ability to strike distant targets by hitting the headquarters of the PLO terrorist organization in Tunisia, about 1200 miles from Tel Aviv. In this case the raid was carried out with eight F-15Cs and Ds that made use of inflight refueling and precision weapons, such as the GBU-15 guided bombs.

The third mission that confirmed Chel Ha'Avir's capabilities was the "Peace in Galilee" operation, the invasion of southern Libya designed to counter the actions of Syria in the country and neutralize the PLO terrorists' bases. After Egypt had moved within the Western sphere and the signing of the Camp David agreements in 1978, Syria had become Israel's principal adversary. Israel organized its aerial offensive after the assassination of its ambassador in London, and began it on 4 June 1982, followed two days later by a land-based assault. The main air battles developed over the Beka'a Valley where the Israeli F-4, A-4 and Kfir fighter-bombers struck the Syrian anti-aircraft defenses and troop positions, while the F-15s and F-16s flew support missions and engaged the enemy fighters. On 10 June a ceasefire agreement was reached that in effect brought the aerial operations to a close, although the land war along the Libyan coast and in Beirut continued for some months.

The result of the air battles was stunning: in just two days the Syrians had lost 85 MiG-21 and MiG-23 fighters while the Israeli forces remained completely intact. Syrian losses totaled 92 aircraft against a dozen aircraft and helicopters lost by the Chel Ha'Avir to anti-aircraft fire. The reasons for this crushing victory are once again to be found in the superior quality of the Israeli men and machines. They were able to employ highly efficient tactics in terms of identifying and attacking enemy anti-aircraft systems, thanks in part to the use of remotely piloted drones, electronic warfare pods, and dedicated missile systems. Moreover, Israel's recently acquired Grumman E-2C Hawkeye radar planes (two of which were constantly in flight), and the Boeing 707 electronic warfare and reconnaissance aircraft played a vital role. They provided Israeli fighter pilots with a global, precise and updated view of the situation, identifying Syrian fighters as soon as they took off from their home bases. These American planes also enabled the Israelis to neutralize the enemy's communication and identification systems.

Since 1982 the Chel Ha'Avir has had no further need to fight a true large-scale war. However, the persistence of the motives underlying the Arab-Israeli conflict, which today primarily focus on jurisdiction over Jerusalem and the occupied territories, and previous pretexts for hatred, means that peace is no closer. Israeli aircraft continue to be used in action, frequently for isolated pre-emptive or punitive raids against terrorist forces. In the light of other crisis situations such as the Gulf War of 1991, the enduring hostility of countries such as Syria and Iran, and the appearance of new Islamic governments in the region, the Israeli air force has continued to maintain the level of its training programs and technology. It has updated some of its aircraft (such as its F-16s and F-15s) and acquired new cutting edge aircraft and weaponry. These are the F-15I multi-purpose fighters (derived from the USAF F-15E), which entered into service in 1998, and the Lockheed Martin F-16I, the latest model of the successful American fighter; in service since 2003, 102 units of this latter were acquired. Moreover, Boeing AH-64 Apache attack helicopters have been in service since 1990, including the advanced D model, which arrived to Israel in 2005, and a whole series of new precision weaponry, primarily of Israeli construction, like the Python, Derby, Delilah, and Popeye missiles, and the Spice bombs. There are also sophisticated unpiloted aircraft, like the Heron and the Hermes, and the aircraft for war and electronic reconnaissance, like the Gulfstream G550 CAEW and the G.5. The latest jewel in the Israeli Air Force's crown is the Lockheed F-35I Adir, which entered service in 2016 and has already been successfully deployed in real operations.

Owing to its advanced training and these new weapons systems, the Israeli Chel Ha'Avir continues to be one of the world's strongest and best prepared air forces.

253 top An Israeli AH-64A Apache during a flight demonstration. This helicopter has undergone several updates and is now even used in counter-terrorism operations.

253 bottom The model F16I is the most advanced model of the F-16 that Israel uses: deployed in 4 squadrons, they have acquired 102 units since 2003.

The early 1960s also saw the birth of two great aeronautical projects that were to produce contrasting planes. Both planes, however, were destined to achieve a place in history as exceptional aircraft of their time. Once again, economic and technological rivalry existed between Europe and America. France and Great Britain placed the emphasis on speed and decided to build a supersonic airliner; the United States instead placed faith in load-carrying capacity.

Back in 1962, the two European nations were working on similar projects for a high technology supersonic passenger jet that would be capable of breaking the American hegemony in the air transportation sector. It soon became evident that only with a joint venture in which expenses were shared would such an onerous project be possible and have any realistic chance of success.

The agreement (specifically, the Anglo-French Supersonic Aircraft Agreement) was signed by the two governments, which also committed themselves to financing the entire project. The participating manufacturing companies, Sud Aviation in France and BAC in the UK, would not be required to risk any capital. After the announcement was made, the American manufacturers began to worry. They already knew, thanks to the feasibility studies they had conducted, that because of the costs involved, the design and construction of a supersonic airliner could not be carried out by a single company. The major firms had thus erased the concept from their future plans.

The new aircraft was given the name "Concorde," which was particularly appropriate, given the traditional rivalry between France and Britain. Work on the two prototypes, one for each company, began in 1963.

From a technical point of view, the Concorde was a major challenge. The chosen format featured a tapering, 223-foot long fuselage and a delta wing that optimized performance both in supersonic flight and at lower take-off and landing speeds. Nonetheless, the Concorde would still be landing at about 177 mph, roughly the speed of an F-104 fighter! In order to allow an optimum angle of attack, a tilting nosecone or "droop-snoot" was fitted to provide the pilots with forward vision when the plane was in a nose-up attitude. Among the Concorde's other notable characteristics was its top speed, which had to be restricted to within Mach 2.2 or 1,354 mph at a cruising altitude of about 51,000 feet. This allowed for the airframe to be built in aluminum rather than the far more expensive stainless steel or titanium that would have otherwise been necessary with higher speeds, owing to the heat generated by air friction. Concorde was to be powered by four afterburner-equipped turbojets mounted in pairs beneath the wing.

256-257 The Concorde, whose slim nose, extreme lines and extraordinary speed was a revolution in commercial air travel in 1969.

256 bottom left A British Airways Concorde taking off from Heathrow in June 2001. The Concorde made its maiden flight in 1969. To date, only 20 have been built.

The Rolls-Royce/SNECMA Olympus 593 Mk. 610 units were specially built; each produced an exceptional 38,000 lb of thrust. The aircraft had a maximum take-off weight of just over 200 tons, and had a range of over 3,850 miles, thanks to the 34,000 gallons of fuel carried in its tanks.

While the development work continued, a number of shadows began to be cast over the project. The airlines felt that the aircraft was too small, but this was a relatively minor factor and the designer stretched the fuselage by 6 feet 6 inches to allow the total number of passengers to be increased to 136. There were two other significant problems, however. The first was the take-off noise and sonic boom that would accompany the Concorde's every flight, with the risk that many cities would refuse permission for the plane to land, citing reasons of public health. The second problem concerned sales. Initial projections had identified a market for between 160 and 400 examples to be sold over a period of about 20 years. By March 1967, however, only 74 options had been taken out by 16 airlines, including, of course, Air France and BOAC. The future appeared to be uncertain, but the Concorde was also a symbol of national pride for the two countries, and despite a degree of wavering (above all on the British side), the project went ahead. The French prototype successfully completed its subsonic maiden flight on 2 March 1969. In the meantime, two interesting things had happened in the United States. First, the news that the Soviet Union was also working on a supersonic civil airliner (the Tupolev 144) rekindled American interest – and pride. Congress approved a resolution to finance 75% of the costs of an American SST (Supersonic Transport) project, and this was all it took for the US to enter the arena once again. Second, after responding to a USAF requirement for a new large transport aircraft (the C-5), Boeing, had lost out to Lockheed. Boeing then decided (under pressure from Pan Am) to derive some advantage from its proposed plane by transforming it into a huge commercial aircraft, an aircraft that would carry two and half times as many passengers as the 707. Boeing, Lockheed and North American committed themselves to the SST project and in December 1966 the government chose the Boeing design. However, the program soon ran into serious technical problems concerning the aircraft's excessive overall weight, while its exorbitant costs aroused public opposition. These problems led to the government's inevitable cancellation of the SST project on 20 May 1971.

257 bottom The Tupolev Tu-144, extraordinarily similar to the Concorde, was the Soviet reply to its rival. Shown here is the presentation of the project in Moscow, May 1969. The second Tu-144 prototype flew at the Paris Air Show on 4 June 1973 but crashed shortly after, killing the six people on board. The Tu-144 was faster than Concorde but was never perfected.

Meanwhile, Boeing's giant airliner project, designated Model 747, had been having an easier ride. It was a vast machine, with a length of over 229 feet, a height of over 65 feet (the equivalent of a six-story building), a weight of well over 300 tons, and a surface area of over an acre. It was powered by four Pratt & Whitney JT9D engines; each developed 20,400 lb of thrust. The 747 was not dissimilar in layout to an enlarged 707, but its distinguishing feature was its fuselage, which was over 19 feet 6 inches wide and over 180 feet long, unblocked from nose to tail because the pilots' cabin and the first-class section were located on an upper deck. The width of the fuselage led to the coining of the term "wide body," a format that since then has characterized all large commercial airliners because of its characteristics of greater comfort and capacity.

Having received sufficient orders from Pan Am, Lufthansa and Japan Air Lines, on 25 July 1966, Boeing announced that construction of the aircraft was about to begin. Here too, however, a number of problems had to be tackled. The further the project advanced, the heavier the plane became; moreover, Boeing lacked a sufficiently large factory in which to assemble the 747 and had to build one ex novo at Everett, in the state of Washington. These difficulties were overcome, however, and the new Boeing 747 advanced toward its maiden flight, an event that took place on 9 February 1969, a month earlier than the first Concorde flight. Pan Am inaugurated scheduled 747 service on the New York-London route on 22 January 1970.

The 747's design had been slightly modified before reaching definitive form, thanks in part to suggestions from Pan Am. The definitive version of the 747 had a greater wingspan, a different undercarriage configuration, and a maximum take-off weight that had been increased to almost 340 tons. The 747's physical dimensions led to its becoming known universally as the "Jumbo Jet." The plane was by no means free of teething troubles; first and foremost, the engines were not sufficiently powerful and pilots were frequently obliged to use excessive rpm, causing overheating which in turn translated into more frequent repairs and higher maintenance costs. The passengers were initially unenthusiastic about the 747 because the engine problems caused delays and the on-board services (primarily meals) and amenities (specifically toilets) were inadequate, given that almost 400 people could be carried. Finally, the world's airports were not yet equipped for such large aircraft, and the baggage handling services in particular were put under severe strain. The problems were gradually resolved, but those relating to the engines persisted for a couple of years until the introduction of the JT9D-3A. In the same period, Boeing was facing another predicament: in 1970 the firm had accepted 190 orders from 28 different companies, but these were insufficient to cover the plane's development costs, while the onset of economic depression and the oil crisis meant that all the airlines

258-259 The "Jumbo Jet," with the pilots' cabin on the upper deck, has an unmistakable front profile. The 747-400 model can carry 568 passengers.

259 top With over 1,500 units built,, the B.747 was also developed into modern versions, such as the 747-8I and the 747-8F.

259 bottom A Virgin Atlantic Boeing 747-400. Nicknamed the Jumbo owing to its enormous size, the 747 first flew in 1969.

were facing severe difficulties. It was not until the end of the oil crisis and the renewed expansion of the economy that orders for the 747 started coming in and Boeing could breath a sigh of relief. By 1983, Boeing had delivered 588 of the 747s. The basic design was subjected to continuous improvement, and Boeing had introduced the 747-200B, powered by JT9D-7FW engines producing 50,000 lb of thrust, with a maximum take-off weight of over 400 tons; the shortened 747SP version suitable for extremely long-range flights; and the 747-300. Then came the 747-400, currently in production, and equipped with PW4056 engines producing 56,550 lb of thrust (or similar General Electric or Rolls-Royce units). This version has a maximum take-off weight of over 437 tons, a range of 8,370 miles, and a maximum capacity of 568 passengers. The latest versions placed on the market are the 747-8I, for passengers, and the 747-8F, for freight, certified in 2011. A total of 1,574 Jumbo Jets were built, and production ended in 2023.

In Europe in the meantime, the cancellation of the SST program and in the Soviet Union the failure of the Tupolev 144 had lent new confidence to the Concorde executives: despite the difficulties still facing the program, they at least knew that they no longer had any competition. What they had to do, however, was to start to sell the airplane. In 1971, five confirmed orders came in from BOAC and four from Air France. This was better than nothing, but it has to be said that the state airlines of Great Britain and France were virtually obliged to purchase the supersonic jet. It was therefore necessary for one of the world's major airlines to order the plane so that others would be convinced of its potential. Concorde executives decided to concentrate on Pan Am, but the airline then withdrew. Despite the projections presented, Pan Am decided not to take the risk: the Concorde was too expensive (at $65 million, it was already three times the price of the already expensive 747), and tickets would be too expensive for the vast majority of passengers.

In order to publicize the Concorde's capabilities, demonstration tours were organized throughout the world and it was decided to start up scheduled service as soon as possible.

Regular service in fact began on 21 January 1976, with an Air France flight to Buenos Aires and a British Airways (the successor to BOAC) flight from London to Bahrain. Difficulties regarding landing permits for the United States were finally overcome and flights to New York and Washington began within a few months. Unfortunately, from the outset the two companies found it very difficult to fill their planes, and even the transatlantic service was running at a loss.

In the end, none of the companies who had taken options to purchase the Concorde

260-261 An Iberian McDonnell-Douglas DC-10 landing. In 1977, the USAF selected this airplane as its future Transport/Tanker. They bought 60, and designated the plane the KC-10A Extender.

260 bottom An Air Canada Lockheed L-1011 TriStar as it lands. This aircraft was built at the end of the 1960s at the specific request of American Airlines, which wanted a large capacity, wide-bodied plane for medium- and short-range flights.

261 bottom A Japan Air Lines McDonnell Douglas MD-11 landing. This model was introduced in 1990 as the improved version of the McDonnell Douglas DC-10; it had new engines, new avionics and a longer fuselage.

actually took them up; only France and Britain, the two countries that had constructed the aircraft, operate it. Only 16 Concordes have been built, with each one eventually costing no less than $500 million. During the 1980s and 1990s the situation improved, but Concorde was doomed to remain little more than an exceptional symbol of prestige for the airlines and for the relatively few privileged passengers who have used it. A terrible accident involving an Air France Concorde in July 2000 decreed the beginning of the end for this splendid aircraft, which was withdrawn from service definitively on 26 November 2003.

In the meantime, the Jumbo Jet had opened the way for a new generation of wide-body jets. As result of the increase in air traffic, the airlines needed more capacious planes for their medium- and short-range services. At the behest of American Airlines, two new aircraft came into being. McDonnell-Douglas (formed by the St. Louis-based McDonnell's acquisition of Douglas) proposed the DC-10, which first flew on 29 August 1970. Work on the Lockheed L-1011 Tristar began in 1967, and the plane took to the air for the first time on 16 November 1970. Both were tri-engined airliners with a short landing capability and the possibility of carrying about 300 passengers in a wide-body fuselage. Despite initial sales successes, both planes then suffered a period of commercial stagnation, partly as a result of the mid-1970s' air transport crisis. A total of 250 Tristars were built, but this was insufficient to cover its development costs. The DC-10 was more successful, and in 1990 it was replaced by the MD-11, an improved version with new avionics, a lengthened fuselage and new engines. The aircraft remained in production until 2000, and a total of 646 units were produced, a figure that also includes the military KC-10 model.

In Europe, however, a new chapter in commercial aviation has opened. Airbus Industrie (a consortium France, Great Britain, and Germany set up in 1970 to meet European air transportation needs) has devoted itself to the creation of the A300, a wide-body jet for medium- and short-range routes. This aircraft has proved to the first of an exceptional range of airliners.

Chapter 21

Fighters of the Seventies

The 1970s saw the appearance of a new generation of combat aircraft. They were the result of lessons learned and experience gained during the most recent wars (the Vietnam and Yom Kippur wars above all), and also the availability of new technologies and the advances made in the computing sciences. In the 1970s, Cold War strategies and the enormous numerical threat that the Warsaw Pact forces posed obliged the United States to concentrate on the quality of its defense forces. The US had to exploit the technological gap between the West and the USSR in order to guarantee that US military forces were capable of facing up to and defeating any attack from the East.

The United States, the world's leader in aviation, introduced a number of fighters during the 1970s that, after appropriate improvements, still represent the backbone of its own military air forces and those of many allied countries. The first to appear was the new air superiority fighter designed for the US Navy for which the Navy announced a request for proposals in 1968 with the VFX program, aimed at replacement of the F-4 Phantom II following cancellation of the General Dynamics-Grumman F.111B. In 1969, the Grumman proposal was declared the winner of a design competition. The Grumman design reprised a number of the F.111B's features including the variable-geometry wing, the engines, and the weapons system that hinged on the AWG-9 long-range radar and AIM-54 Phoenix missiles. The first example of the aircraft, designated the F-14A Tomcat, took to the air on 21 December 1970. It was a two-seater with a tandem cockpit and was distinguished by a very wide fuselage with the two engines set well apart, twin rudders and fully mobile tailplanes. It was powered by two afterburner-equipped Pratt & Whitney turbofan engines each delivering about 20,900 lb of thrust, sufficient to propel the plane to a top speed of Mach 2.34 (about 1,545 mph) at high altitude. It had a wingspan of between 62 feet (minimum sweep) and 33 feet (maximum sweep), a length of 62 feet, and a maximum take-off weight of about 70,500 lb.

262 An F-14A Tomcat at low altitude over its aircraft carrier. In total, Grumman built 712 of these aircraft. The only ones exported were sold to Iran, which bought 80 in 1976-1978.

263 top A pair of Grumman F-14B Tomcats of the VF-103 squadron flying over the ocean. The fighter entered service in 1972 and is still considered an exceptional aircraft despite being close to retirement.

The weapons system comprised a Vulcan 20-mm cannon and eight AIM-9 Sidewinder, AIM-7 Sparrow or AIM-54 Sparrowhawk missiles. The principal combination was, however, the AWG-9 radar system with a range of over 180 miles and the Phoenix active guidance missile (fitted with its own radar), which had a range of over 80 miles. With this armament the Tomcat could provide the fleet with a degree of protection significantly greater than that offered by any other aircraft. In fact, the Tomcat took US Navy aviation to a new level of excellence.

During the development phase, three prototypes were lost, including the first prototype, which crashed nine days after its maiden flight. Nonetheless, the Tomcat demonstrated its worth, and deliveries to front-line units began in October 1972. The plane undertook its first operational cruise two years later, with fighter squadrons VF-1 and VF-2 being embarked on the USS Enterprise. The only country to which the Tomcat was exported was Iran, which between 1976 and 1978 (when the Shah still governed) acquired 79 planes. New versions of the aircraft were developed during the 1980s to resolve deficiencies in power output and to update its avionics.

The first F-14B (originally known as the F-14A Plus) flew in 1987; it was powered by new General Electric F110-GE-400 engines delivering almost 27,000 lb of thrust and fitted with new avionics. In 1990 the F-14D was introduced, fitted with the same engines and digital avionics, along with a new APG-71 radar system. Planes in these two versions were built either from scratch or by converting older F-14A airframes.

From 1997 on, part of the Tomcat fleet could also be used for bombing missions, thanks to the development of LANTIRN (Low-Altitude Navigation and Targeting Infra-Red for Night) pods for navigation and attack and laser-guided weaponry; these planes were nicknamed Bombcats.

In all, 712 Tomcats were produced, and the last remaining units were retired from service with the US Navy in September 2006, performing consistently well including in the two wars in Afghanistan and in Iraq.

In the mid-1960s, the USAF also turned its attention to a replacement for the F-4, and established the F-X program. McDonnell-Douglas was declared the winner of the competition in 1969, with an innovative design that was radically different from the

263 bottom Gunner loading 550-lb GBU-12 bombs under an F-14D of the VF-213 squadron in November 2001, during Operation Enduring Freedom.

In 1971, the USAF launched another competition, this time for an aircraft designated the LWF (Light-Weight Fighter). As a partner for the F-15, the USAF wanted a smaller multi-role, single-seat, single-engine fighter that would also be less expensive and capable of replacing part of the F-4 and A-7 fleets still in service. The competition was narrowed down to the General Dynamics YF-16 project and the Northrop YF-17, and in January 1975 the former was declared the winner. The General Dynamics fighter had already flown on 20 January 1974 and, although small, it immediately appeared to be highly advanced and have great potential. The plane's aerodynamics were futuristic, with the oval air intake placed beneath the belly and the fuselage making a significant contribution to the lift provided by the wing.

For the first time fly-by-wire computerized controls were used, while the cockpit featured a lateral joystick, a seat inclined at 30° to counter g-force and a fully glazed, high-visibility canopy. The thrust-weight ratio was again better than 1:1, and in flight the F-16 was extremely fast and maneuverable, to the extent that the first cases were even recorded of pilots losing consciousness (referred to as the g-LOC phenomenon) because of abrupt or extreme maneuvers.

The F-16A made its first flight in 1976, and the first operational unit received the aircraft in 1978, at Luke Air Force Base. In the meantime, in 1975, the F-16 had already earned a place in the history books thanks to the signing of the so-called "contract of the century" whereby four NATO countries, Belgium, Holland, Denmark and Norway, agreed to adopt the aircraft. In fact, 500 F-16s were, ordered to replace the old F-104Gs.

The F-16A had a wingspan of 30 feet, a length of 46 feet, and a maximum take-off weight of almost 35,400 lb, with external loads accounting for up to 15,200 lb. It was powered by the same engine as the F-15A, a single Pratt & Whitney F100-PW-100 unit delivering around 23,900 lb of thrust that was good for a top speed of Mach 2. The F-15C version (D was the two-seater) was introduced in 1983 and featured a new APG-68 radar system, new avionics, and a reinforced airframe that allowed the maximum take-off weight to be increased to about 37,500 lb. The F-15C Block 30 was the first version to introduce the possibility of selecting a General Electric F110-GE-110 engine (producing 28,000 lb of thrust) or a Pratt & Whitney F100-PW22o unit (23,900 lb), as well as improved avionics and weaponry such as the introduction of AIM-120 AMRAAM air-to-air missiles. In 1986 General Dynamics received a contract for the updating of the National Guard's 270 F-16As to the ADF specification that featured a new radio system, new APG-66 radar capabilities, and the ability to launch AIM-7 and Aim-120 missiles. The most recent model currently in service for the USAF is the Block 50/52, featuring advanced avionics, an F100-GE-129 engine delivering 29,000 lb of thrust, a maximum take-off weight increased to 42,295 lb, and the possibility of using the most modern weapons and navigation and attack systems. The F-16, later named the Fighting Falcon, which since 1993 has been a Lockheed-Martin project, has been subject

266 top A pair of F-16CJs of the 35th FW display their main weaponry, an AGM-88 HARM antiradar missile. The F-16C Block 50 with HTS pod specializes in taking out enemy air defenses.

266 bottom The Lockheed-Martin F-16 Fighting Falcon is the most widely used modern fighter in the world. Shown here is the rear cockpit of an F-16D belonging to South Carolina's National Guard.

266-267 The F-16 is a superb multi-role fighter for both aerial combat and attack missions. This USAF F-16C is flying over the 'no-fly-zone' in Iraq; it is armed with laser-guided GBU-12 bombs and AIM-120 and AIM-9M air-to-air missiles.

267 bottom The Falcon is highly maneuverable. This F-16C from the USAF 52nd Wing has two smoke containers that underline the aircraft's dramatic handling during an air show.

to continual improvements and has been exported to 26 countries around the world. The latest models in the series are the F-16E/F Block 60 and the F-16I Block 70/72, whereas in 2012 the firm presented the F-16V version, with AESA radar and last-generation avionics. Over 4,700 F-16s have now been produced, and it can safely be said that the Falcon is one of the most successful and widely used jet fighters of all time.

Another American plane developed in the same period was the Fairchild-Republic A-10 Thunderbolt Ii, which made its maiden flight on 10 May 1972. This was a plane developed by the company in response to the USAF's A-X request for a new close air support plane designed primarily to neutralize the threat of Soviet armored forces in Europe.

Early in 1973, the YA-10 was declared the winner at the expense of Northrop's YA-9, and the first production A-10As entered service in 1976. This was an aerodynamically innovative plane that placed firepower and survival capabilities above traditional fighter characteristics. It featured a large straight wing, two engines paired above the fuselage and shielded by the wing and the dual-rudder tailplanes. The titanium cockpit was sufficiently armored to resist 23-mm shells, and its fixed weaponry consisted of an enormous Gatling GAU-8a cannon with seven 30-mm rotating barrels, capable of perforating tank armor. Powered by two General electric TF34-GE-100 turbofan engines, each producing around 9,000 lb of thrust, the A-10 was a top speed of about 435 mph. The plane had a wingspan of 53 feet, a length of 53 feet, and a maximum take-off weight of 52,000 lb; it could carry a weapons load of up to 16,600 lb on eleven belly fittings.

The A-10A (with 716 planes built, all for the USAF) was considered to be virtually obsolete following the collapse of the threat posed by the Warsaw Pact and the Soviet Union. It was re-evaluated after its use in the Gulf War in 1991 and in NATO's aerial operations in the former Yugoslavia during the period between 1993 and 1999, as well as after its use in Operation Enduring Freedom in Afghanistan. Even though many planes have been withdrawn, the A-10A (designated as the OA-10A for forward air control missions) has been improved thanks to the Precision Engagement Upgrade Program (PEU) modernization program, beginning in 2005, that has transformed 356 units of the model A-10C, equipping them with new wings, new digital avionics, target designating pods, and new high precision weaponry.

268-269 Two A-10As launching heat flares designed to draw heat-seeking infra-red missiles off track. The A-10 can carry a varied payload of up to 16,000 lb.

269 top The A-10A's design was based on tactical considerations: high engines shielded by the dual-rudder tailplanes, a raised cockpit and armor-plating on the nose.

269 bottom The Fairchild-Republic (today Northrop-Grumman) A-10A Thunderbolt II was introduced at the end of the 1960s as a tactical support craft for the USAF. Its main characteristic was its 30-mm Gatling anti-tank cannon with seven rotating barrels.

Naturally, the Soviet Union was also proceeding with modernization of its aircraft in the1970s, introducing new planes that drew on the wartime experience and demands of the Soviet air force. The new planes had to match the operational capabilities of the United States and NATO. Even though the MiG-21 was an aircraft with excellent specifications, it lacked a weapons system capable of competing with those of the West's aircraft; moreover, the need to improve its maneuverability was recognized. Program 23 was thus launched, featuring a series of new components: Sapfir-23 radar, an AVM-23 analog computer, R-23 air-to-air missiles, and so on. Two prototypes were constructed, the 23-01 with a delta wing and the 23-11 with a variable-geometry wing, which (both) flew for the first time on 10 June 1967.

The second version was chosen for serial production, and 28t May 1969 saw the maiden flight of the MiG-23S (known to NATO as the "Flogger-A") that was, however, still retained theMiG-21's avionics. This aircraft was powered by an R27F2M-300 engine delivering 21,825 lb of thrust and was capable of a top speed of Mach 2.27. The airframe was particularly interesting as it broke with MiG traditions. The variable-geometry air intakes were located either side of the fuselage (like those of the Phantom) while beneath the tail was a stabilizing fin that folded away when the undercarriage was lowered. The variable-geometry wing (inspired by the F-111 design) improved take-off and landing characteristics while maintaining high-speed capabilities. The ground-attack version, designated the MiG-23B, flew for the first time on 20 August 1970; this model had a new nose section stripped of radar equipment, fixed air intakes, a Lyulka AL-21F3 engine, a reinforced airframe with larger tires, simplified avionics, and various other minor modifications.

The first true MiG-23 was the M version, which flew for the first time in June 1972, and immediately entered large-scale serial production. It had a wing span of between 43 feet (minimum sweep) and 23 feet (maximum sweep), a length of 53 feet, and was powered by a Soyuz/Khachaturov R29-300 engine producing about 27,550 lb of thrust and good for a speed of Mach 2.35. The MiG-23M's maximum take-off weight was about 45,550 lb, and it could carry weaponry weighing up to 4,400 lb, principally composed of six R-23 and R-60 missiles. In 1973, the strike version was renamed the MiG-27 and in 1983, after 910 had been built, the production line was transferred to India. Various versions of the fighter model were developed, some for export to friendly countries. The series culminated in the MiG-23MLD (or "Flogger–K"), which had an R-35-300 engine producing 28,660 lb of thrust, a radar system with close combat functions, a Shchel 3-U missile-guidance visor fitted to the pilot's helmet and R-73 missiles. By 1985, 5,047 MiG-23s had been produced, and it remained in service in the USSR/Russia until 1994. A total of 1,075 MiG-27s were built. The Mig23MLD and the Mig-27 were used by the air forces of over 20 countries, and they saw combat in the Middle East, Afghanistan, Angola and India, where they were reasonably successful.

Development of the MiG-25, a project initiated in the early 1960s before that of the MiG-23, involved very different plane, designed to respond to the threat posed by the American's XB-70 intercontinental supersonic bomber then in development. Exceptional performance was required of the new super-interceptor: a maximum speed of 1,865 mph, a maximum ceiling of around 65,500 feet and a range of 2,509 miles. In order to meet these criteria, the MiG design department came up with a large single-seat aircraft equipped with enormous turbofan engines, twin rudders at the rear and a high-mounted, moderately swept wing. The first prototype E-115R-1 (the reconnaissance model) completed its maiden

270 top The Russian MiG-23 fighter first flew in June 1967; 5,000 of the interceptor version have been built until 1985.

270 bottom The MiG-25RBF, the electronic recognition version of the trisonic MiG-25P fighter, appeared in 1981. Shown here is one from the Soviet airbase at Finow, East Germany.

flight on 6 March 1964, while the E-155P-1 interceptor took off six months later. Even though not all the performance requirements were achieved, the plane met the principal criteria and went into service on 13 June 1972 as the MiG-25P interceptor ("Foxbat-A" to NATO). This version was characterized by two Soyuz/Tumanski R-15B-300 engines, each producing almost 25,000 lb of thrust on afterburn, sufficient to power it to a top speed of Mach 2.83 (about 1,865 mph). It had a wingspan of 43 feet, a length of 69 feet, and a maximum take-off weight of about 90,400 lb. The weapons system was based on a Smerch-A radar and four R-40 missiles.

December 1972 saw the introduction of the MiG25-PD fitted with Sapfir-24 radar, R-40D missiles, and with further improvements to the avionics. The MiG-25BM version, dedicated to destroy enemy anti-aircraft defenses, was also introduced that year. Its principal armaments were four Kh-58 air-to-surface missiles. Production of the MiG-25 ended in 1985 with the 1,186th plane. Interceptor and reconnaissance versions were delivered to certain friendly countries: Algeria, Libya, Iraq, India and Syria. Only a few units remain in service throughout the world today, and these are primarily of the reconnaissance version. The MiG-25 also formed the basis for a new two-seat interceptor designed to provide greater range and independent capability to engage multiple targets.

270-271 The MiG-23B attack model was derived from the MiG-23S interceptor and first flew in 1970. It was a simple aircraft without nose radar. In 1973 it was renamed the MiG-27 and in 1983 production was transferred to India. Shown here are two Indian Air Force MiG-27s.

271 bottom right When the MiG-25 appeared in 1964, it was the fastest interceptor fighter ever built. It was retired in the Soviet Union in 1994. This is a two-seat training MiG-25RU during landing.

The first to be built, designated the MiG-25MP, completed its maiden flight on 16 September 1975, while December 1976 saw the first appearance of the pre-production MiG-31 ("Foxhound"). This was a heavier, more powerful fighter than the Foxbat. It featured a two-seat tandem cockpit, a reinforced undercarriage, a Zaslon radar system capable of tracking ten targets and attacking four simultaneously, and a weapons system composed of four-33 semi-active guidance missiles with a range of 75 miles, and two R-40TD missiles. It was powered by two Aviadvigatel/Solovyov D-30F-6 engines, each producing over 34,000 lb of thrust with afterburner, and could attain a top speed of around 1,865 mph. The MiG-31 had a wingspan of 43 feet, and a length of 72 feet, and a maximum take-off weight of about 102,000 lb. It went into service in December 1981, and since then over 500 have been built. The MiG-31M version was introduced in 1986, and features R-37 missiles, Zaslon-M radar, a maximum take-off weight increased to more than 114,500 lb, a greater internal fuel capacity, and more powerful engines. Only a few have been built because of the crisis the Russian aviation industry after the breakdown of the Soviet Union. Recently, with the improvement of Russia's economic situation, there has appeared a willingness to revitalize this powerful interceptor. In 2011, the Ministry of Defense signed a contract to upgrade 60 units to the MiG-31BM standard (conceived in 1999 and updated in 2005), which includes new avionics, the new Zaslon-AM radar, and new K-37 and K-77 missiles. Recently, these aircraft were updated with Kh-47M2 Kinzhal hypersonic missile capabilities.

272 A heavy two-seat MiG-31 fighter climbing in Soviet skies. The fastest interceptor plane in the world, it can fly at 1,740 mph.

272 bottom A direct derivative of the two-seater Su-27UB, the Sukhoi Su-30MK multipurpose fighter looks very similar to it externally.

The MiG-23's successor as the Soviet Union's principal air defense fighter was being discussed as early as 1969, when the government issued a request for a new fighter in the same class as the American F-15. Subsequently, the request was divided into two parts, and the MiG engineers concentrated on the creation of a lightweight fighter, the preliminary design for which was accepted in 1974. The plane was heavily influenced by the American designs, featuring twin rudders, a wing integrated with the fuselage, and square-cut belly air intakes. The first prototype, the 9-01, took to the air on 6 October 1977. Its official designation was the MiG-29, but NATO identified it as the Ram-L. The two-seat MiG-29UB flew in 1981, and serial production of the MIG-29B began the following year.

Deliveries to operational squadrons began in August 1983. The MiG-29B, finally designated the "Fulcrum-A' by NATO, was powered by a pair of Klimov/Sakisov RD-33 engines each producing 18,300 lb thrust on afterburn and good for a top speed of Mach 2.3 (close to 1,500 mph). The plane had a maximum take-off weight of 48,500 lb, of which external armaments in the form of up to six R-27, R-73 and R-60 air-to-air missiles and a 30-mm cannon accounted for about 4,400 lb. The MiG-29 could carry bombs and ground attack rockets if required. The plane's RLPK-29 radar system had a range of over 40 miles and was capable of tracking ten targets and attacking one.

The Fulcrum soon proved to be an excellent fighter, agile and capable of achieving extreme angles of attack. It was particularly deadly in dogfights, thanks in part to the capability of guiding missiles by means of a visor built into the pilot's helmet. In 1992, production concentrated on the MiG-29S, which was equipped with an improved radar system and flight commands, and with a weapons load increased to 8,800 lb. The MiG-29 has been built in a further 20 other successive variants, like the MiG-29M, the MiG-33 and MiG-35 for exportation, the naval MiG-29K, and the more recent multirole MiG-29SM and SMT versions. To date over 1700 units have been produced and are in use by 34 countries throughout the world. A direct descendant of the MiG-29 is the MiG-35, which completed its maiden flight in December 2016 and went into operational service in 2023.

The Sukhoi industry has also produced a number of outstanding aircraft, such as the Su-24 fighter-bombers, which entered into service in 1974, and the Su-25, adopted beginning in 1980. The most interesting aircraft of those years, however, was the Su-27, the fruit of the 1969 PFI program for a new high-performance fighter to provide air superiority. The Sukhoi project was designated T-10 and took its maiden flight on 20 May 1977. It was a large aircraft with aerodynamic solutions vaguely resembling those of the American F-14 and F-15. With a wingspan of 48 feet and an overall length of 72 feet, the series production aircraft, designated the Su-27, had a maximum take-off weight of 72,750 lb. It was powered by two Saturn AL-31F turbofans providing 27,560 lb thrust, for a top speed of about 1,550 mph. The payload could include up to 17,600 lb of exterior load and up to 10 air-to-air missiles. The Su-27 (coded "Flanker" by NATO) entered into service in 1984 and was followed by the improved Su-27S model, by the two-seater trainer Su-27UB, and by the Su-27K, the version used for aircraft carriers, equipped with folding wings and landing hook. The Su-27 has been exported, as the base model Su-27SK/UBK, to some dozen countries, with more than 600 total units having been built. The Su-27 has also been acquired by China and is produced under license as the J-11.

273 top A MiG-29A fighter belonging to the Slovak air force. More than 1,500 were built and used by more than 25 countries around the world.

273 bottom right A MiG-31B being rolled in after landing, with its parachute-brake still attached. More than 500 of these planes were built; they can carry up to 10 missiles.

274-275 An RAF Tornado ADV F Mk.3 interceptor. This plane was developed from the IDS attack model to meet RAF needs. If first flew in 1979.

274 bottom Three Tornado ECRs of the German wing JaBoG.32 flying in formation in the skies over Bavaria. The Luftwaffe has been employing this aircraft since 1990.

275 top The Italian Aeronautica Militare took 100 Tornado IDS into service after 1982. This is the line of aircraft used in the Desert Shield and Desert Storm operations before they left Italy.

During this period, another important combat aircraft was developed, but not by either of the two superpowers. Instead, Panavia, a European multinational corporation established by the principal aeronautical companies of Great Britain, Germany, and Italy on 26 March 1969, developed and built it. Panavia's Multi-Role Combat Aircraft (MRCA) project involved not only a new combat aircraft (to be constructed by BAC, MBB and Aeritalia), but also a new turbofan engine (entrusted to the tri-national Turbo Union corporation). The aircraft was essentially an all-weather, low-level fighter-bomber whose layout featured a two-seat tandem cockpit for pilot and navigator, a high-mounted variable-geometry wing, twin engines, large tailplanes with a single rudder, and sophisticated radar and avionics. Named the PA-200 Tornado, the German prototype P.01 completed its maiden flight on 14 August 1974. Pre-production planes began to fly in 1977, and in June 1979 the first of the interdiction-and-strike (IDS) production planes was delivered to the Royal Air Force.

The Tornado had a wingspan of between 43 feet (minimum sweep) and 26 feet (maximum sweep), a length of 53 feet, and a maximum take-off weight of about 61,700 lb, of which external loads accounted for about 19,850 lb. It was powered by two RB.199 Mk.101 turbofans, each producing 16,000 lb of thrust on afterburn, and could attain a top speed of 920 mph at low altitude.

The basic IDS version was subsequently optimized for reconnaissance and anti-shipping

275 center and bottom
An RAF Tornado ADV passing at high speed. The wings are swept back as much as possible; the humidity of the air creates a cloud where aerodynamic pressure is greatest.

strike missions through addition of pods and dedicated weaponry. In total, the RAF received 229 IDS planes, Germany 332, and Italy 100.

In order to satisfy British requirements for an interceptor, Panavia then developed the ADV (Air Defense Variant), which made its first flight on 27 October 1979. The main differences were sleeker aerodynamics resulting from lengthening the fuselage to 59 feet, addition of the Foxhunter long-range radar system and new avionics, removal of one of the two 27-mm cannons, adoption of RB.199 Mk.104 engines producing about 16,500 lb of thrust.

The new version had a weapons system composed of four medium-range Sky Flash missiles and four AIM-9 short-range missiles. The RAF eventually acquired 173 Tornado ADVs, with which it replaced its F-4 Phantoms. Italy also benefitted from use of the Tornado ADV, having leased 24 units from the RAF during the period 1995-2004. The last version of the Tornado was the ECR (electronic combat reconnaissance) variant, dedicated primarily to suppression of anti-aircraft defenses. A total of 35 Tornado ECRs were built for the German Luftwaffe (entering service in 1990), and 15 for the Italian air force, which has used them since 1998.

The ECR is fitted with sophisticated electronic equipment for identification of enemy signals (ELS and RHWR), and dedicated anti-radar weapons such as AGM-88 HARM air-surface missiles; in addition, it is fitted with the latest version of the RB.199 engine, the Mk.105, producing 16,800 lb of thrust. A total of 993 Tornados have been built in various versions, including 96 of the IDS and 24 of the ADV, which were exported to Saudi Arabia.

The aircraft still in service have been subjected to updating programs aimed at keeping them operational until 2027 lthough Great Britain withdrew the model completely in 2019.

The Dassault Mirage 2000 was last important fighter of the 1970s. The plane, approved by the French government in December 1975, was designed to provide the Armée de l'Air with a replacement for the Mirage F.1. The first prototype of the new fighter completed its maiden flight on 10 March 1978; it marked a return to the delta- wing layout that once again offered advantages, thanks to the introduction of fly-by-wire controls, new aerodynamic features, and a thrust-weight ratio close to parity. In November 1982, following intensive development involving five prototypes, the French air force took delivery of the first production Mirage 2000Cs, optimized for interception and air superiority roles. The aircraft was 46 feet long, with a wingspan of 30 feet, and had a maximum take-off weight of about 24,000 lb. Its SNECMA M53-5 engines produced 18,850 lb of thrust and gave the aircraft a top speed of Mach 2.2 (about 1,430 mph). The weapons system was based on a mission computer, a Thomson-CSF RDM radar system, and up to four Matra Super 530D and 550 Magic 2 missiles, along with two 30-mm cannons. After the 38th plane had been built, all further ones were fitted with the RDI radar system and M53-P2 engine, which produced about 21,400 lb of thrust; both modifications added to the plane's performance.

The year 1983 saw the maiden flight of the 2000N. This version was designed for nuclear strike roles. It had a tandem cockpit with a second seat for a co-pilot dealing with the weapons and also radar and avionics systems optimized for low-level operations. The 2000N formed the basis for the 2000D version, which was designed for attacks using conventional weapons; it was fitted with new avionics and the capability for using sophisticated precision weaponry.

In 1984, the Mirage 2000E was introduced as the export version of the 2000C, but with multi-role capabilities. Next, in October 1990, the 2000-5 model completed its maiden flight. This version incorporated a redesigned cockpit with multifunction screens and digital instrumentation, a new RDY multi-mode radar system, and up to six new radar- and infrared-guided Mica missiles, with multi-role capabilities. The first 2000-5s were delivered in 1996, and they are currently in use by the air forces of France (37 units designated as 2000-5F), Qatar, the UAE, Taiwan, and Greece. The Mirage 2000 was built up until 2007, for a total of 601 units in use by 10 nations, including Ukraine, which has received an unspecified number of second-hand units from France beginning in February 2025.

276 top Four Mirage 2000-9DADs of the United Arab Emirates Air Force flying in formation. This model is equipped with advanced multi-role capabilities.

276 bottom The Mirage 2000N, the nuclear attack version of the Mirage 2000 family, now modified for conventional weapons missions. Shown here is a rocket-carrying Mirage of the French air force's 4th Wing.

277 top A Mirage 2000C interceptor of the French air force refueling in flight. The 2000-5 version in part resolved the problem created by the 2000C's short flight range.

277 bottom A Mirage 2000-5F operated by the French Air Force's Escadron de Chasse 1/2 Cigcognes seen landing at the Orange base in 2010.

Chapter 22

The Third Level

The growth in commercial air traffic after the Second World War, and in particular its expansion during the 1950s, created a marked divide. The divide separated the major state or international airlines that served international capitals and major cities from the medium-to-small companies that concentrated on shorter routes, frequently within individual countries and geographical areas. This second type of business came to be defined as third level or regional air transport, and the aircraft used were grouped under the umbrella term "commuter planes." The first generation of this type of aircraft came into being during the second half of the 1960s.

From the outset, the principal characteristics of commuter aircraft have been their reduced dimensions, with restricted numbers of seats (usually between 10-20 and 30-40), propeller or turboprop propulsion, relatively short take-off and landing capabilities (STOL), and low operating costs. Regional transport aircraft were used, and still are used, primarily to link secondary airfields with major airports (the so-called hubs) and vice versa. Passengers can reach the most important cities by air and then take advantage of connections to international and intercontinental flights departing from and arriving at the hubs. The United States was the first nation to distinguish and develop this sector. Its vast territory naturally demanded the creation of a dense network of air traffic routes linking the principal cities within a state to each other and linking the various states to one another and to the major metropolises.

Oddly enough, however, despite recognition of the third level market, the American aviation industry remained focused on the production of jet airliners with over 100 seats and took relatively little interest in the third level's equipment needs. While Fairchild, Beechcraft and de Havilland Canada did produce some significant third-level aircraft in North America, many of the smaller airlines relied on planes built abroad.

The Swearingen Merlin IIA, an eight-seat turboprop designed in 1964, was the first of a successful range of American commuter planes. Five years later, the Merlin was used as the basis for the 226TC Metro; it was fitted with new engines and a stretched fuselage that could accommodate up to 20 passengers, the Metro first flew on 26 August 1969 and was immediately sold to Air Wisconsin and Mississippi Valley Airlines. Later Fairchild acquired the Swearingen project, and introduced the structurally modified Metro II fitted with 950-shp engines. Then came the Metro III, with a

278 top The de Havilland Canada DHC 8, known as the Dash 8, made its maiden flight in 1983. It was developed to provide a quiet, economical plane for regional flights.

greater wingspan, 1,100-shp engines, and a higher maximum take-off weight. The Metro 23, certified in June 1990, is the latest version; it has a wingspan of 55 feet 8 inches, a length of 59 feet, and is powered by two AlliedSignal TPE-12UHR engines, each delivering 1,100 shp. The plane has a maximum take-off weight of 16,500 lb, and can reach a top speed of 348 mph. Staffed by a crew of two pilots and a flight assistant, the Metro 23 can carry up to 19 passengers. The cargo version, known as the Expediter, can carry a load of 5,500 lb. The Metro remained in production until 2001, with over 600 units built.

Beechcraft meanwhile developed its Model 99. It mated a fuselage capable of accommodating 15 passengers to a wing design based on the 65/80 Queen Air, and fitted it with two 557-shp turboprop engines. The prototype first flew in December 1965, and deliveries to clients began three years later. The C99 Commuter, the last model in the series, boasted two 774-shp turboprop engines and was produced until 1987. In the meantime, the Beechcraft 1900 had already appeared; it made its maiden flight on 3 September 1983. This was a larger aircraft a T-tail similar to that of the Super King Air, two Pratt & Whitney Canada PT6A turboprops delivering 1,110 shp and a cruising speed of 295 mph; maximum take-off weight was 16,600 lb. The fuselage could accommodate up to 19 passengers or 4,500 lb of cargo. The improved 1900D version appeared in March 1990, equipped with 1,280-shp engines and four-bladed propellers that provided a top cruising speed of about 330 mph. This version had a maximum take-off weight of 17,550 lb. Over 680 1900Ds have been built. Production was terminated in 2002, with 695 built.

In the mid-1960s de Havilland Canada (DHC) introduced an aircraft designed specifically for regional operations. It flew for the first time on 20 May 1965, and was named the DHC 6 Twin Otter. Designed with STOL capabilities, this rugged plane featured a fixed undercarriage, a high wing, a particularly robust airframe and twin turboprop engines. The Twin Otter had two PT6A engines that gave it a top speed of about 217 mph; the cabin accommodated 18 passengers, and the plane had a maximum take-off weight of 12,345 lb. With its specifications and the possibility of fitting snow skis or floats to turn it into a snow surface or seaplane, the Twin Otter was particularly attractive to companies operating in difficult terrain such as the American Great North, and it was produced until the late 1980s. Given the Twin Otter's success, DHC decided to proceed with the design of a larger, more capacious aircraft with the same characteristics of robustness and economy as its predecessor.

The DHC 7 Dash-7 completed its maiden flight on 27 March 1975. It was an all-new plane with a fuselage capable of accommodating up to 54 passengers or 11,680 lb of cargo. The Dash-7 had four 1,130-shp PT6A engines, a top speed of 280 mph, and a maximum take-off weight of 44,000 lb. The Dash-7's STOL capabilities were enhanced by the use of large double-slot flaps; the plane could operate from runways less than 2,000 feet long. The success of this model was limited, however, and the similar but more modern Dash-8 replacement completed its maiden flight on 20 June 1983. This plane featured twin engines and an increased top speed, but had a lower loading capacity. The Dash-8 could plane could carry 39 passengers or about 7,825 lb of cargo. Its two Pratt & Whitney turboprop engines each delivered 2,000 shp and a cruising speed of 310 mph. Responding to market demands for an even better plane, DHC then introduced the enlarged and uprated Dash-8 200, 300 and 400 versions. The latest variant, the 70-seat Q400 flew for the first time on 31 January 1998. To date, over 1100 Dash-8 aircraft in all the various forms have been sold, a fact that places Bombardier (Canadair's successor company) among the leaders in this manufacturing sector.

278 bottom The dual turboprop Beech 99 has a conventional design. It remained in production from 1968 till 1983.

279 top This de Havilland Canada DHC 6 Twin Otter is the hydroplane version. It was designed in 1964 to provide a regional aircraft capable of dealing with Canada's difficult terrain.

279 center left The Fairchild Metro II is one of the many light commuter planes derived from the Swearingen Merlin II, which first flew in 1965.

279 center right A four-engined de Havilland Canada DHC 7 (Dash 7) flown by the British Brymon company. It was designed to carry passengers and goods, and has the advantages of short take-off and landing capabilities.

279 bottom The Beechcraft 1900D, introduced in 1990, improved on the 1900 model with more powerful engines, a four-bladed propeller, and greater carrying capacity.

The South American manufacturer Brasiliana Embraer is another producer of successful commuter aircraft. This company, founded at the behest of the Brazilian government in the late 1960s, concentrated on the production of an original aircraft, a twin-engined light transport destined for the Brazilian air force. The aircraft was designed by the Frenchman Max Holste and was a fairly traditional, rather unsophisticated but robust design with a low wing, two PT6A turboprop engines and conventional tailplanes.

The prototype, designated YC-95, first flew on 26 October 1968, and the aircraft entered military service as the C-95 in transport, search-and-rescue, and maritime patrol roles. The civil version with a stretched fuselage was designated the EMB-110P Bandeirante, and was converted to seat up to 18 passengers. The EMB-110P had a maximum take-off weight of 12,500 lb, and was powered by two 750-shp engines providing a top speed of 286 mph. By 1994, 469 Bandeirantes had been built. The EMB-120 Brasilia model followed the 120, making its maiden flight on 27 June 1983, with the first production deliveries made in 1985 to the America Atlantic Southeast company. The new Brasilia represents a considerable all-round improvement over the earlier plane, and is equipped with a T-tail and two Pratt & Whitney Canada PW118 engines producing 1,800 shp. It is capable of a cruising speed of 342 mph and can carry up to 30 passengers or just over 7,700 lb of cargo. In all, seven different versions of the Bandeirante have been developed to respond to diverse requirements and in all more than 350 planes have been sold. A military version is also offered. Production was terminated in 2007.

The greatest proliferation of third-level transport aircraft occurred in Europe, where the aeronautical industries of Great Britain, Holland, Germany, France, Spain, Italy and Sweden have produced a number of designs. The first was the Fokker F27 Friendship, a design conceived as early as 1950 and developed two years later, thanks to the backing of the Dutch government. The prototype first flew on 24 November 1955, and presented features that were to become familiar characteristics of commuter planes in general: a high wing, turboprop engines, traditional tailplanes, a retractable undercarriage and a pressurized fuselage with a maximum capacity of 48 seats. The initial Mk.100 series was equipped with Rolls-Royce Dart 514 engines, but they were replaced in the Mk.200 by more powerful Dart 532-7R engines that developed 2,285 shp. The Mk.200 had a maximum take-off weight of 45,000 lb, and a cruising speed of about 300 mph. Production got under way in 1958 in both the Netherlands and the United States, where Fairchild had acquired a license to manufacture it. The Fokker 27 finally went out of production in 1987, with about 785 planes having been built

Prior to this, the Dutch company had launched the Fokker F50 in 1983. In effect, the F50 was a complete revision of the earlier model, with 80 percent of the components being replaced. The first F50 flew on 28 December 1985, and deliveries began in 1987, with Lufthansa Cityline being the first purchaser.

280 top left The Britten Norman BN2 Islander is a successful commuter plane also used for search and rescue operations. More than 1,200 have been built.

280 center The Brazilian Embraer company was also busy at the end of the 1960s producing commuter planes. The EMB-110 Bandeirante shown here entered production in 1972.

280 bottom The Fokker F-27 Friendship was one of the first regional turboprop planes. The development project, which was launched in 1950, benefited from grants from the Dutch government.

The plane was equipped with two 2,500-shp Pratt & Whitney Canada PW125B turboprop engines driving six-bladed propellers and providing a top cruising speed of 330 mph. A maximum of 58 passengers could be carried.

A total of 319 of the F50s (and of the stretch F60 version) were built, until 1997. The F27 and the F50 were sold throughout the world, and the former in particular was a true leader in its market sector. Fokker later produced the models Fokker 70 and 100, derived from the Fokker 28. The 100 was produced from 1986 to 1997, with 283 units built. Fokker declared bankruptcy in 1996, and its aviation construction activities were taken up by Stork, now renamed Fokker Technologies.

The British Britten-Norman BN 2 Islander was another of the earliest true European commuter planes, and was specially designed for short-haul flights. The aircraft completed its maiden flight on 13 June 1965 and went into service two years later with Aurigny and Loganair. The Islander was a small plane with twin piston engines, a high wing, a fixed undercarriage, a square-section fuselage, STOL capabilities, and room for nine passengers. The improved BN 2A version went into production in 1969, while 1970 saw the introduction of a tri-motor variant designated the BN 2A Mk.III Trislander. This version had a longer fuselage that accommodated more passengers and a third engine installed on the rudder. It never enjoyed significant sales success, and went out

version was powered by the ubiquitous 1,215-shp PT6A engine, and had a maximum take-off weight of 23,000 lb and a cruising speed of about 217 mph. The 330 formed the basis for the Sherpa cargo plane and, above all, the model 360 that flew for the first time on 1 June 1981. The 360 was a larger aircraft, with 36 seats, new traditional tailplanes and 1,440-shp PT6A engines. The maximum weight at take-off was increased to 26,450 lb and the cruising speed to about 245 mph. Modified versions of the Shorts 360 were subsequently developed and featured new six-bladed propellers.

On 28 March 1981 the British Aerospace Jetstream 31 made its maiden flight. It was a slim twin-turboprop plane with a low wing and a retractable undercarriage derived from the Handley Page HP.137 Jetstream, an aircraft developed in the late 1960s. While the first Jetstream had not been particularly successful, the Jetstream 31 enjoyed far better fortunes. It introduced new Garret TPR331 engines producing 950 shp and equipped with four-bladed propellers. These power units brought a considerable increase in performance, the top speed rising to 298 mph and the maximum take-off weight to about 15,325 lb, with a payload of up to 19 passengers or about 3,950 lb of cargo. In the 1980s, this aircraft was virtually the market leader, winning, for example, 60 percent of the difficult American market's commuter segment. In 1988, the Super 31 flew for the first time; it was equipped with 1,100-shp AlliedSignal TPE331-12UAR engines, which provided a top speed of 304 mph and permitted a maximum take-off weight of 16,200 lb. In 1991, the Jetstream 41 appeared, fitted with more powerful engines and with room for up to 29 passengers, but its production was terminated already in 1997.

280-281 The Embraer EMB-120RT Brasilia, introduced in 1983, was a natural successor to the Bandeirante. The slender twin turboprop is shown here displaying Flight West Airlines' colors.

281 center The British Aerospace Jetstream 31 of 1980 derived from the same-name aircraft built by Handley Page at the end of the 1960s. The Jetstream Super has been in service since 1988.

281 bottom The Shorts 360 made its first flight in 1974 as the SD3-30. It is used to carry passengers, cargo and for military use.

of production in 1984. However, the original Islander is still in production, in both piston-engined form and in a version equipped with turboprop engines. Over 1,300 have been built. Since 1977 the company has been owned by the Swiss firm. The BN 2T model is powered by two 320-shp Allison 250-B17C turboprop engines and can maintain a top cruising speed of about 195 mph. It has a maximum take-off weight of 7,000 lb, and can take-off from runways just 835 feet long.

The Northern Irish Shorts company also entered the sector, first in 1963 with the Skyvan, a rustic twin-turboprop aircraft with a large square-section fuselage, suitable for both civil and military purposes, and then with the 330, which flew for the first time on 22 August 1974. This aircraft, which also had STOL capabilities, was developed from the earlier plane, but had a greater wingspan and a new stretched fuselage equipped with a retractable undercarriage. It could seat up to 30 passengers. The 330-200

282 top The CASA-Nurtanio CN-235 resulted from a Spanish-Indonesian cooperation on construction of a civil and military transport plane.

282 bottom left ATR (Avions de Transport Regional) is the company formed by Alenia and Aerospatiale to produce a third level turboprop transport plane. The ATR-42 made its first flight in 1984; since then it has been a best-selling plane in its class.

282 bottom right The SAAB 2000 twin turboprop, developed from the 340 model at the end of the 1980s, was designed to be a faster, more comfortable plane.

282-283 A SAAB flown by British Airways. At the start of the 1980s, the American Fairchild and Swedish SAAB companies joined forces to produce a new regional transport plane. The resulting 430 first flew in 1983 and has met with commercial success.

In Sweden, Saab created a family of highly successful regional air transport planes originally born out of a joint venture with Fairchild in America. On 25 January 1983, the Saab-Fairchild SF.340A, which became the Metro II's successor, completed its maiden flight. It was a fairly large plane, with a low wing, a retractable undercarriage and sleek lines, and was powered by two General Electric CT7-5A2 engines producing 1,750 shp, good for a cruising speed of 320 mph and a maximum take-off weight of 27,270 lb. The 340A could carry 35 passengers or up to 7,580 lb of cargo. Fairchild withdrew from the series 1985, but Saab went on to introduce the Saab 340B in 1989. This version had a higher maximum take-off weight of 29,200 lb, allowing for more passengers or cargo. The Swiss airline Crossair was the 340's first purchaser, as it would be for its successor, the Saab 2000.

The Saab 2000 completed its maiden flight on 26 March 1992, earning the nickname "Concordette" because at the time it was the world's fastest turboprop. The 2000 represented a redesign of the 340 and shared numerous components with it. However, the 2000 had a greater wingspan, a longer fuselage capable of seating up to 58 passengers, 4,150-shp Allison GMA2001 engines, a maximum take-off weigh of 48,500 lb, and a top speed of 422 mph. Despite predictions to the contrary, the 2000 never achieved the hoped-for commercial success, perhaps owing to cost. It went out of production in 1999 after 64 planes had been built. Production of the 340 also ceased that year after the 461st plane had been delivered. A military version of the 340, the S110B Argus, was also produced and used by the Swedish air force as radar plane.

Spain was also active in the commuter sector, initially with an aircraft derived from a military design, The CAA C-212 Aviocar was designed in the early 1970s as a transport plane for the Spanish armed forces. It was a rugged STOL design characterized by a square-section fuselage with a high wing and traditional tailplanes over the rear cargo door. The C-212 completed its maiden flight on 26 March 1971, and enjoyed reasonable sales in both military and civilian versions, the latter optimized for passenger- and cargo-carrying duties. The most recent model in the series is the 212-300, capable of seating 26 passengers or carrying about 6,170 lb of cargo, with a maximum take-off weight of 16,975 lb. This version is powered by two 900-shp AlliedSignal TPE331 turboprops, and has a cruising speed of 220 mph. The 212M military version is certified for slightly higher take-off weight and can perform various duties such as parachute-drops, air-ambulance, reconnaissance and observation.

In 1979, CASA signed an agreement with the Indonesian company Nurtanio for the design and construction of a new, second-generation civilian transport plane for the third level market, with possible military applications. Designated as the CN-235 and built by the Airtech consortium, the plane had a similar layout to the previous model, but was larger and sleeker. The prototype flew for the first time on 11 November 1983. The initial series 10 version was followed by the 100/110 series equipped with two 1,750-shp General Electric CT7-9C turboprops driving four-bladed propellers and guaranteeing a top speed of 298 mph and a maximum take-off weight of 33,289 lb (34,833 lb for the 200 series). Planes in the 100/110 series could carry a payload of up to 44 passengers or 8,800 lb of cargo (9,480 lb for the 200 series). Over 220 CN-235s were built, with the majority sold to military operators. The plane also formed the basis for the CN-295, an enlarged version that is currently under development.

The early 1980s saw the proliferation of new projects for regional air transport, a sector then in rapid expansion. The development of a new aircraft for this sector led to the establishment in 1982 of the ATR (Avions de Transport Régional) consortium, founded by the French firm Aérospatiale and the Italian firm Alenia. The result was the ATR-42, which completed its maiden flight on 16 August 1984. The first production plane was delivered to the French Air Littoral company in December 1983. The aircraft reflected fairly traditional design, with a high wing, a T-tail, and two 2,000-shp Pratt & Whitney Canada PW120 turboprop engines driving four-bladed propellers. These engines were used throughout the ATR-42-300 series. Later, the more powerful PW121 engines were adopted, and finally the PW127Es, which were used in the ATR-42-500 series. The ATR-42-500 has a power output of 2,750-shp per engine (restricted to 2,160-shp), offering a cruising speed of 348 mph and a maximum take-off weight of 41,000 lb. ATR-42-500 aircraft could carry up to 50 passengers, the equivalent of a payload of 1,190 lb.

In 1985, ATR announced the introduction of a stretched 72-seater version of the plane, appropriately designated the ATR-72, to satisfy operators' demands for greater capacity. The first prototype took to the air on 27 October 1988, and in 1989 the first aircraft were already being delivered to the Finnish carrier, KarAir. The most powerful version currently in production is the ATR-72-600, launched in 2007 and equipped with 2,750-shp PW127M engines that permit a cruising speed of over 370 mph and a maximum take-off weight of 49,600 lb. The ATR series has sold over 1800 units to date, including the military versions, and the planes have won a leading position in the third level sector, with deliveries to commercial operators throughout the world.

283 center right The CASA 212 resulted from the Spanish air force's need in the early 1970s to replace some of its transport planes. The military model was followed by several civilian passenger and cargo versions.

283 bottom The ATR-72 of 1988 was developed from the basic ATR-42. It has more powerful engines and a lengthened fuselage to take more passengers.

Following the airline deregulation of the 1990s, European regional air transportation boomed. The new frontier in commuter planes became represented by jet aircraft, better suited to satisfying the demands of operators on the numbers of passengers carried and the reduction of flight times.

The Hawker Siddeley HS.146 was first aircraft of this new generation. It was actually designed and developed back in 1973, and completed its maiden flight on 3 September 1981, named the British Aerospace 146. The 146 had a high wing and a fairly traditional T-tail, but was characterized by its four particularly quiet turbofan engines that made it suitable for operation from inner-city airports. The BAe 146 100 was fitted with Avco Lycoming ALF502R-5 engines producing 7,000 lb of thrust; it had a top cruising speed of 441 mph, a maximum take-off weight of about 93,000 lb, and could carry up to 93 passengers. The first customer was the British DanAir company, which began operating the 146 in 1983. Subsequent development led to the introduction of the 200 and 300 series with stretched fuselages seating up to 109 passengers. In 1993 the series passed into the hands of Avro International, a BAe subsidiary. In 1994, after the construction of the 222nd series 146 plane, production concentrated on the improved Avro RJ70, 85, 100 and 115 versions, the names of which indicate the number of passengers carried. The most successful model in the series, fitted with electronically governed AlliedSignal engines and digital avionics, is the RJ85, of which over 80 have been sold. This model flew for the first time on 23 March 1992 and is equipped with four LF507-1F turbofans producing 7,000 lb of thrust, providing with a top speed of 497 mph, and a maximum take-off weight of 97,000 lb.

In 1989, Canadair (now Bombardier) developed its Regional Jet in Canada, basing design on the Challenger business jet. The new Regional Jet completed its maiden flight on 10 May 1991. The plane is powered by two jet engines mounted at the tail; it features a T-tail, a low wing and seats for 50 passengers. The first deliveries were made to Lufthansa CityLine in October 1992. In 1995 Bombardier introduced the RJ200 series, powered by General Electric CF34B1 engines producing 9,225 lb of thrust, good for a cruising speed of 534 mph and a maximum take-off weight of 47,443 lb. The RJ200 could seat up to 54 passengers, the equivalent of 12,000 lb of cargo. In 1997 Bombardier announced the RJ700 model, characterized by a fuselage stretched by just over 13 feet to bring the passenger capacity up to 70, and fitted with CF34-8C1 engines producing 12,675 lb of thrust, and maximum take-off weight of 72,500 lb. The Bombardier RJ series has enjoyed notable success, owing to its models 900 and 1000, with production exceeding 1900 units.

In the late 1980s, following its success with turboprops, the Brazilian firm Embraer launched a regional jet project. The firm's internal problems and redesign needs delayed the project, but the first EMB-145 prototype, a plane with clean, conventional lines, took to the air on 11 August 1995. Continental Express, which had been involved in launching the project, was taking the its first deliveries in 1996. In October 1997, the EMB-145 was redesignated the ERJ-145 (Embraer Regional Jet 145) in order to emphasize its precise market position. The first version was 50-seat ERJ-145ER, which was equipped with two Allison AE3007A turbofans, each producing 70,54 lb of thrust and ensuring a cruising speed of 516 mph and a maximum take-off weight of 45,415 lb. Next came the 70-seat ERJ-170 and the ERJ-135, introduced in 1997 with shortened fuselage, and the ERJ-195, up to 122 seats. Over 1000 ERJ series planes have been sold, and the aircraft is still in production since it appears that the regional air transport market is likely to expand still further in the future.

One of the most recent regional jets is the Russian Sukhoi Superjet 100, or SSJ100, an aircraft seating between 70-110 passengers and built by Sukhoi, sold in Western countries by the Superjet International consortium created by AleniaAermacchi (51% owner) and Sukhoi. The aircraft took its maiden flight on 19 May 2008 and already has more than 360 orders from around the world. It entered into service with Aeroflot as its launch customer in June 2011.

The SSJ100/95B is powered by two Power Jet SaM145 turbofan engines providing 16,120 lb thrust for a top cruising speed of Mach 0.78 and maximum range of 1890 mi, or 2,840 miles for the LR version. Wingspan is 91 feet and overall length is 98 feet. Other aircraft in this same class are the Ukrainian Antonov An-148, which took its maiden flight in 2004 (it is also produced in the stretch version, the An-158); the Russian Yak-42, production of which terminated in 2003; and the Russian Tupolev Tu-334, the project for which was canceled in 2009.

284 top The Sukhoi Superjet 100 is an interesting aircraft in the world of third-generation machines. Entering into service in 2011, it is marketed by the Superjet International, of which Alenia Aermacchi is majority holder.

285 top The Bombardier RJ700 is a twin jet that can carry up to 70 passengers, shown here bearing the colors of the United States carrier American Eagle.

285 center The BAe 146 project was transferred in 1993 to the company's Avro International subsidiary, which produced the RJ series with different load capacities. This Lufthansa RJ85 is shown landing.

285 bottom The 1995 Embraer RJ-145, seating 50 passengers, reflected a regional transport shift to jet planes that began at the start of the 1990s.

288 top A formation of F-117 fighters in flight. The aircraft in the foreground clearly shows the squarish arrow forms developed as a result of early research into radar invisibility.

288 bottom An F-117 parked, with its bomb bay open, showing the bombs (laser-guided GBU-24s) extracted by means of special compasses.

Spearheading the American air combat force was the Lockheed F-117A Nighthawk fighter-bomber, the famous yet top secret invisible "stealth" fighter, around which many rumors and legends had grown. The origins of this remarkable aircraft date back to 1974, the year in which the United States government issued a requirement for a revolutionary new stealth fighter with a very low radar footprint. In 1976, the government selected the design proposed by Lockheed's secret "Skunk Works" group and authorized the Have Blue research program calling for the construction of two prototypes. The first of these completed its maiden flight in December 1977, and production of the definitive F-117 commenced the following year, with a batch of 20 aircraft.

The F-117 was highly revolutionary; its shape was almost exclusively the result of research into the refraction of electromagnetic waves. The skin was faceted like a cut diamond, with the plane's fuselage almost integral with the V-shaped wing. The two butterfly tailplanes were fully mobile. The F-117's handling left much to be desired, and it was largely controllable in flight thanks to the application of computerized flight controls. The single-seat aircraft was powered by two General electric F404-GE-F1D2 turbofans producing 10,800 lb of thrust with no afterburners. Its top speed was no higher than 646 mph, and maximum take-off weight was about 52,470 lb. The weapons load (a maximum of 5,000 lb) was carried in two internal bays and was typically composed of two laser-guided 2,000-lb Paveway bombs. In practice the F-117 was a night bomber which, thanks to sophisticated navigation and strike avionics, was capable of carrying out surprise attacks without being picked up by enemy radar. The first F-117s went into service in 1982, but only in 1988 was the plane's existence officially acknowledged, and it made no public appearances until 1990. A total of 64 expensive and sophisticated F-117s have been built, and 42 of them were employed in the Gulf War. The F-117 was retired from service in 2008, but subsequently, it would seem from 2016, a certain number have silently returned to operations for training purposes.

The F-117 Nighthawks were the first Allied planes to see active combat, attacking command centers in Baghdad during the night of 17 January 1991. This date marked the commencement of Operation Desert Storm and the beginning of the aerial offensive against Iraqi forces that followed the United Nations' the vain formal requests for Iraq's withdrawal from Kuwait.

The aerial assault was carried out primarily with F-117, F-16, F-15E, F-4G, F-111, A-10, Tornado IDA, Buccaneer Mirage F.1 and Jaguar fighter-bombers, along with F-15, Tornado ADV and Mirage 2000 air-superiority fighters, backed up by imposing support services for in-flight refueling, surveillance and electronic warfare. The carrier-borne forces were composed of F-14, A-6 and A-7 fighters, but primarily by F/A-18s and AV-8Bs, two new planes that represented the backbone of the US Navy and Marines' strike force. The McDonnell-Douglas F/A-18 Hornet could trace its origins back to the Northrop YF-17 project, the plane that together with the YF-16 had participated in the USAF's Light-Weight Fighter (LWF) competition in 1971. With the aim of optimizing resources, in 1974 the American government had decided that the new fighter produced by the US Navy's VFAX program (a replacement for the Navy and the Marines' A-7s and F-4s in strike and aerial combat roles) should also be chosen from the LWF contenders. McDonnell-Douglas proposed an alliance with Northrop in order to exploit its experience with carrier-borne aircraft, and in 1976 the Navy chose the project jointly presented by the two companies. This project involved the construction of two almost identical planes, the F-18 air combat fighter and the A-18 strike version. Subsequently, thanks to technological advances and the Navy's logistical requirements, a single multi-role model was defined, the F/A-18, which completed its maiden flight on 18 November 1978.

The F/A-18 was a twin-engined plane with a high wing and dual rudders that benefited from the recent aerodynamic research into flight

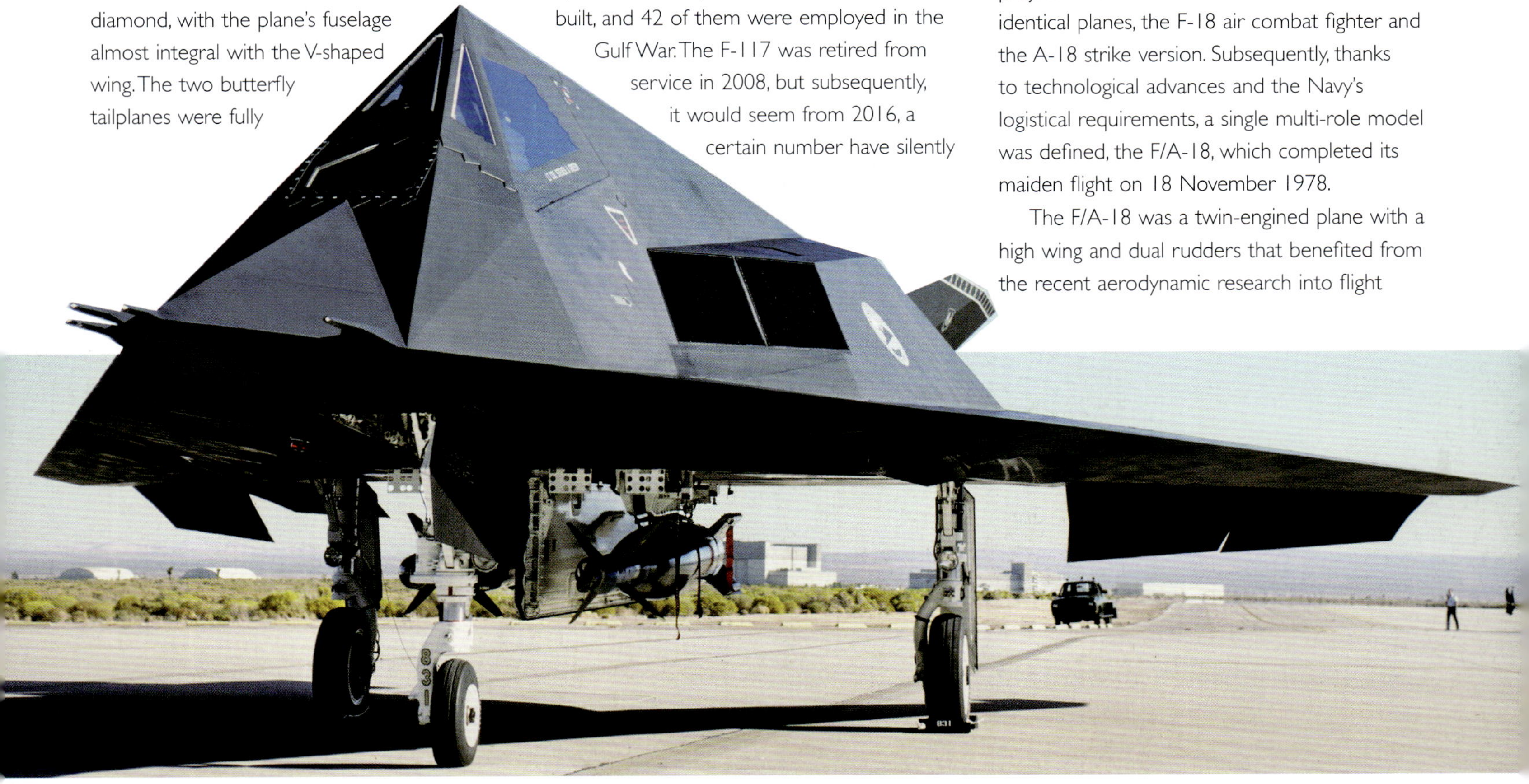

at high angles of attack. The plane was highly maneuverable, especially at low speeds. Thanks to its mission computer and radar systems' capabilities the F/A-18 could carry air-to-air and air-to-surface weapons, taking on the role required by the various phases and demands of its mission. Deliveries of the F/A-18A (and the two-seater B) to operational units began in May 1980. By June1982 exports had begun, first to Canada and then to Australia and Spain. In 1987, the F/A-18C/D model appeared, featuring new avionics that supported the possibility of using new weapons (such as the ATM-20 missiles), night-fighter capabilities, and subsequently improved engines and radar. The F/A-18C had a wingspan of 36 feet and a length of 55 feet 9 inches, and was powered by two General Electric F404-GE-402 turbofans, each producing about 17,700 lb of thrust and guaranteeing a top speed of about 1,180 mph (Mach 1.8). The plane had a maximum take-off weight of 56,000 lb; its weapons system comprised a 20-mm Vulcan cannon and up to 18,000 lb of external load. About 1,480 F/A-18s were built, and the aircraft was exported to Kuwait, Finland, Switzerland and Malaysia. The US Navy definitively retired the model in 2019, with Australia following suit in 2021.

The early 1990s saw the introduction of the F/A-18D Night Attack, a multi-role two-seater version developed for the Marines to replace the A-6E as an all-weather bomber. The F/A-18E/F super Hornet project was launched in 1992, aimed at developing the Hornet into an aircraft capable of replacing both the F-14 as the navy's defense fighter and also the earliest F/A-18s. The Super Hornet is a new aircraft, far larger and more versatile machine. Its wingspan was extended to 42 feet 7.5 inches and its length to 59 feet; maximum take-off weight was increased to 66,000 lb, and the internal fuel tanks' capacity was increased by 3,600 lb in order to extend the plane's range. Two extra underwing pylons were added to carry weapons, and F414-GE-400 engines delivering 22,000 lb of thrust were adopted: top speed is around Mach 1.8. The first F/A-18E flew on 29 November 1995, and the plane went into service with the first training unit in 2000. The US Navy ordered 520 more units of the Super Hornet, including the model EA-18G Growler for electronic warfare (this entered into service in 2008) and production continues. For the time, Australia is the sole foreign customer, having acquired 24 units. During the Gulf War, the Hornets effected a majority of the missions entrusted to the Navy, including ground attack sorties, attacks of air defenses, and air superiority.

289 top A pair of F/A-18Cs above the USS Kitty Hawk in the Pacific Ocean. The model C Hornet, incorporating improvements over the model A, first flew in 1987.

289 bottom A Boeing F/A-18F Super Hornet of the VFA-102 squadron, armed with conventional bombs, shown taking off from an aircraft carrier during Operation Enduring Freedom.

The Marines also made good use of the Hornets during the conflict, using it together with four groups of V/STOL McDonnell-Douglas AV-8B Harrier II fighter-bombers. This plane was born in 1976 out of the collaboration between British Aerospace and McDonnell-Douglas, which had resulted from a program designed to produce an improved version of the Harrier fighter then in use with the Marines in its AV-8A version. The YAV-8B, the prototype of the new plane, first flew on 9 November 1978. In practical terms, the YAV-8B was an all-new aircraft, even though it reflected the general layout of its predecessor. It had a new, larger wing offering greater lift and more weapon-mounting points; the engine was new, and the weapons load and range had been increased.

New avionics were fitted and the airframe, largely composed of composite materials, had a new cockpit. The AV-8B (and the TAV-8B two-seater) entered service with the Marines in January 1984, and it was delivered to the RAF (as the Harrier GR.Mk.5) in July 1987. The AV-8B had a wingspan of 29 feet 6 inches and a length of 45 feet 11 inches; its Rolls-Royce Pegasus F402-RR-408A engine provided 23,800 lb of thrust, good for a top speed of 652 mph. Its maximum take-off weight was 31,000 lb, and the plane could carry a weapons load of up to 13,225 lb. Subsequently, in its AV-8B Night Attack and Harrier GR.Mk.7 versions, the plane was equipped with night-strike capabilities. In 1987 the Harrier II Plus program was launched to transform the fighter into a multi-role plane also suitable for aerial combat.

The Plus version is in fact equipped with an APG-65 radar system (the same as that fitted to the early F/A-18s) and air-to-air weaponry such as AIM-9 and AIM-120 missiles.

In 1993, the Plus version went into service with the U.S. Marines and the Italian and Spanish navies. Around 480 Harrier IIs were built, and the RAF continued to operate its Harriers, uprated to GR.9 specification, through to 2010..

290 top An AV-8B Harrier II fighter-bomber lands on the USS Essex' flight deck during Operation Enduring Freedom, Afghanistan, October 2001.

290-291 A Harrier II armed with laser-guided GBU-12 bombs weighing 500 lb each taking off from the helicopter-carrier USS Bataan in January 2002 during Operation Enduring Freedom.

291 top F-14D and F/A-18C fighters, an E-2C radar plane, an EA-6B electronic warplane, an S-3B anti-submarine plane and an A-6E bomber flying over the USS Carl Vinson. The photograph was taken in 1996 during Operation Southern Watch in the Persian Gulf.

As mentioned earlier, electronic warfare aircraft played a key role in the Gulf War. Although not strictly belonging to this category, the Boeing E-3 AWACS radar plane was one of the Allies' crucial aircraft in the their complex aerial operations. Development of the E-3A began in 1970. The design derived from the civilian 707-320B model, and drew on the experience gained in Vietnam by the USAF's first C-121 radar planes. The new aircraft (then designated as the EC-137D) made its maiden flight on 9 February 1972, and deliveries of the production version began on 24 March 1977. In essence, the E-3A Sentry was a flying container for the Westinghouse AN/APY-1 radar system characterized by the enormous antenna located in a rotating disc above the fuselage, inside which were the radar operators' work stations. The system could be used not only as a flying radar station (to track other aircraft at distances of up to 250 miles), but also as a true airborne command post. The E-3s were subsequently updated to B and C specifications through installation of the APY-2 radar system, new tactical evaluation command and control computers, new electronic support systems (ESM), and new secure communications and data exchange equipment. A total of 68 E-33s have been built, and are currently in service with the United States, Great Britain, France, Saudi Arabia and a NATO unit.

Other electronic warfare aircraft employed in the Gulf included the EF-111A and the EA-6B for the electronic neutralization of anti-aircraft defenses; the EC-130E/H for tactical command-post operations, electronic spying and psychological warfare; and the RC-135U/V/W for the interception of communications and electronic signals. During Gulf War operations, the RF-4C, Mirage F.1CR

292 bottom The Boeing E-6A Mercury is based on model E-3. The US Navy uses to the plane maintain contact with its fleet of nuclear submarines. The plane has been in service since 1989.

293 top A Grumman EA-6B Prowler on the USS Theodore Roosevelt's flight deck during Operation Enduring Freedom, December 2001.

and U-2R reconnaissance planes gave these electronic warfare planes invaluable support.

The war against Iraq was virtually one-way, especially in the air. Apart from a few dogfights (in which Coalition fighters shot down 41 enemy aircraft without loss) the greatest threat to Coalition aircraft came from the anti-aircraft missile and artillery systems. These, however, managed to shoot down only 43 Allied aircraft in 110,000 combat missions. Much of the Allies' success in the air was due to the work of the SEAD (Suppression of Enemy Air Defences) planes that attacked the enemy anti-aircraft defenses. The F-4G, F-16, EA-6B, A-7, F/A-18 and Tornado aircraft, armed with HARMs (High-speed Anti-Radiation Missile) and ALARMs, and employing increasingly sophisticated tactics, were able to pinpoint the Iraqi positions; they frequently managed to silence them either by force or by their mere presence, obliging the Iraqis to shut down their radar systems.

After almost 40 days of air attacks, the Coalition's land forces (540,000 men and 3,700 tanks) finally attacked on 24 February, and in just 100 hours overran the relatively untrained and demoralized Iraqi troops. The Gulf War ended on 28 February 1991.

292-293 The Grumman EF-111A Raven, developed from the F-111 fighter-bomber, was the USAF's equivalent of the US Navy's Prowler. It remained in service from 1981 until 1998.

293 center The Boeing E-3 Sentry is the AWACS version of the Boeing 707-320 airliner. The main system (with the antenna installed in the large rotating disk) is the Westinghouse AN/APY-2 radar.

293 bottom The Prowler is a ship-based electronic warplane used to blind enemy air defenses electronically.
It first flew in 1968, and has a four-man crew.

At much the same time, however, another major center of tension and subsequently war was developing in the Balkans as Slovenia and Croatia decided to abandon the Yugoslav federation and proclaim their status as independent nations. The war between Yugoslavia (Serbia) and Croatia expanded into Bosnia, provoking reaction from the United Nations and NATO. These organizations decided upon the formation of a military and aerial force that would re-establish and maintain peace. Airborne operation began in summer 1992, with the airlift that brought relief to Sarajevo and in spring 1993, with Operation Deny Flight that was intended to impose control over the region's airspace and prevent military actions. These activities gradually intensified until winter 1998-99 when Serbian troops began a series of criminal "ethnic cleansing" operations against people of Albanian origin in the province of Kosovo. This led to a new UN and NATO initiative that, given the failure of negotiations, took the form of military intervention begun during the night of 24 March 1999.

with the remaining 25 percent supplied by eleven NATO countries including Italy, which provided abut 50 aircraft and offered the use of some 20 airfields. The Yugoslav air force was able to sustain opposition only in the conflict's very earliest phases, when five of its MiG-29s were shot down. Thereafter, Serbia's coordinated and constantly mobile and anti-aircraft defenses posed the main threat to NATO – one that was greater than that faced in the Gulf War.

The Balkan conflict saw three very different types of aircraft in action for the first time. One was the AMX fighter-bomber, an aircraft developed in Italy in the early 1980s by a consortium formed by Aeritalia, Aermacchi and the Brazilian company Embraer. The AMX designed to replace the Italian air force's G.91s and F-104Gs. The first prototype flew on 15 May 1984, and the first production planes went into service in 1989. The AMX is a single-seat, single-engined plane with traditional lines but a moderately swept high wing, developed from the profile

294 The Russian MiG-29, shown here in 1998, before the Kosovo conflict began, is the Yugoslav air force's most important fighter.

294-295 The AMX resulted from a 1980 Italian-Brazilian project to build a light tactical fighter-bomber; 192 were built. The Italian air force used the AMX with success during the 1999 war in Kosovo.

Operation Allied Force saw the onset of aerial attacks on Serbian military and strategic targets. The aerial offensive against Yugoslavia (the first war NATO had fought in its fifty-year existence), was initially conducted with a limited number of aircraft (about 500 combat and support planes). However, given the opposition's strength and tenacity, the Coalition intensified its effort both in terms of numbers (over 1,000 aircraft, including those embarked on five aircraft carriers) and the frequency of the missions that were carried out night and day. About 75 percent of the aircraft were supplied by the United States (operating from bases in Italy, France, Germany, Spain and Great Britain),

of the Tornado's wing. It has a wingspan of 26 feet 3 inches and a length of 32 feet 7.5 inches. Powered by a Rolls-Royce (FIAT Avio) Spey RB-168-807 engine delivering 11,000 lb of thrust with no afterburner, it has a top speed of Mach 0.86. (The RB-168-107 is an old engine with an unexceptional power output but moderate fuel consumption; it was to cause considerable problems in terms of reliability.) The AMX has a maximum take-off weight of 28,660 lb, and can carry an external weapons load of up to 8,377 lb, together with a 20-mm Vulcan cannon. The two-seater AMX-T version appeared in 1990. Over the years the plane matured into an effective fighter-bomber and tactical

reconnaissance plane of consistent reliability (diurnal capabilities), equipped with generally advanced avionics and precision weaponry such as laser and infra-red guided bombs. From 2007 to 2010, 52 units were upgraded to the new ACOL standard, which introduced improvements in avionics and weaponry, with the integration of JDAM bombs, RecceLite reconnaissance pods, and Litening targeting pods. In Kosovo, the Italian AMXs were utilized with a certain continuity in support missions, giving an excellent show of their capabilities as was the case in later operations in Afghanistan, Libya, and Kuwait. The AMX was retired from service with the Italian Air Force in 2024.

295 bottom left Puma helicopters take French paratroopers to Kosovka Mitrovica during NATO peace force operations in Kosovo, September 1999.

295 bottom right A Puma helicopter transports French commando paratroopers over Macedonia during Kosovo war operations in April 1999.

The second plane that made its operational début in the Balkans was the American strategic bomber, the Rockwell B-1B Lancer, which operated from Great Britain. For various reasons, the United States did not used the B-1B Lancer in the Gulf War. It is a large, four-engined aircraft with a variable-geometry wing and intercontinental range. It dates back to 1970, to the development project launched to find a replacement for the Boeing B-52. The first prototype of the supersonic B-1A completed its maiden flight on 23 December 1974, but then in 1977 President Carter canceled the project. Four years later, President Reagan authorized the revival of the project in B1-B form, for a slightly simplified and subsonic version of the original design but one featuring improved avionics and structural technology. Plans called for building 100 B1-Bs rather than 244 B-1As. The B1-B has a wingspan of between 75 feet 6 inches (maximum sweep) and 134 feet 6 inches (minimum sweep) and a length of 144 feet 4 inches. It is powered by four General Electric F101-GE-102 engines, each delivering 30,800 lb of thrust on afterburn, good for a top speed of 746 mph. Maximum take-off weight is 238 tons, of which the weapons load accounts for up to 61.5 tons. The plane has a crew of four and is capable of dropping both nuclear and conventional weapons. Deliveries of the B-1B began in 1985.

The Northrop-Grumman B-2A Spirit was the third plane to receive baptism of fire during the Kosovo war. This was the first "invisible" strategic bomber in the history of aviation. The origins of the project date back to the second half of the 1970s, following the launch of the Have Blue program and the F-117. The development program for the B-2 was signed off in 1981.

The plane revived Jack Northrop's original ideas from the 1940s for a flying-wing bomber, but was adapted and improved to benefit from the technological advances available in the 1980s. An aircraft with highly futuristic lines, the B-2A has no aerodynamic line-breaking appendages whatsoever; its only protuberances are the cockpit and the four engines on the top of the airframe, all very carefully faired. The first B-2 completed its maiden flight on 17 July 1989 and the first production version was delivered to the USAF in 1993. The plane has a wingspan of 170 feet 7 inches and a length of 68 feet 11, and is powered by four General Electric F118-GE-100 engines, each providing 19,000 lb thrust, for a top speed of 528 mph. Maximum take-off weight is over 167 tons, and a weapons load of 40,000 lb can be carried. The flight crew numbers two persons.

Thanks to the combined use of sophisticated materials, coatings and shapes, the B-2A is practically invisible to enemy radar and can strike targets of great strategic importance on a global scale, with either nuclear or conventional weapons.

Its performance in the Kosovo War confirmed the plane's potential; taking advantage of in-flight refueling, the B-2As flew a number of missions against targets in Serbia, with the planes taking off and landing from their home base of Whiteman, Missouri.

Operation Allied Force officially concluded with the ceasefire of 20 June 1999. The objectives set by the political leaders of the NATO countries had been achieved, in particular the withdrawal of Serbian troops from Kosovo.

Of notable importance in this conflict was the fact that, despite the immense military and economic expenditure, the Allies attained a successful result with the loss of only two combat aircraft and without having to make recourse to land forces.

Given the particular political and ethnic scenario in the Balkan theater, land-based operations could have resulted only in expansion of the conflict and new international tensions, as well as the indefinite prolongation of the war with further suffering for the civilian population and significant casualty numbers on both sides.

296 top A B-1B Lancer creating aerodynamic compression visible in the humid air as it flies at high speed. It has a top speed of 750 mph.

296-297 bottom The B-2A Spirit's unconventional and science-fiction forms show to advantage in this ground-based photograph.

297 top A B-1B Lancer refueling. This aircraft first flew in 1974 as the B-1A but the project was put on hold; President Reagan revitalized it in 1981.

297 bottom right The Spirit bomber's radar-absorbent paint accounts in part for its invisibility to radar.

24

With the arrival of the 21st century, the international situation does not seem to have changed character and unfortunately, the crises have not stopped, showing, rather, an increased frequency. The date of September 11, 2001 is one that changed the world. On that day, 19 Al Qaeda terrorists, guided by the Saudi sheikh Osama Bin Laden, hijacked four airliners. They succeeded in crashing two against the Twin Towers of the World Trade Center in New York City, and another crashed into the Pentagon in Washington, D. C. The dead numbered nearly 3000. Besides the technical and regulatory implications subsequently established for civilian flight security as a result of that day, this attack upon the United States provoked a response within a few weeks: a military intervention in Afghanistan, where Al Qaeda had been operating undisturbed, with the support of the local Taliban regime. As part of the Global War on Terror as decided by the United States administration, Operation Enduring Freedom was launched on 7 October 2001. The initial attack against the Al Qaeda training camps and infrastructure was conducted by Tomahawk missiles, B-1, B-2, and B-52 bombers, and carrier-based F-14 and F/A-18 fighters. Due to the difficulties imposed by the type of terrain and the characteristics of asymmetric warfare against elusive guerrilla units, the conflict quickly spread, resulting in a massive intervention of ground forces, not only American, but also of NATO through the ISAF mission (International Security Assistance Force), established by mandate from the United Nations Security Council on 20 December 2001. Its numbers grew to the point of including over 58,000 troops from 40 countries. To consider the aircraft, the operations in Afghanistan saw a broad use of transport and support components, such as reconnaissance and surveillance aircraft, designed for identifying concentrations of insurgent troops within the territory. Among the combat means available, the majority of those engaged were fighter aircraft, attack helicopters, and unmanned aircraft, designed to provide the vital missions of Close Air Support (CAS) for ground troops. Despite undeniable tactical successes, the war operations in Afghanistan did not lead to resolution of the country's endemic problems, and the 2008 outbreak of a severe global economic downturn and financial crisis induced the United States and the countries participating in the ISAF mission to begin a gradual military disengagement from the theater, which went hand in hand with the attempt to create Afghan police forces and armed forces capable of taking on and maintaining control of the territory. The complete withdrawal of ISAF forces is slated to take place within 2014. While fighting was underway in Afghanistan, on 19 March 2003 the United States opened a second war front by attacking Iraq, with the declared purpose of "Depriving the country of weapons of mass destruction, ending terrorist support from the regime of Saddam Hussein, and liberating the Iraqi people." Operation Iraqi Freedom was conducted by the United States military, with support by the United Kingdom, Australia, and Poland. The campaign concluded on 9 April, with the taking of the capital of Baghdad and the fall of Saddam Hussein's regime. Nevertheless, the conflict was prolonged, in the form of guerrilla warfare, and was conducted with the participation of contingents sent from 45 other countries. The last United States forces completed their withdrawal from Iraq in December 2011. This second war against Iraq saw the air forces in action using means and aircraft similar to those already used in preceding conflicts, but, similarly to what was going on at the same time in Afghanistan, with a much more massive use of unmanned aircraft and of Close Air Support than in the past.

298 A Marine Corps' CH-53E Super Stallion helicopter taking off from the USS Bataan during Operation Enduring Freedom. The carrier is en route toward Afghanistan.

299 top An AgustaWestland AW.129C combat helicopter of the Italian Army shown in Afghanistan during joint operations with paratroopers of the "Folgore" (meaning "Thunderbolt") Brigade.

299 center A Boeing F-15E Strike Eagle of the USAF, armed with laser guided bombs, shown after inflight refueling during the course of Operation Iraqi Freedom in 2003.

299 bottom A Dassault Rafale B of the French Air Force flying over the mountains of Afghanistan in March 2007. The aircraft is equipped with two additional fuel tanks and GBU-12 laser guided bombs.

The last conflict, chronologically speaking, that has affected the world was the one in Libya, which erupted on March 19, 2011 and lasted until 31 October. In this case, the motives for the crisis are to be found in the popular uprising that broke out in February against the dictatorship of Muammar Gaddafi, which was grafted into a broader situation of open rebellion against many highly restrictive Arab governments in power in the Middle East, known as the Arab Spring. In order to protect the civilian population from the loyalist armed forces, and (unofficially) to safeguard the enormous wealth in oil and raw materials in Libya, the United Nations issued a resolution authorizing member states to take "all necessary measures to protect civilians and areas of civilian populations from threat of attack by Libyan armed forces," in addition to establishing a No Fly Zone over the country. French, British, and United States troops attacked major Libyan military objectives beginning from the night of 19 March, triggering Operation Odyssey Dawn. Within a few days, a massive naval air system was established in attempting to reduce military forces still loyal to Gaddafi to a state of harmlessness. The air forces of 18 countries (including 3 Arab states) responded to the appeal, redeploying their assets (over 390 aircraft) primarily in Italy and Greece. In order to better organize and support the coalition from the political point of view, the operation went under NATO command on March 31, taking on the name Unified Protector, whereas Odyssey Dawn now referred solely to United States activities. A primary characteristic of this war, conducted almost exclusively by means of aircraft (operating from land bases or aircraft carriers), was utilization by the aircraft of only the most modern precision weaponry, designed to avoid at all costs any collateral damage, and thus of involvement of civilians, their property, and infrastructures of the country. Although

300 top A pair of CF-188 fighters, the Canadian version of the Boeing F/A-18A Hornet, in flight over Libya during Operation Unified Protector.

300 center The crew of a Mirage 2000D fighter-bomber of the French Air Force during Operation Unified Protector.

300 bottom A Eurofighter EF.2000 Typhoon fighter from III Squadron, Royal Air Force. The aircraft is armed with laser-guided Paveway II bombs for a precision ground attack mission during the 2011 Libyan war.

elements of the air forces were engaged in air-to-air missions of patrolling the No Fly Zone, the actual operations were conducted by aircraft for ISTAR missions (Intelligence, Surveillance, Targeting, and Reconnaissance) and by fighter-bombers, which made use of the most modern systems of recognition and identification of targets, supported by radar aircraft and inflight refueling. This conflict saw a baptism under fire for various weapons systems, including the fighters Eurofighter Typhoon, Dassault Rafale, JAS Gripen, Lockheed Martin F-16E/F, and Boeing EA-18G Growler, as well as certain weapons systems, like the MBDA Storm Shadow cruise missiles and the French AASM bombs. The experience gained in the conflict proved very useful for the ISTAR component, which has already been engaged for some time in Afghanistan. Nevertheless, with the lack of effective opposition forces, it was not possible to obtain data toward assessing the real capacities of the last-generation combat aircraft, among which, notably, the Lockheed Martin F-22A was absent, the most advanced Western-built fighter, which the United States did not feel appropriate to use in the engagement.

301 top A Boeing EA-18G Growler electronic warfare aircraft of the US Navy's VAQ-141 Squadron, shown landing on aircraft carrier USS George H. W. Bush in 2010.

300-301 An F-16AM of the Norwegian Air Force over the base at Souda Bay in Crete during Operation Unified Protector on Libya. The fighter is armed with AIM-120B and IRIS-T air-to-air missiles, with a GPS-guided GBU-54 bomb. It has a Sniper target designation pod and, in the ventral position, an AN/ALQ-131 electronic warfare pod, as well as two additional fuel tanks.

Chapter 24

Airbus and Boeing

In the 1960s, the global market for large airliners was in effect an American prerogative, with Boeing, Douglas and Lockheed sharing the vast majority of orders. However, during that period a new wind began to blow through Europe. Thanks to the intervention of national governments, the foundations were laid for the rebirth of the European aeronautical industry. A number of international projects, such as the Anglo-French Concorde and Jaguar and the tri-national Tornado, came into being. But the initiative that most influenced the global industrial scene was the Airbus project, which began in 1965 as the result of collaboration by France, Germany and Great Britain's governments and aviation industries.

The need for a twin-engined, wide-body, 300-passenger jet for medium-range operations was identified. The agreement that initiated the development of the aircraft, designated the Airbus A300, was signed on 26 September 1967. According to its terms, France and Great Britain were each to contribute 37.5% of the cost of the development and construction of the airframe, while Germany would provide the remaining 25%. With regard to the engines, Great Britain was to contribute 75% of the cost of developing the Rolls-Royce RB207 turbofan, while France and Germany were each to contribute 12.5%. The primary contractor for the airframe was to be Sud Aviation (later Aérospatiale). Airbus Industrie was officially constituted on 18 December 1970, to oversee the development, production and marketing of the A300. The Spanish firm CASA subsequently joined the consortium. Rolls-Royce announced that it would be unable to develop the RB207 engine in time, and this unit was replaced by the proven General Electric CF6-50 delivering 49,000 lb of thrust. Another significant modification involved the reduction of seats from 300 to 250, as requested by the principal potential clients.

Construction of the prototype A300B1 began in September 1969, and its maiden flight took place on 28 October 1972 at Toulouse. In 1971, the consortium received its first order, with Air France requesting six planes plus options for a further ten. Two years later, in 1973, Lufthansa ordered three planes and placed options on a further four. After the third plane had been built, design modifications were made; the plane's fuselage was lengthened by 6 feet 6 inches and the more powerful CF6-50C engine, providing 51,000 lb of thrust, was adopted. This modified model, designated the A300B2, could carry a greater payload and had a greater range .In March 1974, the European and American authorities certified the aircraft, and the fifth plane flew for the first time that year. It was the first to be delivered to a client, going to Air France on 11 May and entering service on 23 May on the Paris-London route.

The A300 had yet to win the confidence of the market, however, and sales were slow; in fact in 1976 the consortium failed to win a single order for the A300. It received a massive boost in May 1977, when the American company Eastern Airlines agreed to lease

302 top The four-jet A340 that first flew in 1991 is the largest plane in the Airbus family. This one belongs to Virgin Atlantic.

302 bottom The nose of an American Airlines' Airbus A300-600 in close-up. This model can seat 360 and was designed for medium- and long-range flights.

four A300B4s: this was the first contract to be won in the United States, and in breaking the monopoly American manufacturers enjoyed on their home turf, Airbus had taken a fundamental step forward. Nonetheless, by the end of 1978, the production total had almost reached 50 planes, and the break-even point appeared to be little closer. Airlines and their pilots considered the A300 to be an excellent aircraft and in 1979 and thereafter, thanks in part to the publicity provided by the Eastern Airlines deal, Airbus's stock began to rise. In the meantime, new versions of the original design were in mind that would eventually materialize as new projects. In 1983 the A300-600, which was to become the principal production variant, made its maiden flight. The A300-600 a wingspan of 144 feet 4 inches, a length of 177 feet 2 inches and a maximum take-off weight of about 189 tons. It could be ordered with two Pratt & Whitney PW4156 engines (delivering 56,000 lb of thrust) or two PW4158s (58,000 lb thrust) or alternatively with General Electric CF6-80C2A1/A3/A5 units delivering similar outputs. The plane had a top cruising speed of Mach 0.82 and in standard configuration could carry 266 passengers, with up to 361 in high-density configuration. The A310 model, with a shorter fuselage seating 210-234 passengers (255 at high density), a new wing and smaller tailplanes, had gone into development in 1978. The prototype flew for the first time on 3 April 1982 and the first delivery (to Lufthansa) took place in March 1984. The early 1980s saw development of the A320 model, an aircraft for medium- and short-haul work that could seat up to 179 passengers; it incorporated a number of innovative features such as fly-by-wire controls and offered reduced operating costs. Air France was the first airline to order the 320, signing as early as 1981 for 25 planes. The prototype made its maiden flight on 22 February 1987. This base model was followed by a further two variants, the stretched A321 and the shortened A319. The A321 first flew on 11 March 1993 and was delivered to Lufthansa the following year. The A319 first flew on 29 August 1995, with the first delivery being made to Swissair the following year.

In the late 1980s Airbus was working on completion of its range with a large aircraft for medium- and long-haul flights. The A330/340 project was launched in 1987 and resulted in an average-capacity wide-body plane with a four-engine configuration (A340) or a-two-engine configuration (A330). The A340-300 prototype flew for the first time on 25 October 1991, followed by the short-fuselage A340-200 version. The base model had a wingspan of 196 feet 10 inches and a length of 206 feet 8 inches, with a maximum take-off weight of about 283 tons. It was powered by four CFM International CFM-56-5C engines, each delivering 31,000 lb of thrust, and was capable of a top speed of Mach 0.83 and had an operational range of about 6,800 miles. The standard configuration provided between 250 and 350 seats, but 440 could be squeezed in a maximum density format. The A330, which flew for the first time on 2 November 1991, retained the same dimensions, but was powered by two General Electric, Pratt & Whitney or Rolls-Royce engines (as specified by the client), with power outputs ranging from 64,000 to 67,000 lb of thrust. It had a maximum take-off weight of about 233 tons and passenger capacities similar to those of its four-engined sister plane. The first A340 was delivered to Lufthansa in January 1993.

303 top The A320 developed by the Airbus consortium for short- and medium-range distances, seen here in Air Jamaica colors. The A320 first flew in 1987; it can seat 180 passengers.

302-303 An Alitalia Airbus A300 landing. The A300 was the first Airbus produced; it made its maiden flight in 1972.

304 left The Boeing 757 was designed to repeat the success of the 727. It is a short- and medium-range plane that can carry 239 passengers, though rather tightly packed.

304-305 One of a Boeing 767's two engines. At the customer's request, the plane can be equipped with Rolls-Royce, Pratt & Whitney or General Electric engines, which provide 23-24 tons of thrust.

305 top right A British Airways Boeing 777. Boeing developed the 777 to compete with the Airbus 330 and 340, and the plane made its maiden flight in 1994. It can seat 440 passengers.

305 bottom Production line for the new B.787 Dreamliner at Boeing's plants in Everett. Entered into service in 2011, this aircraft is a concentration of high technology. About 80% of the airframe is produced from composite materials.

The American aircraft manufacturing industry was by no means idle in the face of the Airbus' growing technological and commercial importance. Boeing, now the sole remaining major company active in the civil airliner sector, began developing two new planes in the late 1970s that became operational in the 1980s. These planes marked the opening of a new era for Boeing and were intended to maintain the firm's advantage over its European rival.

The first of the two new Boeing planes was the 767, which flew for the first time on 26 September 1981.

In appearance the 767 was similar in appearance to the Airbus A300, with a wide-body fuselage, two engines slung beneath the wing and traditional tailplanes. It had a wingspan of 154 feet 3 inches, a length of 157 feet 6 inches and a maximum take-off weight of 345,000 lb. It was powered by two turbofan engines, which could be chosen by the client from a list of twenty units built by General Electric, Pratt & Whitney and Rolls-Royce that ranged in power outputfrom 52,000 to 62,000 lb of thrust. The standard configuration provided between 180 and 224 seats (285 at maximum density) and the 767-200 could maintain a cruising speed of Mach 0.8. The first production 767-200 was delivered to United Airlines in 1982.

In 1986 Boeing introduced the 767-300 with a fuselage stretched to just over 180 feet and a standard configuration offering 218 seats, which could be increased to 350 in a maximum density format. The 767-300ER is the long-haul version; it has a maximum take-off weight of 204 tons and a range of more than 7,000 miles. The latest version is the 767-400ER, an aircraft capable of carrying about 290 passengers, and the first 400-ER was delivered to Delta Airlines on 29 August 2000.

The heir to the Boeing 727, this aircraft was designed for use on medium-range routes as a direct competitor to the A300. While it retained the same fuselage section as the 727, the 757 was an all-new plane that resembled a scaled-down version of the 767. The first 757 went into service with Eastern Air Lines in January 1983. It had a wingspan of 124 feet 8 inches and a length of 154 feet 2 inches. Two Rolls-Royce RB211 or Pratt & Whitney PW2037 engines (delivering 40,000 and 38,000 lb of thrust respectively) provided the 757 with a cruising speed of Mach 0.86 and a maximum take-off weight of 254,850 lb. The 757 could carry between 150 and 178 passengers in standard configuration and 239 in a high-density format.

Naturally, the existing 747 "Jumbo Jet" series was exploited to the full, and in 1988 the 747-400 was introduced, featuring updated aerodynamics, engines and avionics. The typical configuration seated 420 passengers, but a high-density format for short-haul domestic flights allowed a maximum of 568 to be squeezed in.

In order to meet its clients' demands, Boeing also introduced the 777 version, which completed its maiden flight on 12 June 1994. Externally, the new aircraft is very

similar to the smaller 767, but is in fact an all-new design incorporating the most advanced technology. The launch customer for the 777 was United Airlines, which acquired the first one in May 1995. The 777 is a truly impressive aircraft, despite the fact the twin-engine layout makes it appear smaller than it really is: it has a wingspan of 196 feet 10 inches, a length of 206 feet 7 inches, and a maximum take-off weight of over 534,800 lb. The engines (chosen by the client from units produced by the usual big three of General Electric, Pratt & Whitney and Rolls-Royce) have power outputs ranging from 74,000 lb to 86,500 lb of thrust. Passenger capacity ranges from 320 in a standard seating format to 440 in a high-

density format. The basic 777 was followed in 1995 by the stretched 777-300.

This variant has a length of 239 feet 6 inches (it is, in fact, the world's longest airplane) and with its increased maximum weight can carry between 350 and 550 passengers, depending on the seating format and the flight range required by the operating airline.

The 777 holds the record for longest nonstop flight by an airline aircraft (November 2005, with 22 hours and 42 minutes of flying). In 2009, the version 777F appeared for freight transport. Overall, more than 1,800 orders have been placed for the 777 family.

In the 1990s, McDonnell-Douglas decided to exploit the positive trends dominating the air transport markets by modernizing its MD-80 series. The first plane of the MD-90 series made its maiden flight on 22 February 1993. The plane featured electronically controlled engines, improved avionics and structural features, and a stretched fuselage that allowed for ten additional seats. The introduction of the MD-95 followed in 1995; this was an airliner in the 100-passenger class that, following Boeing's acquisition of McDonnell-Douglas in 1997, was redesignated the Boeing 717, joining the smallest 737 models at the lower end of the market.

Airbus has also continued to expand its aircraft range in the lower end of the sector, launching the A318 on 26 April 1999. This is a slightly smaller version of the 319, equipped with two Pratt & Whitney PW6000 engines and carrying about 100 passengers. The maiden flight of the new aircraft was scheduled for the end of 2001, but orders for over 120 planes have already been accepted.

In order to compete with Airbus's diversified range in the medium- and short-haul sector (where the 3 existing Airbus models had been joined by the A218), Boeing decided to revitalize its best-seller in this sector, the now antiquated 737. In the 1990s Boeing launched the 737 Next-Generation, once it could spare enough energy from its 767, 757 and 777 projects. The new project aimed to get the best out of the 737 design by equipping it with the most advanced technology available (engines, aerodynamics, avionics), in order to keep it competitive with the Airbus A319/320/321 in the important 100-200 seat sector.

The first 737 Next-Generation to fly was the 737-700, on 9 February 1997. This plane, which was positioned in the 128/149-seat segment, had a maximum take-off weight of about 77 tons and two engines producing between 20,000-22,000 lb of thrust. Other versions followed the 737-700. First was the 737-800, which flew for the first time on 31 July 1997. It was a 160/189-seat plane with a maximum take-off weight of 88 tons. Next came the 737-600 (maiden flight on 22 January 1998), 108-132 seats, maximum takeoff weight of about 71 tons), and lastly, the 737-900 and 900ER and the first unit was delivered to Lion Air in April 2007. The 737-900ER is the largest of the 737 family, with a capacity for 215 passengers. Its maximum takeoff weight is 187,700 lb, with a maximum range of 3750 miles. The 737 family's upcoming model is called MAX, a model equipped with improved engines and aerodynamics, for lower fuel consumption and further reduction of harmful emissions. The 737 MAX first flew on 29 January 2016 (MAX 8 model), followed by a further three versions, the MAX 7, 9 and 10 and to date more than 1,450 examples have been produced. Output was affected by two accidents that occurred in 2018 and 2019 and led to operation of this family of planes being suspended. Production restarted in May 2020. Further quality control issues were encountered in 2024, with production getting under way again in December of that year. These aircraft share a notable slice of the commercial market with the Airbus. From its foundation through to October 2021, Airbus delivered more than 13,600 machines, compared with the 14,800 examples of the four models currently in production by Boeing (737, 767, 777, 787). The Airbus A320 is currently the most successful commercial aircraft in history, with more than 12,200 examples delivered and orders through to August 2025 taking that number to more than 19,200.

With the world market in expansion and now split virtually fifty-fifty between Boeing and Airbus, the two industrial colossi have also plunged into the latest sector for new aircraft, that of the long-range super-giants. Their approaches to this market are quite different. After careful evaluation and research, Airbus has opted for the construction of the largest airplane ever built, while Boeing has initially chosen to go for the fastest (apart from Concorde).

The Airbus aircraft – the A380, initially designated A3XX – was officially launched on 19 December 2000 when an order from Virgin Airlines brought the order book to the fateful total of 50, which had been fixed as the minimum number for Airbus to begin the program. The first flight was recorded on 27 April 2005, and the first operator to introduce it into service was Singapore Airlines, in October 2007. The basic model A380-800 has a wingspan of 261.7 feet, an overall length of 238.6 feet, and a maximum takeoff weight of 620 tons. The standard layout has both upper and lower passenger decks capable of seating 525 in three classes, allowing also for a maximum density of 853 passengers. The four engines are Rolls-Royce Trent 970/Bs or 972/Bs, or General Electric GP7270s, all in the class of 70,100 lb of thrust. The aircraft's range with standard maximum load is 9,570 miles. Also planned were the A380-900 version, designed to seat 650 passengers in the standard layout (for a maximum density of 900), and the A380F version for freight transport, with a maximum takeoff weight of 650 tons. However, neither model was actually produced, a fate that has also befallen the Neo and Plus versions. The A380 went out of production in 2021 after just 251 examples had been completed. The aircraft proved to be too large and too expensive, the fruit of inaccurate market analyses that had estimated a market for Very Large Aircraft of 1,283 planes through to 2027. The A380 is still flown by several companies, the largest fleet operated by Emirates Airlines, which acquired 116 examples.

Boeing, on the other hand, is following a different path: that of building a smaller but faster long-range plane, the Sonic Cruiser. The project was announced in March 2001, and revolves around a transport airplane with canard twin-jets with a dual delta-wing, able to cruise at an altitude of between 39,000 and 48,750 feet at a speed of Mach 0.98 so as to reduce current flying times by 15 to 20% (two or more hours on Atlantic routes).

It should be able to seat around 300 passengers. Nevertheless, the market has not displayed an appreciation for aircraft that are fast when the operating costs are high, and Boeing terminated the program in 2002. Research thus turned toward a more efficient aircraft with reduced costs, designated the 7E7, which nonetheless drew advantage from the research carried out for the Sonic Cruiser, such as the widespread use of carbon fibers for

fuselage and wings, and high technologies for the avionics systems. In 2005 the project was renamed 787 Dreamliner and the aircraft took its maiden flight on 15 December 2009. Some of the primary characteristics of the plane are that 80% of the construction, as mentioned, is of composite material, including the airframe (equal to a weight of 70,400 lb carbon fiber reinforced with plastic material); high-efficiency engines and wings; cockpit equipped with large color LCD screens, Head-Up Display for both pilots, and broad use of electrical systems architecture, eliminating the weight and complexity of hydraulically operated systems. The two engines are new-generation turbofans, with reduced noise levels, and the customer can choose between the Rolls-Royce Trent 1000 or the General Electric GEnx, both of which deliver about 63,900 lb of thrust. The Dreamliner has a wingspan of 197 feet, a length of 186 feet, and a maximum takeoff weight of 251 tons; it can accept 242 passengers on board in the typical three-class configuration, whereas top cruising speed is Mach 0.85 (567 mph) at 35,000 feet, with a range of 9,445 miles. The model in production, the 787-8, has already reached a total of more than 1,100 deliveries, with the first unit entering service with All Nippon Airways in September 2011. The stretch version, the model 787-9, with 250-290 seats and a range of up to 9,785 miles, appeared in September 2013, while the 787-10, capable of carrying up to 350 passengers, replacing the B.777-200ER and of competing with the Airbus 350 first flew in 2017. This last aircraft now represents the most advanced program of the European colossus; it was officially released in December 2004 to respond to the Boeing 787. Initially the A350 was a simple development of the model 330, but airlines did not care for this, and requested instead a completely redesigned aircraft that would be capable of introducing the necessary new technologies. In 2006, a new project was thus begun, and was presented at the Farnborough Air Show as the A350XWB (eXtra Wide Body). The 350XWB, also, is built largely of composite materials, and makes use of two Rolls-Royce Trent XWB engines, an engine model deriving from the A380 and the B.787 engines, delivering 79,000 lb of thrust. The base model 350XWB-800 has been canceled, while the standard production model A350-900 has a wingspan of 212 feet, a length of 199 feet, and a maximum takeoff weight of 285 tons. It can carry 270 passengers in the three-class layout or 440 passengers for high-density travel. The maximum cruising speed is Mach 0.88 (675 mph) at 40,000 feet, with a maximum range of 9,320 miles. The 350-900 model entered service on 15 January 2015 with Qatar Airways as its launch customer. There are already more than 600 orders for the new aircraft, including those for the lengthened 350-1000 version with a range of 10,000 miles that entered service in 2018.

The most recent Airbus product is the A220, a 100-seat aircraft conceived in 2004 as the Bombardier CSeries, which from July 2018 has been marketed by Airbus. This aircraft flew for the first time on 16 September 2013 and entered service in July 2016 with Swiss Global Air Lines. To date, around 320 examples have been produced in the A220-100 and larger A220-300 versions, this last carrying up to 145 passengers.

The "war" between Boeing and Airbus is ongoing, reflecting trends in the air transport market and the ability of the world's skies to absorb increased airline traffic.

306 top In service since 2007, the Airbus A380 can accommodate up to 853 passengers for high-density travel.

307 top The Boeing 787 Dreamliner represents the direct competitor for the Airbus 350; the most recent version first flew in 2017.

307 center The Airbus A350, seen here in the -1000 version, is the French company's product for intercontinental flights and boasts a range of 10,000 miles.

307 bottom The Airbus A330neo, here in -900 form, is the largest version of this model and has a range of over 8,250 miles.

Chapter 25

Unmanned aircraft

In the history of aviation, unmanned aircraft are not new: in fact, their use was being pioneered even before the Second World War. After the war, the rudimentary technology for controlling what were then called drones was primarily limited to target drones, small aircraft used in the military field for development of air-based and land-based weapons systems. However, the possibility of using unmanned aircraft (Remotely Piloted Vehicles, or RPVs) for other missions as well, primarily for reconnaissance, dates back to the mid-1950s, when the U.S. Army drone MQM-33 was transformed into a photographic reconnaissance instrument, becoming the RB-71, and later the MQM-57 Falconer. The first operational uses of the reconnaissance RPV occurred between 1964 and 1975, during the Vietnam War era, when the 100th Strategic Reconnaissance Wing of the USAF made use of Ryan Model 147 Lightning Bug jet drones, designated as AQM-34. In total, 3,435 launches were made, and 554 aircraft were lost through various causes. In the early 1970s, the USAF also evaluated the use of eight BGM-34Bs as attack drones, armed with MK-82 bombs and Maverick missiles, while others also adopted laser designators on the nose to guide other RPVs armed with laser-guided Paveway bombs. It should be pointed out here that remotely piloted aircraft all require a dedicated Ground Control Station (GCS) where the pilot and system operator are generally located, with control of the entire mission taking place via satellite links. Given the duration of many missions, crews take shifts. One further step along the road leading to the current UASs (Unmanned Aerial Systems) took place in 1982, during "Operation Peace for Galilee," the invasion of Lebanon by the Israeli Armed Forces. On that occasion, small propeller-driven aircraft equipped with video cameras were successfully used, the Tadiran Mastiff and the IAI Scout, which contributed significantly to the discovery and destruction of Syrian antiaircraft missile systems deployed in the Bekaa Valley.

This success attracted the attention of the U.S. Armed Forces, and in particular, in 1984, the U.S. Navy requested of the Israeli industry a tactical UAV (Unmanned Aerial Vehicle) for the battlefield, which materialized in the form of the RQ-2 Pioneer, a derivative of the Mastiff. In the 1990s, Israel was at the forefront of UAV sector, with the IAI Heron, which flew for the first time in 1994. This was a straight-winged aircraft with a wingspan of 54 feet, a length of 28 feet, and an aerodynamic formula with thrusting propeller and double tail boom. The motor is a four-cylinder Rotax 914 piston engine with 115 hp, providing a maximum velocity of 143 mph and maximum takeoff weight of 2,535 lb, 550 lb of which form the so-called "payload," namely, the mission system for observation and reconnaissance. A fundamental characteristic of the aircraft is its ability to fly to an altitude of 35,100 feet with a range of fully 52 hours. The payload may include various types of sensors, including infrared or daylight television cameras, electronic reconnaissance systems for ELINT (Electronic Intelligence) and COMINT (Communication Intelligence), and radar systems. The sensors transmit in real-time the data received via a ground station, not only making use of direct data-link transmissions (LOS, or Line of Sight), but also making use of satellites or other aircraft as a bridge. The aircraft can operate in a completely automatic mode, if so programmed, in manual mode, or even in mixed mode. The crew is made up of at least two persons, including pilot and system operator, who are positioned in the ground control station, usually located at the starting base, but which may also be located in other places. These operating principles have been employed by all other UASs that have appeared on the market. The Heron, which falls into the category of MALE (Medium-Altitude Long-Endurance) was even exported (as the Harfang) to France, Turkey, India, Germany, Australia, and seven other countries. From the Heron, the IAI even developed a version with the turboprop engine, called the Eitan, or Heron TP. Completing its maiden flight in July 2006, it is equipped with a 1200-shp Pratt & Whitney PT6A engine, giving it a range of over 5,000 miles, or 70 hours of flight, at altitudes greater than 45,900, while the payload is increased to 4400 lb and maximum takeoff weight, to 10,250 lb. The wingspan is 85 feet, and its length is 43 feet. The Eitan has been in service with the Israeli air force since 2009, but other countries as well have shown interest in the system with orders being placed by Greece and Germany.

308 top A US Navy AQM-34 target drone. In Vietnam, between 1964 and 1975, the USAF engaged this sort of machine in reconnaissance activities.

309 top A General Atomics MQ-1B Predator of the US Air Force shown landing. Under the half wings, note the two Hellfire missiles for ground attack.

309 bottom The Israeli IAI Heron is a MALE-type (Medium-Altitude Long-Endurance) reconnaissance UAV that was adopted by France, under the name Harfang.

The United States is also obviously dedicated to developing UAS technologies, which have significant advantages in comparison with manned platforms: they allow construction of smaller aircraft that are lightweight, economical, and silent, making them ideal for ISTAR missions (Intelligence, Surveillance, Target Acquisition, and Reconnaissance), because they can remain airborne in operations areas for a very long time, remaining virtually invisible to ground-based enemy eyes.

Back in the 1980s, both the CIA and the U.S. Department of Defense had begun experimental work with next-generation reconnaissance drones. At the invitation of the CIA, Abraham Karem's Leading Systems developed the Gnat, a small aircraft powered by a very quiet engine, and which was designated the Predator. In January of 1994, General Atomics of San Diego was commissioned to develop the Predator, and the Advanced Concept Technology Demonstration phase lasted until June 1996, when the program was finally passed to the U.S. Air Force by a mixed Army/Navy team. At that time, Predators were also being used in exercises and in real operations, in the Balkans, flying from Gjader, in Albania. The Predator series was officially introduced into service under the designation RQ-1B Predator A, although this abbreviation identified the entire operational system, including four aircraft, a ground control station, and a group of antennas for transmitting and receiving data. The individual aircraft was denominated RQ-1L. The core of the machine's capacity was the AN/AAS-52 system, which included an entire series of electro-optical sensors for target identification and acquisition, plus a Synthetic Aperture Radar system (SAR) AN/ZPQ-1, which however was subsequently uninstalled, given its low use. The RQ-1B system was unarmed, but it was soon evident that the possibility of arming the aircraft would have greatly increased its operational capabilities. Thus was born, in 2002, the MQ-1A, which could be equipped with two AGM-114 Hellfire air-to-ground missiles or four AIM-92 Stinger air-to-air missiles. This was followed by the MQ-1B, and then by the B Block 10/15 (also known as the Predator A Plus), which introduced several improvements, including a greater wingspan, modified tailplanes, and an improved AN/AAS-52 system. This latter aircraft has a wing span of 55 feet, a length of 27 feet, a height of 6.8 feet, a maximum takeoff weight of 2,250 lb, an operational cruising speed between 80 and 103 mph, a range of 684 miles, or more than 24 hours, a maximum altitude of 4,735 feet, and is powered by a 115-hp Rotax 914F piston engine, connected to a two-bladed propeller. More than 360 units of the Predator A/A Plus were produced through to 2018 and have been used by Italy, Turkey, and the air forces of a further five countries.

310 top The Northrop Grumman RQ-4 Global Hawk took its maiden flight in 1998. The USAF intends to use it to replace the U-2 piloted reconnaissance aircraft.

310 bottom In 2019 Leonardo presented the Falco Xplorer, suitable for surveillance missions, but which can also be armed.

In 2004, the MQ-1C Gray Eagle also appeared, entering into service with the U.S. Army in 2009. This model has a 165-hp Thielert Centurion 1.7 diesel engine with a range of 30 hours, and it can carry four AGM-114 missiles or four GBU-44 bombs. The U.S. Army plans to acquire a total of 200 Gray Eagles.

The experiences of General Atomics lead this California-based company to develop its own, more capable version of the Predator, so anticipating the requests of the USAF. Thus was born the Predator B, in 2001, coming in three different versions, one with a jet engine, and two with turboprop engines. In October 2001, the USAF signed a contract for the development of the B-003 prototype, which was eventually introduced into operational service on May 1, 2007, with the official designation of MQ-9A Reaper. This machine is quite a bit larger and more capable than the Predator A, with a wingspan of 66 feet, length of 36 feet, and height of 12 feet; it is powered by a 900-shp Honeywell TPE331-10 engine, coupled with a three-bladed propeller. Its maximum takeoff weight is 10,494 lb, while the payload is subdivided as 794 lb of mission sensors (an AN/APY-8 radar and an AN/DAS-1 MTS-B target acquisition and designation system) plus 3,086 lb of armament, which may be coupled to six underwing joists. Cruising speed is 360 mph, maximum flight altitude is 49,212 feet, and the range is nearly 1,150 miles, or 14 hours at full load, but this may be extended to 24 hours. The armament consists of AGM-114 missiles, 507-lb laser-guided GBU-12 bombs, and 507-lb JDAM GBU-38 bombs. More than 60 specimens of the MQ-9A have been produced, and its production continues. It is used by the USAF (including wartime use, in the Afghan theater), but also by NASA (under the name Altair), U.S. Customs and Border Protection, Royal Air Force, and Italian Air Force. In 2014, development began on the improved MQ-9B Sky Guardian and Sea Guardian model, which boasts a range of over 40 hours and was acquired by the Royal Air Force as the Protector RG.1, entering service in 2023. Finally, in 2021, General Atomics flew a short take-off and landing (STOL) version of the Predator, designated Mojave, which attracted the interest of several operators.

The Northrop Grumman RQ-4 Global Hawk falls, instead, into the HALE (High-Altitude Long Endurance) category. This aircraft flew for the first time in February 1998 and was designed to have a UAS that could replace the strategic U-2 scouts. The aircraft had already been employed, during its testing and development stage, in the war in Afghanistan, and then in Iraq. The Global Hawk has a wingspan of 116 feet, a length of 44 feet, a height of 15 feet, and a maximum takeoff weight of 22,900 lb, with a payload of 1984 lb. It is powered by a 7055-lb Allison Rolls-Royce AE3007H Turbofan engine, allowing a cruising speed of 400 mph, a ceiling of 65,000 feet, and a range of 15,525 miles, or 36 hours of flight. The RQ-4 is equipped with an Integrated Sensor Suite (ISS) produced by Raytheon, and it includes a Synthetic Aperture Radar, an infrared viewing system, and an electro-optical system. In addition, the aircraft also has an AN/ALR-89 self-defense suite, consisting of an AN/AVR-3 Laser Warning Receiver, an AN/APR-49 Radar Warning Receiver, and electronic noise system, and an AN/ALE-50 towed decoy. The Global Hawk RQ-4B Block 30 can also mount an electronic reconnaissance payload (SIGINT), denominated Advanced Signals Intelligence Payload (ASIP). The B model boasts a more powerful engine, greater dimensions, and maximum takeoff weight of 32,250 lb, but its range is reduced to 8,700 miles or 28 hours flying time. The most advanced model is the Block 40, whereas the older versions did not test well, especially with regard to mission

systems capacities. In addition to the USAF and the U.S. Navy (introduced the MQ-4C Triton maritime patrol version into service in 2018), Germany also acquired the aircraft (as the RQ-4E or Euro Hawk) but was unable to use it as its systems prevented it from being certified for flight in European air space. The Global Hawk was acquired by NASA in 2007 and has also been used by NATO, with the Alliance Ground Surveillance Force based at Sigonella, Sicily from 2019. The Global Hawk has been produced in around 100 examples.

Nowadays there are hundreds of models of UASs in service in the world, from the micro UAVs, the size of a few centimeters, to the large-scale HALE models, since the low-tech construction that characterizes them is within reach of many countries and the transfer of technological information, including through agreements between companies, is continuous. Among the most successful and popular projects is the Israeli Elbit Hermes family, from the small model 90, to the 180, to the 450, and finally up to the 900. The most popular is the 450, which appeared in 1998, currently used by the air forces of 12 different countries. In Italy, the Meteor company was among the first in the world to deal with drones, with the Mirach series of drone targets. Acquired by

311 With much greater performance and capacities, the MQ-9A Reaper is the successor to the General Atomics Predator. It can be used for surveillance and reconnaissance tasks, as well as for attack.

311 bottom In 2005, the Italian firm Alenia Aermacchi released the Sky-X, a jet-propelled high-tech combat drone (UCAV) demonstrator.

Selex Galileo and today part of Leonardo, the company has produced, among others, the Falco (meaning "hawk" in English), a Medium-Altitude Medium-Endurance aircraft which completed its maiden flight in 2003. Over 50 units have been sold to seven international operators, including Pakistan and Saudi Arabia. Production continues, and an improved version with a greater range appeared in 2011, denominated Falco Evo. Moreover, Alenia Aermacchi has recently built two demonstrators for the UCAV and MALE technologies, the Sky-X, with a jet engine, in 2005, and the Sky-Y, width turboprop engine, in 2007. In 2019, Leonardo presented the Falco Xplorer model, a MALE with Intelligence Surveillance and Target Acquisition capabilities that can also be armed. The aircraft is a demonstrator for the European Detect and Avoid System (EUDAAS) and will reach full operational status in 2026.

In the future, it is expected that UASs will become increasingly more widespread, especially for performance of tasks that are riskier or require long flight times. The new category of UCAV (Unmanned Combat Aerial Vehicles) is already in the advanced stages of development and the first unmanned aircraft with strike capability far more significant than those provided by the Predator will be in service within a few years. The United States has shown itself, in this field as well, to be in the forefront, with experimental projects that are moving forward with interesting results. These are the fruits of the work of the DARPA (Defense Advanced Research Projects Agency), dating back to 2000. Initially the programs of the U.S. Air Force and the U.S. Navy progressed in parallel, the first with the X-45 and the second with the X-47. The Boeing X-40 5A was a small-scale demonstrator that completed its maiden flight on May 22, 2002, and two years later it completed its first successful test bombing of a target. The aircraft shape was modeled using advanced stealth technology (meaning using forms that tend to be invisible to radar), and it adopted the all-wing formula, without tailplanes, fins, or empennage of any sort. In 2005, the X-45 completed its series of tests, and the two models produced were transferred to museums. Then the models X-45B and X-45C were produced, larger in size and with greater range, but the USAF decided to terminate the program in 2006. The later derivative from these projects was the Phantom Ray, a UCAV intended to demonstrate the possibility of effecting ISTAR missions, suppression of enemy air defenses, electronic attack, and automatic airborne refueling. This craft completed its maiden voyage on April 27, 2011, and is in practice a technological demonstrator. It is powered by a General Electric F404-GE-102D Turbofan and has a maximum velocity of Mach 0.85, equivalent to over 560 mph. With a wingspan of 50 feet and a length of 36 feet, its maximum takeoff weight is around 36,376 lb. For its UCAS-D (Unmanned Combat Air System Demonstration) program, the U.S. Navy has, however, preferred the Northrop Grumman X-47 Pegasus project, initially realized as the small-scale Model A, and in 2007 it proceeded to Model B, which has the same size of a possible series UCAS. The first prototype flew on February 4, 2011: a classic stealth aircraft, all wing, with the engine on the back, so as to shield the air intake and the form of the exhaust cone from enemy identification systems. One peculiarity of the X-47B is that it is an embarked aircraft, and it therefore it has a reinforced trolley with hooking systems for a catapulted launch and a stopping hook for landing, as well as folding wings. The aircraft is powered by a Pratt & Whitney F100-PW-220U Turbofan with 14,330 lb of thrust and is expected to reach a top speed of over 560 mph, with a maximum ceiling of over 39,370 feet. The wingspan is 62 feet and the length is 38 feet, while maximum takeoff weight is 44,566 lb. The Pegasus has already undergone a good deal of ground testing, and in 2013 flight tests were conducted aboard an aircraft carrier, demonstrating its automated landing capabilities. Further tests in 2015 concerned its capacity for automatic airborne refueling of carrier-borne fighters. Since 2010, the US Navy has been pursuing the Unmanned Carrier-Launched Airborne Surveillance and Strike (UCLASS) program, which in 2018 led to Boeing being awarded the contract for the Carrier-Based Aerial-Refuelling System

(CBARS), a remotely piloted aircraft for in-flight refueling of carrier-borne aircraft. The Boeing aircraft is the MQ-25A Stingray, which flew for the first time on 19 September 2019. It is 51 feet long with a wingspan of 75 feet (31 feet with the wings folded), and is powered by a Rolls-Royce AE3007N turbofan engine delivering approximately 9,920 lb of thrust. It has the capacity to transfer almost 16,000 lb of fuel, but also has two underwing

312 top The MQ-25 Stingray is an unmanned aircraft that can be deployed from aircraft carriers for in-flight refueling missions, but can also be armed.

312-313 With its first flight in 2011, the Boeing Phantom Ray is a stealth technology demonstrator for a UCAV system designed for ground use by the USAF.

313 top The nEUROn, shown here under construction, is the result of a French-led European program for the construction of a UCAV stealth demonstrator. The prototype took its first flight on 1 December 2012.

attachment points for two AGM-158C anti-ship missiles. The US Navy plans to acquire 76 Stingrays, which will be distributed to two flight groups, VUQ-11 and VUQ-12, while the VUQ-10 training group was formed in October 2022.

In Europe, various aircraft manufacturers have since the turn of the 21st century begun to conduct experimental activities with next-generation UCAS and UAS aircraft, and in particular, four projects have attracted the most attention: the nEUROn, Talarion, Barracuda, and Taranis. The first is primarily a technology demonstrator for a future unmanned combat aircraft. The program was launched in 2003 by the French government and Dassault, which is responsible for 50% of total costs, while other governments and companies have joined on to the project: French Thales and EADS-France, Italian Alenia Aermacchi, Swedish SAAB, Spanish EADS-CASA, Greek EAB, and Swiss RUAG. The first aircraft ready for flight testing was presented on January 19, 2012, at the base of Istres, and lengthy in-flight testing has been undertaken, aiming to accumulate data and experience necessary for design of future European combat aircraft, both piloted and unmanned. In competition with the nEUROn, Germany and Spain created the Barracuda program in 2006, a jet engine UCAV capable of reaching Mach 0.85, with a maximum takeoff weight of over 6,600 lb. The program, interrupted until 2008 due to an accident, produced a significant amount of data that will be useful for the development of a future operational UCAV.

The Talarion is a MALE designed and developed by the EADS consortium under the approval of the governments of Germany, France, and Spain, and joined later by Turkey. The aircraft was expected to make its first flight in 2015, and its entry into service was scheduled for 2018. France and Germany each planned to acquire six Talarion systems, and Spain, three; each system is formed of three aircraft and one ground control station. The Talarion was to be equipped with electro-optical and infrared mission sensors, as well as a laser designator; it was furthermore to be equipped with SAR radar, electronic support systems (ESM), and electronic reconnaissance systems (SIGINT). The aircraft is equipped with two jet engines and has a maximum speed of 345 mph. Maximum flying altitude is projected at 50,000 feet, with a range of about 20 hours flying time; maximum takeoff weight should be around 15,400 lb, for a payload of 1,765 lb. The program has, however, encountered several difficulties in its realization, and in March 2012, the company substantially halted the program, pending confirmation of a contract by the governments concerned. Subsequently, the program was canceled as Europe preferred to focus on a configuration with turboprops, giving rise to the EuroMALE program.

In 2006, BAE Systems also launched the British Taranis program for a UCAV technology demonstrator, a remotely piloted aircraft with combat capabilities, weighing 17,636 lb and with an intercontinental range. It is also the first stealth aircraft designed by the U.K., and its maiden flight took place on 10 August 2013. However, this ambitious program also seems to have been shelved. Nonetheless, Great Britain and France have continued to collaborate in this field, since February 2012. In any case, it seems very likely that there will be no operational unmanned UCAS entering into service for a military air force before the next 10 or 15 years, at least in Europe.

As mentioned, European efforts to develop a MALE-class remotely piloted aircraft have focused on the EuroMALE Remotely Piloted Aircraft System (RPAS) program, launched in 2015 as a collaboration between the Franco-German manufacturer Airbus (as prime contractor), the French company Dassault, and the Italian company Leonardo, with the Spanish branch of Airbus coming on-board in 2016. The aircraft was unveiled at the ILA aerospace show in Berlin in 2018. In February 2022, the four nations signed a contract for the purchase of 60 aircraft, at a cost of €7 billion, including maintenance for 15 years. The EuroMALE is powered by two General Electric Catalyst turbines, manufactured by Italy's Avio Aero, each delivering 1,300 shp, and is expected to have a maximum takeoff weight of 24,250 lb, a cruising speed of 310 mph, a range of over 24 hours, and a maximum ceiling of 44,950 feet. In 2023, it was announced that the AMS mission system will be supplied by Leonardo,

314 top The Talarion is a MALE aircraft developed by the EADS consortium, projected to satisfy air forces needs of France, Germany, and Spain.

314 bottom The EuroMALE project was launched in 2015 to create a European unmanned platform and is scheduled to enter service in 2030.

315 top The Wing Loong 2 is an unmanned system produced by the Chinese manufacturer Chengdu that flew for the first time in 2017.

315 bottom The Akinci is a twin-engined drone produced by the Turkish firm Baykar that can be used for both surveillance and assault missions, with a maximum payload of 3,300 lb.

which will also be responsible for integrating the weaponry system, as, in addition to reconnaissance and surveillance tasks, the EuroMALE will also have attack capabilities, with Brimstone missiles and JDAM bombs. It is expected to enter service in 2030, with the first prototype flight scheduled for 2027.

In recent years, remotely piloted systems have become increasingly popular for a variety of uses, with aircraft of widely varying sizes and capabilities. The recent war in Ukraine has clearly demonstrated the importance of these systems, both for reconnaissance and surveillance roles and for combat duties. Virtually all the world's major industrialized countries have begun developing or already produce unmanned aerial systems, to the extent that it is now impossible to draw up a list, especially for those in the under-25-kg category. By way of example, here are just some of the main systems in use today.

Russia uses the Orion family of systems, developed by the Kronstadt Group since 2011, which flew for the first time in 2016 and entered service in 2020. The Orion comes in reconnaissance and assault versions, with four guided bombs or four missiles, for a maximum payload of 441 lb. It was first used operationally in Syria in 2019. Then there is the Helios-RLD (also known as Orion-2), a five-ton High-Altitude Long-Endurance (HALE) aircraft with a wing-span of 98.5 feet and a range of over 30 hours. Another member of the same family is the Sirius, a MALE with similar characteristics to the Helios and a warhead capacity of 992 lb, which completed its maiden flight in 2023.

For its part, China also produces a series UAS systems, including the Chengdu GJ-1, also known as Wing Loong 1, a Predator A-class MALE that flew for the first time in 2009 and entered service in 2011. With a wingspan of 46 feet and a maximum weight of 2,425 lb, the GJ-1 is powered by a 100 shp Rotax 914 turbocharged engine, giving it a range of approximately 20 hours. This aircraft can carry various types of sensors weighing approximately 220 lb, or 440 lb of air-to-surface weapons. It has been exported to 11 countries, including Egypt, Morocco, Pakistan, and Saudi Arabia. The GJ-1 was followed by the GJ-2, also known as Wing Loong 2, a Predator B-class UAS that first appeared at the Zhuhai Airshow in November 2016 and completed its maiden flight in February 2017. This is a larger aircraft, weighing 9,260 lb and with a weapons carrying capacity of 1,058 lb. It has a maximum range of 32 hours, and several versions have been developed, including for export. In addition to China, the GJ-2 is also flown by Algeria, Saudi Arabia, Egypt, the United Arab Emirates, Libya, Morocco, Nigeria, and Pakistan. These aircraft were first used operationally during the second civil war in Libya in 2020. The latest and largest model in the family is the Wing Loong 3, which was unveiled in November 2022 and boasts a range of 6,200 miles (or 40 hours) with a maximum takeoff weight of 13,668 lb and a payload of up to 4,410 lb. It is a multi-role platform that can perform air-to-air, anti-submarine, land, and maritime assault, search-and-rescue, reconnaissance, and surveillance missions.

Turkey has also made an international name for itself with the production of high-quality UAS. The leading company in this sector is Baykar, which manufactures various systems: the Bayraktar TB2, TB3, Akinci, and Kizilelma. The TB2 is a small single-engined UAS with a wingspan of 39 feet, a take-off weight of 1,543 lb, a range of over 20 hours, and the ability to carry a payload of 330 lb, whether reconnaissance systems or weapons. The Akinci, on the other hand, is a twin-engined aircraft equipped with two turboprops delivering a power output of up to 850 shp each, a maximum weight of 13,227 lb, and a payload capacity of 3,306 lb, again including various surveillance sensors or precision missiles and bombs for assault missions. The Kizilelma is very interesting. It is defined by Baykar as a highly maneuverable fighter UAV with stealth capabilities. It is an aircraft with an AI-25TLT jet engine developed since 2013, which flew for the first time in 2022. It has a maximum weight of 18,740 lb and a maximum speed of Mach 0.9 and can carry a weaponry load of up to 3,306 lb in two internal bays, in addition to six underwing pylons. Thanks to its short take-off and landing (STOL) capabilities, it can also be deployed from aircraft carriers. These specifications will also allow the Kizilelma to operate in the role of Loyal Wingman, a mission that is becoming increasingly popular among the most advanced air forces. In 2025, Baykar and Leonardo created a 50-50 joint venture, LBA Systems, which will be based in Italy and will be dedicated to the development and marketing of UAS systems.

As mentioned, battlefield requirements have convinced the major air forces that in the future it will be necessary for combat aircraft to be accompanied by fleets of companion aircraft, either controlled by the former or operating independently to perform various roles, such as attacking the most dangerous targets or saturating enemy defenses. All major manufacturers are engaged in the design of jet-powered UAS systems that can be integrated with fourth, fifth, and sixth generation fighters. In addition to the aforementioned Kizilelma, other systems are emerging, intended to accompany aircraft such as the Eurofighter Typhoon, the Dassault Rafale, the Lockheed Martin F-35 and F-22, and the future F-47, GCAP and FCAS produced in the United States and Europe. Boeing has already produced the MQ-28 Ghost Bat, an aircraft designed in a joint venture with the Royal Australian Air Force, which completed its maiden flight on 27 February 2021. The RAAF appears to be planning to acquire an initial batch of 10 aircraft.

To complement and support the USAF's future sixth-generation fighter, two aircraft are competing in the Collaborative Combat Aircraft (CCA) program, the General Atomics YFQ-42A and the Anduril YFQ-44A Fury. Much more than mere drones, these planes will be true unmanned combat aircraft which, thanks to artificial intelligence, will be capable of fighting independently. They will have an operational range of over 800 miles and will also be able to act as advanced sensors and weapons platforms for the F-47, as well as other aircraft such as the F-35. USAF Chief of Staff General David Allvin has stated that he expects to have nearly 1,000 of these aircraft operational by 2029. The decision on when to start production is due in 2026, but Increment 2 of the program could also include other aircraft.

Also produced in the United States is the XQ-58 Valkyrie, under development by Kratos since 2016, which flew for the first time in 2019. Russia, on the other hand, is developing the Sukhoi S-70 Okhotnik-B, which completed its maiden flight in 2019 and is expected to operate in support of the Su-57 fighter, while France, with Dassault, is developing a wingman based on the nEUROn jet drone design. China is instead working on the Feihong FH-97, which first appeared in 2021 but about which little is known, while India is focusing its efforts on the HAL CATS Warrior, also unveiled at Aero India 2021, whose first flight might well take place before the end of 2025. Finally, mention has to be made of the Airbus Wingman, another Mach 1 jet aircraft that was announced at the 2024 Berlin Air Show and is expected to make its first flight in 2030.

The future of UASs is unlikely to be limited military missions only: they are also planned for use in transportation. To this end, Boeing has been working for some time on the X-48 project, an aircraft that completed its maiden flight in June 2007, in X-48B form. This is a small-scale (8.5%) technology demonstrator for a future civil and military transportation aircraft with an all-wing configuration. This was followed by the C model, with improved engines and

aerodynamics, which first flew in August of 2012. NASA and Boeing have announced that a larger demonstrator will be developed, capable of transonic flight, but no further news about this promising program has been heard since 2013.

In addition to the fixed-wing field, unmanned aircraft have also appeared in the rotary wing sector, which has very many applications, in not only the civil field, but also the military. The first unmanned helicopter, a Kaman HTK-1, was already flown back in 1957, but many years passed before the technology could be perfected to the point of making unmanned helicopters feasible. Noteworthy among the many aircraft developed, and the many and various dimensions, are the unmanned K-MAX by Kaman (the unmanned version of the piloted K-1200), which has been employed by the Marines in Afghanistan, Boeing A-160T Hummingbird, Northrop Grumman RQ/MQ-8 Fire Scout, Boeing Little Bird, and PZL-Swidnik SW-4.

Doubtlessly the UASs represent one of the flight systems with the best prospects and potentials for the future, and their technology is within reach of many aerospace industries worldwide. Apart from their military uses, these aircraft have and will become increasingly important for their roles in surveillance, security and police work, civil defense, territorial control, pollution control, and aerial surveying and photography.

316 A rendering of the two Collaborative Combat Aircraft competing for the USAF tender that will result in the future vehicle capable of accompanying fighter aircraft on their missions: the YFQ-42A (foreground) and the YFQ-44A.

317 top The Boeing A-160T Hummingbird project, for an unmanned helicopter, is planned for autonomy of 24 hours or 2,500 miles.

317 bottom Even in the field of rotary wing aircraft, there are a proliferation of designs for unmanned machines. Shown here, a prototype of the Northrop Grumman RQ-8A Fire Scout, developed for needs of the US Navy.

Chapter 26

Today and tomorrow

The twenty-first century future of aviation has already begun, and the trends and problems likely to affect this avant-garde aspect of modern technology and the global industrial economy can already be seen. In the field of air transport, the research programs initiated by the industrial giants Boeing and Airbus have identified a market in dramatic expansion.

The report into air transport developments over the next 30 years, published by Airports Council International (ACI) in February 2025, once again points to strong growth in the future, albeit with a degree of short-term uncertainty. During the period under consideration (2024–2053), the number of passengers carried worldwide is expected to grow to 17.7 billion by 2043 and rise further to 22.3 billion by 2053, representing an increase of almost 2.4 times with respect to 2024 volumes. The greatest developments are expected in emerging countries, with a greater demand for flights from the middle classes. Similarly, the number of flights is expected to rise, reaching 149 million per year in 2043 and 176 million in 2053. In line with these forecasts, Boeing predicts a market for 43,600 new commercial aircraft over the next 20 years, through to 2044, with the greatest demand coming from Chinese and Southeast Asian airlines. Over this period, the global commercial aircraft fleet is expected to double in size, reaching 49,600 aircraft. Airbus's forecasts are broadly similar, with annual growth expected to be 3.6% and demand for new aircraft over the next 20 years estimated at 43,400 units.

The military sector has not enjoyed such good health as the civil aviation industry, primarily owing to continuing defense budget cuts in the Western countries resulting from the end of the Cold War, along with increases in purchase and operational costs and the economic crisis. Nonetheless, the crises that have developed in various parts of the world, including of course the war in Ukraine from 2022 and the permanent state of tension between Iraq, its allies, and Israel in the Middle East, which led to military operations in June 2025, have led to an inversion of this trend, above all with respect to the NATO countries. Over the next few years, the United States will still constitute a large part of the market in the coming years, followed by the Asian and European markets. The need to modernize many air forces will ensure that the fighter sector in particular will remain active and interesting over the coming years.

The fighter aircraft field has been dominated by a new generation of combat planes that entered into operational service right around the turn of the millennium. One of these, the JAS 39 Gripen, has already been in service with the Swedish air force since

1995. Produced by Industrigruppen JAS (led by Saab), the Gripen was conceived in Sweden in the early 1980s as a multi-role successor to the Viggen and the Draken. Characterized by sophisticated fly-by-wire flight controls, simplified maintenance, advanced avionics, and true multi-role capabilities, the JAS 39 flew for the first time on 9 December 1988. However, owing to delays in the development of the control software (which caused two crashes), deliveries to the air force began only in 1993, with the first military pilots being able to fly the plane two years later.

The Gripen is a small airplane, with a wingspan of 26 feet 3 inches, a length of 45 feet 11 inches, and a maximum take-off weight of about 27,500 lb. Its Volvo RM12 turbofan delivers just over 18,000 lb of thrust on afterburn and can power the Gripen to a top speed of about 1,243 mph. A two-seat combat training version has also been developed. Sweden has ordered 130 units, followed by 50 single-seater JAS 39Cs and 14 two-seater Ds, the new and more advanced version including a new main computer, new weaponry, equipment for inflight refueling, and compatibility with NATO systems. The first model C flew in 2002, and the last 20 machines of the first batch were also completed according to this standard. The aircraft has also enjoyed a number of successes in exports, with 28 units being sold to South Africa, 14 being leased to Hungary, 14 to the Czech Republic, and 12 to Thailand. The Gripen NG prototype took its maiden flight in 2008, a model introducing several improvements, including electronic-scan AESA radar, a more powerful F414 engine, strengthened airframe, new points of attachment for weaponry, and increased maximum takeoff weight. Sweden and Brazil were the first customers for this new version, denominated the JAS 39E, with orders for 70 and 36 units, respectively, and deliveries starting in 2022. In July 2025, Peru also announced that it had acquired 24 examples of the Gripen E..

The Dassault Rafale anticipated the Gripen's maiden flight by two years, but this very advanced multi-role fighter's entrance into service has been postponed more than once for reasons of cost. The Rafale was developed as the successor to the French air force's Jaguars and Mirage 2000 series and the French navy's F-8s and Super Étendard's, and the first Rafale prototype flew on 4 July 1986, with the plane due to enter service in the mid-1990s. The Rafale is a twin-engined design that to ensure maximum maneuverability features canard aerodynamic surfaces and a delta wing. It has a wingspan of 32 feet 9 inches, a length of 49 feet 2 inches and a maximum take-off weight of 48,840 lb. The two SNECMA M88-2 turbofans provide 17,000 lb of thrust with afterburner, for a top speed over Mach 1.8 (1,180 mph). Its weapons system comprises up to 20,950 lb of external armament (including the most modern systems, like the MICA IR and EM missiles, SCALP-EG missiles, and the AASM Hammer precision bombs), along with a 30-mm cannon. Three versions of the aircraft have been developed: the C and B (two-seater) for the air force and the carrier-borne M version and the two-seater MB) for the navy. The M model (version F.1) was actually the first to enter service, in the December of 2000, at the Llandivisiau base, while the first planes for the air force were assigned to the base at Mont-de-Marsan in 2004. The first operational department was Group 01.007, coded "Provence," at the St. Dizier base. The versatile multipurpose F.3 version has been in production since 2008, and France has ordered a total of 180 units of this fighter. In spite of all possible efforts, for many years the Rafale never enjoyed export success, but in January 2012, it finally managed to win the MMRCA competition for the new fighter for India, with delivery of 18 of the aircraft programmed for within 2015, and licensed construction in India of another 108. Subsequently, the agreement was annulled and India signed a contract to receive 36 examples directly from France, with the first delivered in 2019. In 2025, the Indian navy also signed a contract for the acquisition of 26 Rafale Ms to be delivered by 2023. Since then, the aircraft has also been acquired by the air forces of Croatia, Egypt, United Arab Emirates, Greece, Qatar, and shortly Indonesia and Serbia. The Rafale is certainly an expensive as well as sophisticated aircraft, but it represents a valid alternative for those countries unwilling to over-commit themselves to products from the United States, Russia, or China. In October 2012, Dassault delivered to the French Air Force the first unit equipped with RBE2 AESA radar, the 137th produced.

Today the aircraft are being produced to F.4 specification, while the company is working on the development of the F.5, which will be fitted with the T-Rex engine, offering 30% more power, a new radar system, and the ability to operate with accompanying drones. Total production has exceeded 300 examples.

In the United States, on 29 September 1990, the first prototype for the USAF's future air-superiority fighter made its maiden flight. The plane is the YF-22, a technological demonstrator built by Lockheed-Martin and Boeing to compete with the Northrop-McDonnell-Douglas YF-23 in the USAF's ATF competition, designed to produce a successor to the F-15 Eagle. In April 1991, the YF-22 was officially declared the winner, and six years passed before the maiden flight of the first pre-production F-22A Raptor on 7 September 1997.

318 top The Dassault Rafale fighter, a multipurpose aircraft, which entered into service in 2000 with the French Navy.

318 bottom A pair of Lockheed Martin F-22A Raptors of the USAF's 1st Fighter Wing. Bearing highly advanced features, this fighter came into service in 2003. With production through 2011, 187 units were built.

319 A pair of JAS39D Gripens belonging to the South African Air Force, one of the foreign customers for this Swedish multipurpose fighter.

maneuverability. The aircraft is 722 feet long with a wingspan of 42 feet, a maximum takeoff weight of 77,200 lb, and a top speed of around Mach 2. The avionics are, obviously, advanced, and the weaponry, probably 8 missiles, is contained in the internal holds. The J20, designated as the Mighty Dragon (Fagin for NATO) entered service in March 2017 and is thought to have been built in more than 250 examples to date. September 2021 saw the presentation of the J-20 equipped with Chinese WS-10C engines providing around 39,680 lb of thrust each. This fifth generation fighter has brought with it new scenarios in the Indo-Pacific region, further affirming the People's Republic of China's military supremacy in the area. On 31 October 2012, another new Chinese combat aircraft completed its maiden flight, apparently with the designation J-31. This fighter appears to be medium-sized, quite similar in form to the United States F-22 and F-35, and therefore presumably with stealth capabilities. With two prototypes built by Shenyang, the J-31 competed with the J-20 in the J-XX program, which was ultimately won by the latter. It should therefore be an aircraft intended for export, to compete with the American F-35, but in 2018 the program received government funding and it appears that a carrier-borne version of the aircraft is being developed for maritime operations. The aircraft attracted considerable interest at the 2024 World Defense Show in Saudi Arabia, and Pakistan is expected to be the first foreign customer, with deliveries due shortly.

Along the same lines, Russia's Sukhoi has developed the Su-75 Checkmate, a single-engined light fighter intended for the Russian air force, but also for export. It is expected to be a low-cost stealth aircraft with a speed of Mach 1.8, but it has yet to complete its maiden flight, originally scheduled for 2023.

Finally, the latest Chinese combat aircraft is the Chengdu JH-36, a large three-engined delta-wing design that was officially unveiled in

324 top One of the top new-generation tanker aircraft is the Airbus A330MRTT, shown landing here, bearing the colors of the British Royal Air Force.

324 top center A CASA/EADS C.295MP for maritime patrol, dressed out in the colors of the Chilean Navy.

324 bottom center A prototype of a new version of airborne radar control, designated C.295AEW, which was extrapolated from the model C.295.

324 bottom The Boeing P-8A Poseidon, a derivative from the commercial aircraft B-737NG, is a new maritime patrol airplane developed for the US Navy.

December 2024, but about which very little is known. Although described by many as a fighter, it could be a long-range interceptor or a bomber, its size suggests that it has a long range, while its three internal bays can accommodate weapons of any type, including large YJ-12 anti-ship missiles.

South Korea has also tackled a new-generation fighter project, the KAI KF-21 Boramae, a 4.5 generation fighter that was unveiled in 2021 and completed its maiden flight on 19 July 2022. This is an ambitious multi-role aircraft, developed in collaboration with Indonesia, which has acquired a 20% stake. It boasts stealth capabilities, a maximum takeoff weight of 56,000 lb, and a speed expected to reach Mach 1.8, thanks to the thrust provided by two General Electric F414-GE-400K turbofan engines delivering approximately 22,000 lb of thrust each. The payload totals 7,000 kg of air-to-air and air-to-ground weapons. South Korea has ordered 40 Block I production aircraft for the time being, with deliveries due to start 2026; another 80 Block II units are expected to be ordered and delivered by 2032.

Looking to the future, mention has to be made of the projects concerning the next aerial superiority aircraft in Europe and the United States.

In Europe, 2014 saw France and Great Britain launch the Future Combat Air System/ Système de Combat Aérien du Futur (FCAS/ SCAF) program regarding a sixth-generation fighter, but it was soon abandoned.

In 2018, at the Farnborough Airshow in the United Kingdom, the British government presented the Tempest project, a super-fighter jet intended to replace the Eurofighter Typhoon from 2035 onwards. In 2019, Italy joined the program, which in 2022, with the arrival of Japan as the third prestige partner, took the name Global Combat Air Programme (GCAP). The program is led by three major companies, BAE Systems, Leonardo, and Mitsubishi. An initial technology demonstrator is expected to fly in 2027, while the first production models are currently scheduled for completion in 2035. The aircraft will be a large twin-engine fighter with a maximum speed greater than Mach 2, stealth capabilities, internal weapon bays, and highly sophisticated avionics, capable of interacting with assets in all domains and controlling support drones, in what has been described as a system of systems. In parallel with this program, France and Germany launched SCAF/FCAS in 2017, essentially the French project with a new partner, with Spain also joining in 2019. This program, with characteristics and capabilities similar to GCAP, is led by Dassault and Airbus, but is encountering significant difficulties and its implementation appears to be in doubt at the moment.

In the United States, on the other hand, the Next Generation Air Dominance (NGAD) program for the Air Force and F/A-XX (or NGAD) for the US Navy have been launched. The demonstrator of what will become the USAF's first sixth-generation fighter, replacing the F-22, should have undergone flight testing as early as 2020, although next to nothing has been leaked about this and it was only in March 2025 that President Trump revealed that the fighter had been designated F-47 and would be built by Boeing. The definitive design's maiden flight is expected to take place by 2029. The Navy program, on the other hand, began in 2008 with the aim of replacing the Super Hornet family of aircraft from 2030 onwards. The designs, of which only renderings have been unveiled, appear very similar.

Other countries are also working on next-generation fighters, Russia with the Mikoyan PAK DP, China with the J-50, and Turkey with the Kaan, but it remains to be seen what capabilities they will actually be able to offer once deployed in service.

325 top The Brazilian Embraer has developed an radar warning and airborne control aircraft from the basis of the EMB-145 aircraft.

325 center The Airbus A400M is a strategic transport aircraft characterizded by the adoption of four turboprop engines rather than jets.

325 bottom The Royal Air Force chose the Sentinel R-1, a version of the executive jet Bombardier Global Express, for airborne ISTAR (Intelligence, Surveillance, Target Acquisition, and Reconnaissance) activities.

The aviation of today and tomorrow is not characterized solely by combat aircraft. Apart from the new category of unmanned aircraft, presented in another chapter, there is another sector counting a great number of advances, and which has witnessed considerable growth in recent times. These are the support aircraft, including, specifically, those for inflight refueling, radar warning, command and control, electronic warfare, maritime patrol, reconnaissance missions, and ISTAR (Intelligence, Surveillance, Target Acquisition and Reconnaissance) missions in general. These are proving increasingly essential in modern military operations, for providing command and engaged units in the battlefield situation with evermore precise and updated information, in real time. In the field of inflight refueling, the two most advanced aircraft in service today are the Boeing KC-46A and the Airbus A330 MRTT (or KC-30), and these are already experiencing a degree of sales success. The United States aircraft is a derivative of the B.767 in airline use, while the European machine is a development of the commercial A330. Another interesting project is that of the Brazilian Embraer KC-390, an aircraft for transport and inflight refueling which flew for the first time on 3 February 2015 and entered service in 2019 with the Brazilian air force. Other contracts have been signed with Austria, South Korea, Holland, Portugal, Czech Republic, Slovakia, Sweden, and Hungary. The radar warning aircraft are always expensive and rare in number; those in service are the well-known Boeing E-3 AWACS and the Ilyushin A-50. Recently, however, other manufacturers have ventured into this field, producing some interesting and effective aircraft, and sometimes at lower costs. We refer to the Boeing E-767 AWACS, adopted by Japan in 2000; to the Boeing 737 AEW&C, introduced into service in 2009, and acquired by Australia, South Korea, and Turkey; to the Embraer EMB145 AEW&C (also called the R-99), in service in Brazil, Greece, India, and Mexico since 1999; and to the Gulfstream G-550 CAEW, acquired by Israel, Singapore, and Italy.

For maritime patrol and surveillance, there are various types of platforms, from the older Lockheed P-3 Orion, Tupelev Tu-142, and Breguet Atlantic; to the ultra-modern Boeing P-8 Poseidon, another military version of the eternal commercial B-737; to the smaller ATR-42/72MP and ASW, and CASA C235MP and 295MP, patrol units with surveillance capabilities.

326 top The only convertiplane designed and developed for the civilian market is currently the AgustaWestland AW609, whose introduction into service is underway with the launch customer, Bristow Group.

326 center Eurocopter tests high-speed possibilities in the helicopter field using the model X3, equipped with a conventional rotor as well as two thrust-providing propellers powered by turbine engines.

Among the more recently developed ISTAR aircraft, lastly, there are airplanes that were created for different purposes, then re-adapted, like the patrol craft and reconnaissance units, while others were specifically developed, like certain derivatives of the twin-engine Beechcraft King Air, like the Shadow R.1 for the RAF, the Desert Owl for the US Army, the various models of RC-12 and RU-21, and above all, the last in the family, the MC-12W Liberty, built to order for the USAF on the model of the King Air 350, for ISR missions in Iraq and Afghanistan. In our roundup of aircraft of today and, especially, of tomorrow, we cannot forget to mention 4 new airplanes, 3 of which have yet to enter into service. The first is the Airbus Military A400M Atlas, the European strategic transport aircraft. This project stems from the Future International Military Airlifter (FIMA) program of 1982, which subsequently became the Future Large Aircraft (FLA) program. Initially, the industries of the UK, Germany, Italy, Spain, France, Belgium, Turkey, and Luxembourg had joined the consortium, but after the withdrawal of Italy, the number of aircraft was fixed at 180, with deliveries to begin in 2009. Denominated A400M, the aircraft was completely new, including the

326 bottom The Bell-Boeing V-22 Osprey is the world's only convertiplane used for military purposes. It entered into service in 2005 and is used by the US Marines and the USAF for special operations.

327 Presented in 2018. The Tempest, produced by British firm BAE Systems, is the basis for the GCAP tri-national program, the first prototype of which is due to fly in 2027.

engines, with cutting-edge technical solutions and an aerodynamic configuration similar to that of other latest-generation military transports, like the United States C-17, but with the difference that it adopted 4 turboprop engines.

The first flight was planned for 2008, but this date was postponed until 11 December 2009 owing to technical difficulties coupled with high costs. With 5 prototypes built, the A400M was intended to replace the C-160 Transall and the C-130 Hercules in many countries. Orders reached a total of 174, subdivided among Germany (53), France (50), Spain (27, albeit only 14 delivered to date), the UK (22), Turkey (10), Belgium (7), Malaysia (4), Luxembourg (1), Indonesia (2) and Kazakistan (2). The A400M has a wingspan of 139.1 feet, a length of 148.0 feet, a maximum takeoff weight of 155.4 tons, and payload of 81,600 lb, equivalent to 116 soldiers

fully equipped. The engines are 4 Europrop TP400-D6 turboprops delivering 11,060 shp each, which operate sophisticated composite 8-blade propellers. Cruising speed is 485 mph at an altitude of 26,000 feet; and with a 20-ton payload, the range is 3,975 miles. Despite various developmental problems, particularly with regard to the engines, the delivery of the first series unit for the French Air Force took place in August 2013. The two new projects mentioned here are the only representatives of the new category of convertiplanes. These aircraft have tilting engines that operate very large (tiltrotor) propellers, more similar to rotors. In normal flight, the engine nacelles, positioned at the ends of the wings, are in horizontal position and the airplane flies like a normal turboprop aircraft. But for takeoff and landing, the engine nacelles are rotated into vertical position, so that the aircraft takes on the flight characteristics of a helicopter.

This type of aircraft was already explored during the 1950s, when the Bell XV- 3 was built, and later the Bell XV-15, a technology demonstrator that took its maiden flight in 1977. It was experiences from these that later made it possible, thanks to new technologies developed in the field of computers and flight controls, to build convertiplanes in series production. Thus was born the Bell Boeing V-22 Osprey, a military machine that resulted from the JVX program of December 1981, aimed toward production of a medium-size tilt-rotor aircraft transport, capable of carrying 24 equipped soldiers for a distance of 230 miles. The first prototype flew on 19 March 1989, but owing to extremely high costs, the program was suspended before yearend by Minister of Defense Cheney. The program was subsequently reactivated and, despite various controversies relating to numerous technical problems, costs, a couple of tragic incidents, the V-22 entered into service in 2005 with the Marine Corps. The Marines ordered 360 examples of the Osprey MV-22B, to replace the older CH-46 and CH-53 helicopters; while the USAF ordered 50 units of the CV-22B to support special operations. It has already been employed in actual operations and the US Navy ordered 48 of the CMV-22B. The MV-22B has a wingspan of 46, a length of 57 feet, and two rotors of diameter 38 feet each. Maximum takeoff weight is 60,400 lb and the propulsion system consists of two Rolls-Royce Allison T406/AE 1107C Liberty turbines delivering 6,150 shp each. Top speed is 316 mph, with a maximum range near 2,230 miles. The CV-22 version also has sophisticated avionics systems for all-weather operations, even in hostile environments, as well as defensive armament and an in-flight refueling probe. The V-22 has been exported to Japan only, with 17 examples delivered. The Osprey's alter ego in the civilian field is the AgustaWestland (today Leonardo) AW609, a tilt-rotor born in 1996 as a project of Bell-Boeing, which became Bell-Agusta in 1998. A smaller machine, it is suitable for use as a passenger aircraft for corporate and business transport; the aircraft took its maiden flight on 6 March 2003. The program suffered from a certain slowness until, in November 2011, the ownership restructuring was concluded, after which the project passed entirely into the hands of AgustaWestland. From that point, the project accelerated, with an aim to achieve civilian certification within the near future, to allow planning deliveries of aircrafts to the first customers already in 2017. In the October of 2015, during a speed test, the second of the two prototypes was destroyed in an accident that killed the two-man crew. In 2025, the AW609 is in the final phase of the American Federal Aviation (FAA) certification, while European EASA certification is also on-going. The AW609 has a wingspan of 38 feet and a length of 43 feet. Powered by two Pratt & Whitney Canada PT6C-67A turbines delivering 1,940 shp, the aircraft has a maximum takeoff weight of 16,750 lb and can transport up to 9 passengers.

Also being researched are models for military and government use, such as for maritime patrol and search and rescue. The future of aviation is not easy to define, given the enormous opportunities offered by the fields of research and new technologies, but also in view of the difficulties arising from the vast development costs required in realizing innovative and revolutionary designs. In the military field the question is open as to whether combat aircraft of the future will be

328 The future sixth-generation fighter for the US Air Force, the replacement for the F-22, has been designated as the F-47, and the production version is due to fly in 2029.

329 The future strategic stealth bomber for the US Air Force that will replace the B-2 is the B-21 Raider, which flew for the first time in November 2023.

unmanned or piloted. In any case, the machines will be ever more audacious from the point of view of electronic capabilities and onboard software, while in the field of armament, research points more and more to weapons of increasing precision capable of generating the least possible amount of collateral damage. In 2011 the United States launched the Long Range Strike Bomber program (LRS-B), with the objective of creating a sophisticated super-bomber, with nuclear, global, and stealth capabilities to replace the B-1, B-2, and later the B-52 fleets. The development contract was assigned in 2015 and the first example of three aircraft, designated as the B-21 Raider, was shown in public on 2 December 2022, while the first flight took place on 10 November 2023. Three examples are being used for the test-and-development program and will later be converted to standard production specification. The USAF intends to acquire at least 100 examples of this model, which is expected to have a maximum takeoff weight of around 397,000 lb, an armaments payload of 20,000 lb, and a maximum speed of Mach 0.8 thanks to two Pratt & Whitney turbofans without afterburners providing around 26,500 lb thrust each.

At the same time, it appears that the Russians are working on a strategic bomber of the same type, denominated PAK-DA, apparently entrusted to Tupolev. In the field of vertical flight, besides the convertiplanes, the future could be represented by the concept of the high-speed helicopter, explored by the experimental aircraft Eurocopter X3, which took its maiden flight on 6 September 2010. Derived from the model EC155, the X3 (also called the X-cubed) is equipped with a conventional rotor, but also two RTM322 turboprops delivering 2270 shp, which operate the principal rotor as well as two 5-blade thrust-providing propellers, installed with the engines at the ends of a short wing. This aircraft proved in 2011 that it can fly at 267 mph, and its technology could be used in the future for a production series helicopter.

Projects for commercial airplanes of the future are numerous, and innovation seems to be concentrating more on greater aerodynamic efficiency and reduced environmental pollution than on speed. For some time now, separate industries have been studying models equipped with a "Box Wing" (or "Joined Wing"), that is to say, a single wing joined to form a rhomboid shape. Others are considering a return to the enormous dirigibles for passenger transport. In 2008, NASA asked the major American air industry players to study projects for aircraft of the future that would be quieter, cleaner, and cheaper. In 2010 the American space agency awarded contracts to three teams—Boeing, Lockheed Martin, and Northrop Grumman—to conceptualize the future airliners for 2025. The designs must be able to burn 50% less fuel and emit 75% less harmful emissions than the aircraft in service in 1998, as well as being safer and quieter. In 2003, the European Commission funded a project (led by Airbus Deutschland) dedicated to developing a new aircraft with hydrogen engines (Cryoplane), whose emissions would consist of simple and harmless water and nitrogen oxide. The study (H2Aircraft) drew positive conclusions, but is still awaiting developments of further research that could lead to concrete results in the future. Of course it is very difficult to predict today what the technological developments will be in 20 to 30 years, considering the incredibly rapid advances made just through the end of the 20th century in the fields of computer science, cybernetics, and wireless communications. Concepts such as spaceports, passenger aircraft capable of flying on a cruise in space, hydrogen motors, and more have now long been discussed and hypothesized. However, it does remain certain that the aeronautics and space industry—always the spearhead at the highest level of the industrialized world—will remain one of the most precious fields of research and development for our future. And perhaps the one that has the most incredible surprises in store for us.

Chapter 27

The Space Shuttle and beyond

The Apollo 11 mission launched by NASA (National Aeronautical and Space Agency) in 1969, which culminated in man's first steps on the moon, also crowned the famous "space race." This is the technological battle the United States and the Soviet Union fought throughout the 1950s and 1960s for what was unrealistically termed the "conquest of space." In fact, Apollo-type missions revealed all too evident limitations: enormous costs for what were effectively throwaway space systems with restricted payloads. By the 1960s it was clear that NASA would have to establish an alternative space program that would permit an enduring and productive presence in space and thus a greater return on the necessary investment. NASA then opted for the Space Transport System (STS), a reusable shuttle capable of carrying large loads into orbit and performing both civil and military duties.

Historically, the idea of the space shuttle dated as far back as the 1940s, as confirmed by the studies Saenger and von Braun conducted in Germany. In the 1950s, German scientists working in the United States revived the concept when NACA (later reorganized as NASA) focused on various research projects that culminated in the design of a piloted space glider for the USAF. While this vehicle had to be launched aboard a missile, it could return to earth and land as an aircraft. The program was named the X-20 Dyna Soar in 1959 and was to have used a Titan III missile as a carrier. In 1963, owing to the high costs and restricted payload, as well as the channeling of resources toward the Apollo program, the X-20 was abandoned. NASA nonetheless continued to support research into "lifting bodies," i.e., gliding aircraft with load-bearing fuselages. Late in 1969, NASA's Integrated Space Program called for a permanent space station with 50 occupants, space shuttles linking it to the earth, and possibly a moon colony and another on Mars, to be operational by 1985. Central to this program was the concept of the reusable spacecraft. On 3 January 1972 President Nixon gave the Space Shuttle Program the go-ahead, by and by year-end NASA and North American Rockwell completed their initial examination of the new vehicle's required technical specifications. The first shuttle launch was planned for March 1979.

The program's underlying concept was the creation of two complementary piloted units: one being a carrier, the other to be used for operational missions. However, the costs involved in this approach proved to be unacceptable, and NASA decided to go ahead with a single piloted mission craft (the Orbiter), with two boosters used to lift it into orbit that would be detachable but still reusable. With regard to the boosters, the choice of a liquid or solid propellant was still problematical. Painstaking research eventually came down in favor of a solid propellant as this made the recovery and reuse of the booster units easier.

The shuttle's final configuration featured a forward section containing the crew module, with the cockpit and all the flight and control avionics; a central bay destined to house the payload; and a rear section containing the engines and their shielding. To this was added the wing and undercarriage and the vertical stabilizer. The airframe was constructed in aluminum or aluminum alloys. The shuttle's nervous system was composed of five IBM AP101 computers. Of these, four were dedicated to flight, navigation, and reciprocal control functions; the fifth governed the on-board systems during orbital operations

330 top John W. Young and Robert L. Crippen, the crew in the first Space Shuttle mission (STS-1). They lifted off on 12 April 1981.

330 bottom The NASA control room for Shuttle missions.

331 top left For moving the Space Shuttle about, NASA used a Boeing 747 that was specially modified to be able to transport the shuttle on its back.

331 bottom left The night-time launch of the STS-93 mission from the Kennedy Space Center on 23 July 1999.

331 right The shuttle Discovery being launched from its support platform.

and was available as an emergency backup during the approach and landing phases. The principal features of the Space Shuttle Program concerned the propulsion system and the thermal shield; both were sources of significant problems and made notable contributions to the delays suffered by the project.

The shuttle engines have different functions and uses. Three main engines make up the SSME (Space Shuttle Main Engines) system, their principal characteristic is that they have been designed for repeated use, the equivalent of around 7.5 hours or 55 shuttle missions; each unit provides around 200 tons of thrust. The main engines can be ignited or shut down a number of times during a mission, thrust delivery can be adjusted (within a range of 65-105% of full power), and they have vectoring nozzles. The shuttle is also equipped with two smaller engines, each providing about 6,150 lb of thrust. These are used to achieve orbit once the main engines have been shut down and to provide thrust for maneuvering and re-entry. Lastly, 42 small rockets developing between 22 and 880 lb of thrust provide fine maneuvering control in space. The main fuel supply is contained in a detachable external tank 154 feet long and just over 26 feet wide, holding 665 tons of liquid oxygen and 111 tons of liquid hydrogen. The external tank is the only non-reusable part of the space shuttle. Two large solid rocket boosters (SRB) 147 feet 8" long provide an extra 3,300 tons of take-off thrust from a solid propellant. With this propulsive system, the shuttle is capable of accelerating from standstill at ground level to Mach 25 (around 15,500 mph) at an altitude of around 77.5 miles in 8.5 minutes.

BIBLIOGRAPHY

BOOKS

AA.VV., Ali italiane, vols.1-4, Milan, 1978.

AA.VV., Guerre in tempo di pace, dal 1945, Novara, 1983.

AA.VV., Ali sul mondo, Florence, 1990.

AA.VV., Russia's Top Guns, New York, 1990.a

AA.VV., Gulf Air War Debrief, London, 1991.

AA.VV., Warplanes of the Luftwaffe, London, 1994.

AA.VV., Brassey's World Aircraft & Systems Directory, London, 1996.

AA.VV., Aerei di tutto il mondo, militari e civili, Novara, 1997.

AA.VV., The Aerospace Encyclopedia of Air Warfare, vols.1-2, London, 1997.

AA.VV., Aerospace Encyclopedia of World Air Forces, London, 1999.

Angelucci, E. & Matricardi, P., Atlante enciclopedico degli aerei militari del mondo, Milan, 1990.

Chant, C., Aviation Record Breaker, London, 1988.

Cohen, E., Israel's Best Defence, Shrewsbury, 1994.

D'Avanzo, G., Evoluzione del caccia NATO, Florence, 1987.

Dorr, R. F. & Bishop, C., Vietnam Air War Debrief, London, 1996.

Francillon, R. J., The United States Air National Guard, London, 1993.

Jarrett, P., Ultimate Aircraft, London, 2000.

Kaminski, T. & Williams, M., The United States Military Aviation Directory, Norwalk, 2000.

Licheri S., Storia del volo e delle operazioni aeree e spaziali da Icaro ai nostri giorni, Rome, 1997.

Luttwak E. & Koehl S. L., La Guerra Moderna, Milan, 1992.

Niccoli R., American Eagles, Novara, 1997.

Niccoli R., Aerei, Istituto Geografico DeAgostini, Novara, 1998.

Niccoli R., Ali Tricolori, Novara, 1998.

Salvadori M. L., Storia dell'età contemporanea, vol. 3, Loescher Editore, 1986.

Steijger C., A History of the United States Air Force Europe, Shrewsbury, 1991.

Taylor M. J. H., Jane's American Fighting Aircraft of the 20th Century, New York, 1991.

Willmott N. & Pimlott J., Strategy & Tactics of War, London, 1983.

Yenne B., The History of the US Air Force, London, 1990.

Zetner C., La guerra del dopoguerra, Milan, 1970.

ENCYCLOPAEDIAS

The Epic of Flight, AA.VV., 23 volumes, 1981.

L'Aviazione, grande enciclopedia illustrata, AA.VV, 12 volumes,

Novara, 1982.

MAGAZINES

Rivista Aeronautica, Stato Maggiore Aeronautica, Italy.

Volare, Editoriale Domus, Italy.

JP-4, mensile di aeronautica, ED.A.I., Italy.

Rivista Italiana Difesa, Giornalistica Riviera Coop., Italy.

Air International, Key Publishing, Great Britain.

Air Forces Monthly, Key Publishing, Great Britain.

Wings of Fame, Aerospace Publishing, Great Britain.

International Air Power Review, AIRtime Publishing, USA-Great Britain.

Combat Aircraft, AIRtime Publishing, USA.

PHOTOGRAPHIC CREDITS

PREFACE
Page 1 Il Dagherrotipo
Pages 2-3 John M. Dibbs/The Plane Picture Company
Pages 4-5 Corbis/Grazia Neri
Page 5 bottom right Mary Evans Picture Library
Pages 6-7 Contrasto

CHAPTER 1
Page 9 right Index
Pages 10-11 Il Dagherrotipo
Page 10 bottom left Il Dagherrotipo
Page 11 top Il Dagherrotipo
Page 11 bottom Il Dagherrotipo
Page 12 top left Photos12
Pages 12-13 Mary Evans Picture Library
Page 13 left Mary Evans Picture Library
Page 13 top Mary Evans Picture Library
Page 13 right Mary Evans Picture Library
Page 13 bottom left Photos12
Page 13 bottom right Photos12
Page 14 top Double's
Page 14 bottom Photos12
Pages 14-15 Photos12
Page 15 bottom Photos12
Page 16 top left Science Photo Library/ Grazia Neri
Page 16 center Otto Lilienthal Museum
Page 16 top right Otto Lilienthal Museum
Page 16 center right top Otto Lilienthal Museum
Page 16 center right bottom Otto Lilienthal Museum
Page 16 bottom right Hulton Achive/ Laura Ronchi
Page 17 top Otto Lilienthal Museum
Page 17 left center left Otto Lilienthal Museum
Page 17 center right Otto Lilienthal Museum
Page 17 bottom right Mary Evans Picture Library
Page 17 bottom center Archivio Privato
Page 18 top right Hulton Archive/ Laura Ronchi
Page 18 bottom Roger Viollet/Contrasto
Page 19 top Hulton Archive/Laura Ronchi
Pages 18-19 Photos12
Pages 20-21 Photos12
Page 20 left Mary Evans Picture Library
Page 20 right Mary Evans Picture Library
Page 21 top Photos12
Page 21 bottom Il Dagherrotipo

CHAPTER 2
Page 22 Photos12
Pages 22-23 top Roger Viollett/Contrasto
Pages 22-23 bottom Il Dagherrotipo
Page 24 left Archiv Luftshiffbau Zeppelin
Pages 24-25 bottom Hulton Archive/ Laura Ronchi
Page 25 top Archiv Luftshiffbau Zeppelin
Page 25 center right Archiv Luftshiffbau Zeppelin
Page 25 bottom Archiv Luftshiffbau Zeppelin
Pages 26-27 Archiv Luftshiffbau Zeppelin
Page 26 bottom right Archiv Luftshiffbau Zeppelin
Page 26 bottom left Archiv Luftshiffbau Zeppelin
Page 27 top Archiv Luftshiffbau Zeppelin
Page 27 bottom left Archiv Luftshiffbau Zeppelin
Page 27 bottom right Archiv Luftshiffbau Zeppelin
Pages 28-29 Hulton Archive/Laura Ronchi
Page 28 bottom left Archiv Luftshiffbau Zeppelin
Page 28 bottom right Archiv Luftshiffbau Zeppelin
Page 29 top Hulton Archive/Laura Ronchi
Page 29 center right Hulton Archive/ Laura Ronchi
Page 29 bottom Hulton Archive/Laura Ronchi
Page 30 top Publifoto Olimpia
Page 30 center Archivio G. Apostolo
Page 30 bottom left Il Dagherrotipo
Page 30 bottom right Il Dagherrotipo
Page 31 top Il Dagherrotipo
Pages 30-31 Hulton Archive/Laura Ronchi
Page 32 top Archivio Privato
Page 32 bottom Il Dagherrotipo
Pages 32-33 Archivio G. Apostolo
Page 33 center right Archivio Privato
Page 33 bottom Il Dagherrotipo
Page 34 top Moffet Field Historical Society
Page 34 bottom Moffet Field Historical Society
Page 35 top left Corbis/Grazia Neri
Page 35 top right Moffet Field Historical Society
Page 35 bottom Moffet Field Historical Society

CHAPTER 3
Pages 36-37 Hulton Archive/Laura Ronchi
Page 36 bottom left Roger Viollet/ Contrasto
Page 37 top left Hulton Archive/ Laura Ronchi
Page 37 top right Hulton Archive/ Laura Ronchi
Page 37 bottom left Popper Foto/Vision
Page 37 bottom right Popper Foto/ Vision
Page 38 top Roger Viollet/Contrasto
Pages 38-39 Il Dagherrotipo
Page 38 bottom Corbis/Grazia Neri
Page 39 bottom Hulton Archive/Laura Ronchi
Page 40 top left Roger Viollet/ Contrasto
Page 40 top right Musee De L'Air
Page 40 center right Roger Viollet/ Contrasto
Page 40 bottom left Roger Viollet/ Contrasto
Page 41 top left Archivio A. Colombo
Page 41 top right Photos12
Page 41 bottom right Hulton Archive/ Laura Ronchi
Page 42 top Musee De L'Air
Page 42 bottom Roger Viollet/Contrasto
Page 43 top left Photos12
Page 43 top right Hulton Archive/Laura Ronchi
Page 43 bottom Publifoto Olimpia
Page 44 Hulton Archive/Laura Ronchi
Page 45 top Musee De L'Air
Page 45 center Popper Foto/Vision
Page 45 bottom Archivio G. Apostolo
Page 46 Popper Foto/Vision
Pages 46-47 Archivio G. Apostolo
Page 47 top Publifoto Olimpia

CHAPTER 4
Page 48 left Il Dagherrotipo
Page 48 right Il Dagherrotipo
Page 48 bottom Il Dagherrotipo
Page 49 top left Denver Public Library
Page 49 center Il Dagherrotipo
Page 49 bottom left Il Dagherrotipo
Page 49 bottom right Il Dagherrotipo
Page 50 top Publifoto Olimpia
Page 50 bottom Il Dagherrotipo
Pages 50-51 Il Dagherrotipo
Page 51 bottom left Il Dagherrotipo
Page 51 bottom right Il Dagherrotipo
Page 52 top Hulton Archive/Laura Ronchi
Page 52 bottom Photos12
Page 53 top left Photos12
Page 53 top right The Bridgeman Art Library
Page 53 bottom Roger Viollet/Contrasto
Page 54 top Roger Viollet/Contrasto
Page 54 top left TRH Pictures
Page 54 center left TRH Pictures
Page 54 bottom Archivio R. Niccoli
Page 54 bottom right Archivio R. Niccoli
Page 55 top Aviation Picture Library
Page 55 bottom left TRH Pictures
Page 55 bottom right Musee De L'Air
Pages 56-57 John M. Dibbs/The Plane Picture Company
Page 57 top right Peter March/ R. Cooper
Page 57 center right Philip Makanna/ Ghosts
Page 57 bottom right Peter March/ R. Cooper
Page 57 bottom TRH Pictures
Page 58 A. J. Jackson Collection
Page 59 top Archivio G. Apostolo
Page 59 center Archivio R. Niccoli
Page 59 bottom Archivio G. Apostolo
Page 60 La Presse
Page 61 top left Hulton Archive/ Laura Ronchi
Page 61 top right Il Dagherrotipo
Page 61 center Il Dagherrotipo
Page 61 bottom right Il Dagherrotipo
Pages 60-61 Il Dagherrotipo

CHAPTER 5
Page 62 top left TRH Pictures
Page 62 bottom left Popper Foto/Vision
Page 62 top right Double's
Page 62 bottom right Double's
Page 63 Aviation Picture Library
Page 63 bottom left Roger Viollet/ Contrasto
Page 63 bottom right TRH Pictures
Page 64 top TRH Pictures
Page 64 bottom left Aviation Picture Library
Page 64 bottom right TRH Pictures
Page 65 top left TRH Pictures
Page 65 center left TRH Pictures
Page 65 center right TRH Pictures
Page 65 bottom left TRH Pictures
Page 66 top Aviation Pictures Library
Page 66 bottom Aviation Pictures Library
Page 67 top left Photos12
Page 67 center left Photos12
Page 67 bottom left Photos12
Page 67 bottom right Aviation Pictures Library
Pages 68-69 Aviation Picture Library
Page 68 bottom Aviation Picture Library
Page 69 top Hulton Archive/Laura Ronchi
Page 69 center A.J. Jackson Collection
Page 69 bottom Aviation Picture Library
Page 70 top Popper Foto/Vision
Pages 70 center left Corbis/ Grazia Neri
Pages 70 center right TRH Pictures
Page 70 bottom Hulton Archive/ Laura Ronchi
Page 71 top left TRH Pictures
Page 71 top right Library Of Congress
Page 71 bottom TRH Pictures
Page 72 top A.J. Jackson Collection
Page 72 left Corbis/Grazia Neri
Page 72 bottom TRH Pictures
Page 73 top Denver Public Library
Page 73 bottom Aviation Picture Library
Pages 74 and 75 TRH Pictures
Page 74 center Archivio R. Niccoli
Page 75 top Hulton Archive/Laura Ronchi
Page 75 center Aviation Picture Library
Page 75 bottom left Aviation Picture Library
Page 76 top TRH Pictures
Page 76 bottom left A.J. Jackson Collection
Page 76 bottom right A.J. Jackson Collection
Page 77 top left TRH Pictures
Page 77 top right Corbis/Grazia Neri
Page 77 bottom Hulton Archive/Laura Ronchi

CHAPTER 6
Page 78 bottom Il Dagherrotipo
Pages 78-79 Photos12
Page 79 top right Roger Viollet/ Contrasto
Page 79 bottom left Aviation Picture Library
Page 79 bottom right Photos12
Pages 80-81 top Roger Viollet/Contrasto
Pages 80-81 bottom Archivio R. Niccoli
Page 81 top Aviation Picture Library
Page 81 center TRH Pictures
Page 81 bottom F. Selinger Collection/ Aviation Picture Library
Page 81 top left Hulton Archive/Laura Ronchi
Page 82 top right Gentile concessione del Museo Gianni Caproni
Pages 82-83 Boeing Co. Archives
Page 83 right TRH Pictures
Page 83 bottom left Musee De L'Air
Page 84 center Il Dagherrotipo
Page 84 bottom left Archivio R. Niccoli
Page 84 bottom right Il Dagherrotipo
Pages 84-85 Musee Municipal Historique de l'Hydraviation
Page 85 top right Il Dagherrotipo
Page 85 center left Aviation Picture Library